Future Directions for Intelligent
Systems and Information Sciences

Studies in Fuzziness and Soft Computing

Editor-in-chief
Prof. Janusz Kacprzyk
Systems Research Institute
Polish Academy of Sciences
ul. Newelska 6
01-447 Warsaw, Poland
E-mail: kacprzyk@ibspan.waw.pl
http://www.springer.de/cgi-bin/search_book.pl?series=2941

Vol. 3. A. Geyer-Schulz
Fuzzy Rule-Based Expert Systems and Genetic Machine Learning, 2nd ed. 1996
ISBN 3-7908-0964-0

Vol. 4. T. Onisawa and J. Kacprzyk (Eds.)
Reliability and Safety Analyses under Fuzziness, 1995
ISBN 3-7908-0837-7

Vol. 5. P. Bosc and J. Kacprzyk (Eds.)
Fuzziness in Database Management Systems, 1995
ISBN 3-7908-0858-X

Vol. 6. E. S. Lee and Q. Zhu
Fuzzy and Evidence Reasoning, 1995
ISBN 3-7908-0880-6

Vol. 7. B. A. Juliano and W. Bandler
Tracing Chains-of-Thought, 1996
ISBN 3-7908-0922-5

Vol. 8. F. Herrera and J. L. Verdegay (Eds.)
Genetic Algorithms and Soft Computing, 1996
ISBN 3-7908-0956-X

Vol. 9. M. Sato et al.
Fuzzy Clustering Models and Applications, 1997
ISBN 3-7908-1026-6

Vol. 10. L. C. Jain (Ed.)
Soft Computing Techniques in Knowledge-based Intelligent Engineering Systems, 1997
ISBN 3-7908-1035-5

Vol. 11. W. Mielczarski (Ed.)
Fuzzy Logic Techniques in Power Systems, 1998
ISBN 3-7908-1044-4

Vol. 12. B. Bouchon-Meunier (Ed.)
Aggregation and Fusion of Imperfect Information, 1998
ISBN 3-7908-1048-7

Vol. 13. E. Orłowska (Ed.)
Incomplete Information: Rough Set Analysis, 1998
ISBN 3-7908-1049-5

Vol. 14. E. Hisdal
Logical Structures for Representation of Knowledge and Uncertainty, 1998
ISBN 3-7908-1056-8

Vol. 15. G. J. Klir and M. J. Wierman
Uncertainty-Based Information, 2nd ed., 1999
ISBN 3-7908-1242-0

Vol. 16. D. Driankov and R. Palm (Eds.)
Advances in Fuzzy Control, 1998
ISBN 3-7908-1090-8

Vol. 17. L. Reznik, V. Dimitrov and J. Kacprzyk (Eds.)
Fuzzy Systems Design, 1998
ISBN 3-7908-1118-1

Vol. 18. L. Polkowski and A. Skowron (Eds.)
Rough Sets in Knowledge Discovery 1, 1998
ISBN 3-7908-1119-X

Vol. 19. L. Polkowski and A. Skowron (Eds.)
Rough Sets in Knowledge Discovery 2, 1998
ISBN 3-7908-1120-3

Vol. 20. J. N. Mordeson and P. S. Nair
Fuzzy Mathematics, 1998
ISBN 3-7908-1121-1

Vol. 21. L. C. Jain and T. Fukuda (Eds.)
Soft Computing for Intelligent Robotic Systems, 1998
ISBN 3-7908-1147-5

Vol. 22. J. Cardoso and H. Camargo (Eds.)
Fuzziness in Petri Nets, 1999
ISBN 3-7908-1158-0

Vol. 23. P. S. Szczepaniak (Ed.)
Computational Intelligence and Applications, 1999
ISBN 3-7908-1161-0

Vol. 24. E. Orłowska (Ed.)
Logic at Work, 1999
ISBN 3-7908-1164-5

continued on page 412

Nikola Kasabov (Editor)

Future Directions for Intelligent Systems and Information Sciences

The Future of Speech and Image Technologies, Brain Computers, WWW, and Bioinformatics

With 116 Figures
and 33 Tables

Physica-Verlag
A Springer-Verlag Company

Prof. Nikola Kasabov
University of Otago
Department of Information Science
P.O. Box 56
Dunedin
New Zealand
Email: nkasabov@otago.ac.nz

ISSN 1434-9922
ISBN 3-7908-1276-5 Physica-Verlag Heidelberg New York

Cataloging-in-Publication Data applied for
Die Deutsche Bibliothek – CIP-Einheitsaufnahme
Future directions for intelligent systems and information sciences: the future of speech and image technologies, brain computers, www, and bioinformatics; with 33 tables/Nikola Kasabov (ed.). – Heidelberg; New York: Physica-Verl., 2000
 (Studies in fuzziness and soft computing; Vol. 45)
 ISBN 3-7908-1276-5

This work is subject to copyright. All rights are reserved, whether the whole or part of the material is concerned, specifically the rights of translation, reprinting, reuse of illustrations, recitation, broadcasting, reproduction on microfilm or in any other way, and storage in data banks. Duplication of this publication or parts thereof is permitted only under the provisions of the German Copyright Law of September 9, 1965, in its current version, and permission for use must always be obtained from Physica-Verlag. Violations are liable for prosecution under the German Copyright Law.

Physica-Verlag is a company in the BertelsmannSpringer publishing group.
© Physica-Verlag Heidelberg 2000
Printed in Germany

The use of general descriptive names, registered names, trademarks, etc. in this publication does not imply, even in the absence of a specific statement, that such names are exempt from the relevant protective laws and regulations and therefore free for general use.

Hardcover Design: Erich Kirchner, Heidelberg

SPIN 10769486   88/2202-5 4 3 2 1 0 – Printed on acid-free paper

Preface

This edited volume comprises invited chapters that cover five areas of the current and the future development of intelligent systems and information sciences. Half of the chapters were presented as invited talks at the Workshop "Future Directions for Intelligent Systems and Information Sciences" held in Dunedin, New Zealand, 22-23 November 1999 after the International Conference on Neuro-Information Processing (ICONIP/ANZIIS/ANNES'99) held in Perth, Australia.

In order to make this volume useful for researchers and academics in the broad area of information sciences I invited prominent researchers to submit materials and present their view about future paradigms, future trends and directions.

Part I contains chapters on adaptive, evolving, learning systems. These are systems that learn in a life-long, on-line mode and in a changing environment. The first chapter, written by the editor, presents briefly the paradigm of Evolving Connectionist Systems (ECOS) and some of their applications. The chapter by Sung-Bae Cho presents the paradigms of artificial life and evolutionary programming in the context of several applications (mobile robots, adaptive agents of the WWW). The following three chapters written by R.Duro, J.Santos and J.A.Becerra (chapter 3), G.Coghill (chapter 4), Y.Maeda (chapter 5) introduce new techniques for building adaptive, learning robots.

Part II contains chapters on intelligent human computer interaction and scientific visualisation. Chapter 6 by J.Taylor and N.Kasabov introduces a new methodology for modeling the acquisition of speech, namely the emergence of the phoneme concept in an unsupervised way. It argues that language can be acquired through learning, which is in contrast to the Chomsky's theory of innate languages. The chapter by Herik and Postma introduces a new approach to building intelligent systems that recognise painters. The two chapters by A.Nijhollt, M.Paulin and co-authors demonstrate the potential of the virtual reality for both scientific visualisation and knowledge discovery.

Part III presents two new connectionist computational paradigms, namely brain-like computing and quantum neural networks. J.G.Taylor points in chapter 10 to new directions for the development of neural networks and intelligent systems based on the recent results from brain imaging. He emphasises on the internal model that the brain builds, manipulates and updates continuously. The chapter by A.Ezhov and Dan Ventura gives an exciting perspective for combining principles from neuro-computing and quantum computing into quantum neural networks. That opens a new horizon for developing new computational models that are both quantitatively and qualitatively superior to any computational models developed so far (including Turing machines, von Neumann computers, neuro-computers, and quantum computers). The chapter by Stocks and Mannella suggests a model of biological sensory neurons.

Part IV includes two chapters on Bio-informatics. Chapter 13 by C.Brown, M.Schreiber, B.Chapman and G.Jacobs is an excellent introduction to the fundamentals, the issues, the problems and to some extent to existing solutions in this area. The second chapter written by V.Bajic and I.Bajic presents a neural network model for discovering promotion regions in DNA sequences, along with a review of existing neural network models for DNA analysis.

Part V is devoted to knowledge representation, knowledge processing and knowledge discovery, and also on some applications. The chapter by W.Pedrycz is a good introduction to the foundations of a new paradigm called granular computing. This paradigm deals with information "granules" represented in different ways (intervals, fuzzy sets, rough sets, shadowed sets, etc). The chapter by J.Kacprzyk reiterates on another new paradigm – computing with words, and illustrates it with examples of information summarization from ordinary numerical databases. The chapter by N.Kasabov, L.Erzegovezi, M.Fedrizzi and co-authors presents a new approach to building intelligent, adaptive decision support systems and illustrates it on the task of analysis and discovery of economic clusters in the European Union. The chapter by D.Yun discusses issues of intelligent resource management through a constrained resource planning model and the last chapter by Ramer and co-authors discusses the efficiency of one class of neuro-fuzzy systems for the task of learning and fuzzy rule extraction. I believe the book will be a useful guide for many researchers and academics who want to embark on new computational models in the exciting 21st century.

Several people helped me with the preparation of this volume. I would like to thank Professor Janusz Kacprzyk and the team of Physica Verlag for accepting this book and for their professional and accurate work for its publication. I am much obliged to Ms.Kitty Ko who helped me a great deal with the organisation of the volume. Brendon Woodford helped in the camera-ready preparation of the volume.

Nik Kasabov
University of Otago
February, 2000

Contents

Part I: Adaptive, evolving, learning systems

Chapter 1. ECOS - Evolving Connectionist Systems – a new/old paradigm for on-line learning and knowledge engineering 3
N.Kasabov

Chapter 2. Artificial life technology for adaptive information processing 13
Sung-Bae Cho

Chapter 3. Evolving ANN controllers for smart mobile robots 34
R.J.Duro, J. Santos and J.A.Becerra

Chapter 4. A simulation environment for the manipulation of naturally variable objects 65
G.Coghill

Chapter 5. Behavior-decision fuzzy algorithm for autonomous mobile robot 75
Y.Maeda

Chapter 6. Modelling the emergence of speech and language through evolving connectionist systems 102
J.Taylor and N.Kasabov

Part II: Intelligent human computer interaction and scientific visualisation

Chapter 7. Discovering the visual signature of painters 129
H.J. van den Herik and E.O.Postma

Chapter 8. Multimodal interactions with agents in virtual worlds 148
A.Nijhollt and J.Hulstijn

Chapter 9. Virtual BioBots 174
M.Paulin and R.Berquist

Part III: New connectionist computational paradigms: Brain-like computing and quantum neural networks

Chapter 10. Future directions for neural networks and intelligent systems from the brain imaging research 191
J.G. Taylor

Chapter 11. Quantum neural networks 213
A.A.Ezhov and D.Ventura

Chapter 12. Suprathreshold stochastic resonance in a neuronal network model: a possible strategy for sensory coding 236
N.G.Stocks and R.Mannella

Part IV: Bioinformatics

Chapter 13. Information science and bioinformatics 251
C.Brown, M.Schreiber, B.Chapman and G.Jacobs

Chapter 14. Neural network system for promoter recognition 288
V.B.Bajic and I.V.Bajic

Part V: Knowledge representation, knowledge processing, knowledge discovery, and some applications

Chapter 15. Granular computing: An introduction 309
W.Pedrycz

Chapter 16. A new paradigm shift from computation on numbers to computation on words on an example of linguistic database summarization .. 329
J.Kacprzyk

Chapter 17. Hybrid intelligent decision support systems and applications for risk analysis and discovery of evolving economic clusters in Europe 347
N.Kasabov, L.Erzegovesi, M.Fedrizzi, A.Beber and D.Deng

Chapter 18. Intelligent resource management through the constrained resource planning model 373
D.Y.Y.Yun

Chapter 19. Evaluative studies of fuzzy knowledge discovery through NF systems 387
A.Ramer, M.do Carmo Nicoletti, S.Y.Sung

Part I

Adaptive, Evolving, Learning Systems

Chapter 1. ECOS - Evolving Connectionist Systems – a New/Old Paradigm for On-line Learning and Knowledge Engineering

Nikola Kasabov

Department of Information Science
University of Otago, P.O Box 56, Dunedin, New Zealand
Phone: +64 3 479 8319, fax: +64 3 479 8311
nkasabov@otago.ac.nz

Abstract. This chapter makes a revision of the major principles, applications and publications on the evolving connecionist systems (ECOS) paradigm. It compares ECOS with other AI models and outlines directions for further research.

Keywords. *Evolving connectionist systems; neural networks; adaptive systems.*

1 On-Line Learning and Knowledge Engineering

The complexity and dynamics of real-world problems, such as adaptive speech recognition and language acquisition, adaptive intelligent prediction and control systems, intelligent agent-based systems; adaptive agents on the Web; mobile robots; visual monitoring systems and multi-modal information processing, and many more require sophisticated methods and tools for building on-line, adaptive, knowledge-based systems (IS). Such systems should be able to: (1) *learn fast* from a large amount of data (using fast training); (2) *adapt incrementally* in an on-line mode; (3) dynamically create new modules – have open structure; (4) *memorise information* that can be used later; (5) *interact continuously* with the environment in a "life-long" learning mode; (6) deal with *knowledge* (e.g. rules), as well as with data; (7) adequately represent in their structure *space and time*.

Several methods and systems have been developed so far that meet some of the criteria above and that have influenced the development of the evolving connectionist systems – ECOS - paradigm and the evolving fuzzy neural networks (EFuNN) [1-38], for example: *adaptive learning* [52,53,39,47]; *incremental learning* [49,55,48]; *local learning* [40,51]; *lifelong learning* [44]; *on-line learning* [42,43,44,46,52-55]; *constructivist structural learning* [55] that is supported by biological facts [49]; *selectivist structural learning*; hybrid constructivist/selectivist structural learning; k*nowledge-based learning neural networks (KBNN)* [56].

The ECOS and the EFuNN models discussed here belong to all the above groups. The models are called evolving because of the nature of the structural growth and structural adaptation of the whole evolving connectionist system (ECOS). In terms of on-line neuron allocation, the EFuNN model is similar to the Resource Allocating Network (RAN) suggested by Platt [52] and improved in other related models [53]. The RAN model allocates a new neuron for a new input example (x, y) if the input vector x is distant in the input space from the already allocated radial basis neurons, and also the output error evaluation y-y', where y' is the produced by the system output for the input vector x, is above another threshold. Otherwise, neurons will be adapted to the example (x,y) through a gradient descent algorithm. In terms of adaptive optimisation of many individual linear units EFuNN is close to the Receptive Field Weight Regression (RFWR) model by Schaal and Atkeson [55]. EFuNNs have also similarities with the Fritzke's Growing Cell Structures and Growing Neural Gas models [43], and with other dynamic radial basis function networks (RBFN) [51,57,39] in terms of separating the unsupervised learning, which is performed first, from the supervised learning, applied next, in a two-tyre structure. The EFuNN learning algorithm though differs from the above in many aspects, mainly in the employment of simpler and faster learning modes, more flexibility when evolving internal structures and representations, knowledge-base orientation. A comparative analysis between EFuNNs and other similar models on benchmark problems shows that while EFuNNs are comparable with the other methods in terms of accuracy of the obtained results, they are much faster, more controllable, and evolve more compact internal representations. EFuNNs suggest a new neuro-fuzzy systemic approach that employs more sophisticated supervised/unsupervised, knowledge-based learning methods. EFuNNs allow for rule insertion and rule extraction at any time of the operation of the system. The functionality of EFuNNs can be fully utilised when EFuNNs are used as elements of the ECOS framework of an adaptive, intelligent, knowledge-based system.

The on-line, knowledge-based type of learning in the ECOS and EFuNNs allow for building dynamic knowledge-based systems where the knowledge-base is built as the system operates. This approach is called here on-line knowledge engineering.

2 Evolving Connectionist Systems (ECOS) – Major Principles

The ECOS paradigm was published first in [15] and then elaborated in [24,16,19]. It allows an intelligent system to be automatically created from incoming data. An ECOS consists of nodes (units) that perform pre-defined functions, and connections between them. The system has a minimal initial structure that includes preliminary input and output nodes and few preliminary connections. Data is allowed to flow into the system so that if an input data vector is associated with a desired output vector, the system stores this association into a new node and creates new connections. Nodes and connections are created automatically to reflect the data distribution. The system's structure is dynamically changing after each data item is introduced.

The number of the input and output variables in ECOS can vary during the learning process thus allowing for more (or less) input and output variables to be introduced at any stage of the learning process. Input and output variables can have 'missing values' at any time of the learning process. If there is no output vector associated with an input vector, the system produces its own output vector (its own solution). If the desired output vector became known afterwards, the system will adjust its structure to produce this output, or one close to it, next time the same input vector is presented. The system continuously and adaptively learns from data to associate inputs to outputs and to cluster the data trough allocating nodes to represent exemplars of data.

The learning process in ECOS is achieved through interaction with the environment which supplies the data flow and reacts to the output produced by the system.

In addition the system provides the knowledge it has learned in the form of IF-THEN rules [16,23,10]. The ECOS framework implies that a system evolves through its operation in an interactive way. The more data are presented to the system, the more the system evolves. The learning process is on-line, life-long.

Nodes and connections in an ECOS system can be created, modified, merged, and pruned, in a self-organising manner, similar to how the human brain learns through creating and wiring neuronal structures. The system's structure grows or shrinks depending on the incoming data distribution and pre-set parameters. Through the process of evolving from data the system learns the rules of its own behaviour. The rules that constitute the system's knowledge can be reported and/or extracted at any time of the system operation.

Here the main principles of ECOS are summarised:

(1) fast learning from a large amount of data, e.g. through one-pass training;
(2) adaptation in an on-line mode, where new data is incrementally accommodated;
(3) 'open' structure, where new features (relevant to the task) can be introduced at any stage of the system's operation, e.g., the system creates "on the fly" new inputs, new outputs, new modules and connections;
(4) memorisation of data exemplars for a further refinement, or for information retrieval;
(5) learn and improve through active interaction with other IS and with the environment in a multi-modular, hierarchical fashion;

(6) adequately represent space and time in their different scales; have parameters that represent short-term and long-term memory, age, forgetting, etc.;
(7) deal with knowledge in its different forms (e.g., rules; probabilities); analyse itself in terms of behaviour, global error and success; "explain" what the system has learned and what it "knows" about the problem it is trained to solve; make decisions for a further improvement.

Biological motivations for ECOS are presented in [14].

3 Evolving Fuzzy Neural Networks (EFuNNs)

One realisation of ECOS is the evolving fuzzy neural network EFuNN. EFuNNs were first introduced in [20]. EFuNNs are models for evolving supervised learning from data that have five-layer structure where nodes and connections are created/connected as data examples are presented. An optional short-term memory layer can be used through a feedback connection from the rule (or 'case') node layer.

The input layer of an EFuNN represents input variables. The second layer of nodes (fuzzy input neurons, or fuzzy inputs) represents fuzzy quantisation of each input variable space. For example, two fuzzy input neurons can be used to represent "small" and "large" fuzzy values. Different membership functions (MF) can be attached to these neurons (triangular, Gaussian, etc.). The number and the type of MF can be dynamically modified in an EFuNN. New neurons can evolve in this layer if, for a given input vector, the corresponding variable value does not belong to any of the existing MF to a degree greater than a membership threshold. A new fuzzy input neuron, or an input neuron, can be created during the adaptation phase of an EFuNN. The task of the fuzzy input nodes is to transfer the input values into membership degrees to which they belong to the MF. The third layer contains rule (case) nodes that evolve through supervised or unsupervised learning. Examples of the latter are given in [33,34]. The rule nodes represent prototypes (exemplars, clusters) of input-output data associations, graphically represented as an association of hyper-spheres from the fuzzy input and fuzzy output spaces.

Each rule node r in an EFuNN is defined by two vectors of connection weights – W1(r) and W2(r), the latter being adjusted through supervised learning based on the output error, and the former being adjusted through unsupervised learning based on similarity measure within a local area of the problem space. The fourth layer of neurons represents fuzzy quantization for the output variables, similar to the input fuzzy neurons representation. The fifth layer represents the real values for the output variables.

The EFuNN evolving algorithm includes also: aggregation of rule nodes (i.e. merging rule nodes); pruning of rule nodes and connections, and other operations.

In the aggregation procedure two thresholds are used T1 and T2, and every two rule nodes for which the distance in the input space is less than T1 and the distance in the output space is less than T2 get aggregated. A normalised distance as a measure for the similarity between two vectors (two rule nodes represented by their connection weight vectors W1 and W2) is used in such a manner that providing the two vectors have been normalised and all their values are inside the interval [0,1], the distance is also inside this interval.

EFuNNs can operate in several modes depending on the number of the rule nodes for which activation is propagated to the output layer:

- One of n (1-of-n) [15];
- Many of n (m-of-n) [13]

A version of EFuNN - dynamic EFuNN (DEFuNN) [13] uses fuzzy rules of Takagi-Sugeno type, rather than the EFuNN Zadeh-Mamdany rules.

The unsupervised version of EFuNN does not use output nodes. Instead, it evolves rule nodes based on similarities between the already evolved nodes and new data [16,33,34]. One version of such systems is the evolving SOM (ESOM) [2].

Simulators of different types of EFuNNs are available from the WWW htp://divcom.otago.ac.nz/infosci/kel/CBIIS.html. EFuNNs are incorporated as models of software environments for knowledge engineering, such as the Repository of the Intelligent Connectionist Base Information Systems (RICBIS) [3] and FuzzyCOPE/4 [37,27,36].

4 Applications of ECOS and EFuNNs

ECOS and EFuNNs have been used in several projects of the Knowledge Engineering Laboratory at the University of Otago as learning tools in the following application oriented projects:

- adaptive and robust speech recognition [18,7,8,9,1];
- modelling speech and language acquisition [33,34];
- robot speech control [5];
- adaptive video mode detection [31];
- multimodal auditory and visual information processing [24];
- image classification [28,30];
- adaptive robot navigation [5,35];
- adaptive robot control [21,25];
- adaptive process control [6,17];
- learning sequences of DNA data [4];
- building adaptive expert systems [38];
- decision making and prediction [11,17,27,29,32];

5 Future Research

The following are directions for future research on ECOS and other adaptive on-line systems for learning and knowledge, manipulation:

(1) Elaboration on the mathematical theory of ECOS (universality; convergence, etc)
(2) Developing different types and modifications of ECOS (e.g. spatial-temporal [12], etc)
(3) Building hybrid models that include ECOS and HMM, for example [1]
(4) Using ECOS for a multi-modal information processing [24].
(5) Building robust to noise, adaptive speech recognisers.
(6) Building on-line learning systems on the WWW.

ECOS combine both known AI features and new features that make them promising techniques for the future development of intelligent systems and information sciences.

Acknowledgements

The development of ECOS and their applications is supported through grant UOO808 funded by the Foundation for Research, Science and Technology, New Zealand.

References

ECOS and EFuNN References:

1. Abdulla, W. and N. Kasabov, Parallel CHMM speech recognition systems, *Proc. of Joint Conference of Information Sciences (JCIS)*, Atlantic City, New Jersey, February 2000
2. Deng, D. and N. Kasabov, Evolving Self-orginizing Map and its Application in Generating a World Macroeconomic Map, In: Emerging Knowledge Engineering and Conectionist-based Systems (Proceedings of the ICONIP/AMZIIS/ANNES'99 Workshop "Future directions for intelligent systems and information sciences, Dunedin, 22-23 Nov.1999), N.Kasabov and K.Ko (eds), 7:12
3. Deng, D., I. Koprinska and N. Kasabov, RICBIS - New Zealand Repository for Intelligent Connectionist-Based Information Systems, In: Emerging Knowledge Engineering and Conectionist-based Systems (Proceedings of the ICONIP/AMZIIS/ANNES'99 Workshop "Future directions for intelligent systems and information sciences, Dunedin, 22-23 Nov.1999), N.Kasabov and K.Ko (eds),182-185

4. Futschik, M; Schreiber, M; Brown, C, and Kasabov, N. (1999) "Comparative Studies of Neural Network Models for mRNA Analysis", in Proc. Int. Conf. Intelligent Systems for Molecular biology, Heidelberg, August 6-10 (1999)
5. Ghobakhlou, A., Q.Song and N. Kasabov, ROKEL: The Interactive learning and Navigating Robot of the Knowledge Engineering laboratory at Otago, In: Emerging Knowledge Engineering and Connectionist-based Systems (Proceedings of the ICONIP/AMZIIS/ANNES'99 Workshop "Future directions for intelligent systems and information sciences, Dunedin, 22-23 Nov.1999), N.Kasabov and K.Ko (eds) 57-59
6. Hegg,D., T. Cohen, N. Kasabov and Q. Song Intelligent Control of Sequencing Batch Reactors (SBRs) for Biological Nitrogen Removal, In: Emerging Knowledge Engineering and Conectionist-based Systems (Proceedings of the ICONIP/AMZIIS/ANNES'99 Workshop "Future directions for intelligent systems and information sciences, Dunedin, 22-23 Nov.1999), N.Kasabov and K.Ko (eds), 152-155
7. Iliev, G. and N. Kasabov Adaptive Filtering with Averaging in Noise Cancellation for Voice and Speech Recognition, (Proceedings of the ICONIP/AMZIIS/ANNES'99 Workshop "Future directions for intelligent systems and information sciences, Dunedin, 22-23 Nov.1999), N.Kasabov and K.Ko (eds) 71-75
8. Iliev, G., Kasabov, N. Adaptive noise cancellation for speech applications, Proc. of ICONIP'99, November 1999, Perth, Australia, IEEE Press (1999)
9. Iliev, G., Kasabov, N. Channel equalisation using adaptive filtering with averaging, in: *Proc. of Joint Conference of Information Sciences (JCIS)*, Atlantic City, New Jersey, February 2000 (accepted)
10. Kasabov, N and B. Woodford, Rule insertion and rule extraction from evolving fuzzy neural networks: algorithms and applications for building adaptive, intelligent expert systems, 1999 IEEE International Fuzzy Systems Conference Proceedings, Seoul, August 1999, v.III, 1406-1409
11. Kasabov, N. and Fedrizzi, M. Fuzzy neural networks and evolving connectionist systems for intelligent decision making, *Proceedings of the Eight International Fuzzy Systems Association World Congress*, Taiwan, August 17-20 (1999)
12. Kasabov, N. and M.Watts. Spatial temporal evolving fuzzy neural networks and applications for adaptive phoneme recognition, *IEEE Transactions on Neural Netwokrs*, (1999) submitted, published as TR99-03, Department of Information Science, University of Otago
13. Kasabov, N. and Q.Song. Dynamic "m-of-n" evolving fuzzy neural networks for on-line learning of dynamic time series, *IEEE Transactions on Fuzzy Systems*, (1999) submitted, published as TR99-04, Department of Information Science, University of Otago
14. Kasabov, N. Brain-like functions in evolving connectionist systems for on-line, knowledge-based learning, in: T. Kitamura (ed) Modeling brain functions for intelligent systems, World Scientific (2000)
15. Kasabov, N. ECOS - A framework for evolving connectionist systems and the 'eco' training method, in: S.Usui and T.Omori (eds) *Proceedings of ICONIP'98 - The Fifth International Conference on Neural Information Processing,* Kitakyushu, Japan, 21-23 October 1998, IOS Press, vol.3, 1232-1235
16. Kasabov, N. Evolving connectionist and fuzzy connectionist systems – theory and applications for adaptive, on-line intelligent systems, In: *Neuro-Fuzzy Techniques for*

Intelligent Information Systems, N. Kasabov and R.Kozma, eds. Heidelberg, Physica Verlag (1999) 111-146
17. Kasabov, N. Evolving connectionist and fuzzy connectionist systems for on-line adaptive decision making and control, in: *Advances in Soft Computing - Engineering Design and Manufacturing,* R. Roy, T. Furuhashi and P.K. Chawdhry (Eds.) Springer-Verlag, London Limited, 1999, 638 pages
18. Kasabov, N. Evolving connectionist systems and applications for adaptive speech recognition, Proc. of IJCNN'99, Washington DC, July 1999, IEEE Press
19. Kasabov, N. Evolving connectionist systems for on-line, knowledge based learning. *IEEE Transactions on Man, Systems and Cybernetics,* (2000), published also as TR99-02, Department of Information Science, University of Otago
20. Kasabov, N. Evolving fuzzy neural networks - algorithms, applications and biological motivation, in: T.Yamakawa and G.Matsumoto (eds) Methodologies for the Conception, Design and Application of Soft Computing, World Scientific, 1998, 271-274
21. Kasabov, N. Evolving fuzzy neural networks for adaptive, on-line intelligent agents and systems, in: O. Kaynak, S.Tosunoglu and M.Ang (eds) *Recent Advances in Mechatronics,* Springer Verlag, Singapore (1999): Proceedings of the international conference, Istanbul, Turkey, 24-26 May 1999, 27-41.
22. Kasabov, N. Evolving fuzzy neural networks: Theory and Applications for on-line adaptive prediction, decision making and control, *Australian Journal of Intelligent Information Processing Systems,* 5 (3): 154-160 (1998
23. Kasabov, N. On-line learning, reasoning, rule extraction and aggregation in locally optimised evolving fuzzy neural networks, *Neurocomputing,* (2000) in print
24. Kasabov, N. The ECOS framework and the 'eco' training method for evolving connectionist systems. *Journal of Advanced Computational Intelligence* (1998) vol.2, No.6, 195-202
25. Kasabov, N. Theory and applications of evolving connectionist agents and systems, *Proceedings of the 1998 international conference on Neural Networks and Brain (NN&B),* Beijing, October 27-30 (1998), Publishing House of Electronics Industry, China, 668-671
26. Kasabov, N., D. Tuck, and M. Watts, Implementing Knowledge and Data Fusion in a Versatile Software Environment for Adaptive Learning and Decision-Making, in: Proc. of the Int. Conference on Data Fusion, San Hose, July 1999 (1999)
27. Kasabov, N., D.Deng, L.Erzegovezi, M.Fedrizzi, A. Beber, D.Deng, Hybrid Intelligent Decision Support Systems and Applications for Risk Analysis and Prediction, International conference on intelligent systems for investment decision making, Bond University, Gold Cost, December (1999)
28. Kasabov, N., Israel, S., and B. Woodford. Hybrid evolving connectionist systems for image classification, *Journal of Advanced Computational Intelligence* (2000)
29. Kasabov, N., L.Erzegovezi, M.Fedrizzi, A. Beber, D.Deng, Hybrid Intelligent Decision Support Systems and Applications for Risk Analysis and Prediction of Evolving Economic Clusters in Europe, in the same volume
30. Kasabov, N., S.Israel, B.Woodford, Methodology and evolving connectionist architecture for image pattern recognition, in: S.Pal, Ghosh and Kundu (eds*) Soft computing and image processing,* Heidelberg, Physica-Verlag (Springer Verlag) (2000)
31. Koprinska, I. and N. Kasabov An Application of Evolving Fuzzy Neural Network for Compressed Video Parsing, In: Emerging Knowledge Engineering and Conectionist-based Systems (Proceedings of the ICONIP/AMZIIS/ANNES'99 Workshop "Future

directions for intelligent systems and information sciences, Dunedin, 22-23 Nov.1999), N.Kasabov and K.Ko (eds), 96-102
32. Swope, J.A., Kasabov, N., Williams, M., Neuro-fuzzy modelling of heart rate signals and applications to diagnostics, in: Szepanjuk ed, Fuzzy Systems in Medicine, Physica Verlag (2000) 519-542
33. Taylor, J. and N.Kasabov Modelling the Emergence of Speech and Language through Evolving Connectionist Systems, in the same volume
34. Taylor, J., N.Kasabov, and R.Kilgour, Modelling the Emergence of Speech Sopund Categories in Evolving Connectionist Systems, Proc. of the Joint Conference on Information Sciences, Atlantic City, 2000
35. Tuck, D., and Q.Song, Demonstration of an adaptive EFuNN application: an on-line robotic arm path planner, (Proceedings of the ICONIP/AMZIIS/ANNES'99 Workshop "Future directions for intelligent systems and information sciences, Dunedin, 22-23 Nov.1999), N.Kasabov and K.Ko (eds),45-48
36. Tuck, D., M. Watts, Q. Song and N. Kasabov, A Practical and Flexible Environment for Adaptive Knowledge and Data Fusion Applications. In: Proc. of Int. Conf. On Applications of Intelligent Systems, Melbourne, Sept. 1999 (1999)
37. Watts,M., B. Woodford, N. Kasabov, FuzzyCOPE - A Software Environment for Building Intelligent Systems - the Past, the Present and the Future, In: Emerging Knowledge Engineering and Conectionist-based Systems (Proceedings of the ICONIP/AMZIIS/ANNES'99 Workshop "Future directions for intelligent systems and information sciences, Dunedin, 22-23 Nov.1999), N.Kasabov and K.Ko (eds) 188-192
38. Woodford, B.J., N. Kasabov and H. Wearing, Fruit Image Analysis using Wavelets, In: Emerging Knowledge Engineering and Conectionist-based Systems (Proceedings of the ICONIP/AMZIIS/ANNES'99 Workshop "Future directions for intelligent systems and information sciences, Dunedin, 22-23 Nov.1999), N.Kasabov and K.Ko (eds), 88-92

Related References:

39. Blanzieri, E., P.Katenkamp, "Learning radial basis function networks on-line", in: Proc. of Intern. Conf. On Machine Learning, 37-45 (1996)
40. Bottu and Vapnik, "Local learning computation", Neural Computation, 4, 888-900 (1992)
41. Edelman, G., Neuronal Darwinism: The theory of neuronal group selection, Basic Books (1992).
42. Freeman, J., D. Saad, "On-line learning in radial basis function networks", Neural Computation vol. 9, No.7 (1997).
43. Fritzke, B. "A growing neural gas network learns topologies", Advances in Neural Information Processing Systems, vol.7 (1995)
44. Gaussier, T., and S. Zrehen, "A topological neural map for on-line learning: Emergence of obstacle avoidance in a mobile robot", In: From Animals to Animats No.3, 282-290, (1994).
45. Hech-Nielsen, R. "Counter-propagation networks", IEEE First int. conference on neural networks, San Diego, vol.2, pp.19-31 (1987)
46. Heskes, T.M., B. Kappen, "On-line learning processes in artificial neural networks", in: Math. foundations of neural networks, Elsevier, Amsterdam, 199-233, (1993).

47. Kohonen, T., Self-Organizing Maps, second edition, Springer Verlag, 1997
48. Mandziuk, J., Shastri, L. "Incremental class learning approach and its application to hand-written digit recognition, Proc. of the fifth int. conf. on neuro-information processing, Kitakyushu, Japan, Oct. 21-23, 1998
49. McClelland, J., B.L. McNaughton, and R.C. Reilly "Why there are Complementary Learning Systems in the Hippocampus and Neo-cortex: Insights from the Successes and Failures of Connectionist Models of Learning and Memory", CMU TR PDP.CNS.94.1, March, (1994).
50. Mitchell, M.T., "Machine Learning", MacGraw-Hill (1997)
51. Moody, J., Darken, C., Fast learning in networks of locally-tuned processing units, Neural Computation, 1(2), 281-294 (1989)
52. Platt, J., "A resource allocating network for function interpolation, Neural Computation, 3, 213-225 (1991)
53. Rosipal,R., M.Koska, I.Farkas, "Prediction of chaotic time-series with a resource-allocating RBF network, Neural Processing Letters, 10:26 (1997)
54. Saad, D. (ed) On-line learning in neural networks, Cambridge University Press, 1999
55. Schaal, S. and C. Atkeson, "Constructive incremental learning from only local information, Neural Computation, 10, 2047-2084 (1998)
56. Towel, G., J. Shavlik, and M. Noordewier, "Refinement of approximate domain theories by knowledge-based neural networks", Proc. of the 8th National Conf. on Artificial Intelligence AAAI'90, Morgan Kaufmann, 861-866 (1990).
57. Topchy, A., O.Lebedko, V.Miagkikh and N.Kasabov, "Adaptive training of radial basis function networks based on co-operative evolution and evolutionary programming", in: Progress in connectionist-based information systems, N.Kasabov et al (eds), Springer, 253-258 (1998)

Chapter 2. Artificial Life Technology for Adaptive Information Processing*

Sung-Bae Cho

Department of Computer Science, Yonsei University, 134 Shinchon-dong, Sudaemoon-gu, Seoul 120-749, Korea

Abstract. Adaptation gives rise to a kind of complexity that greatly hinders our attempts to solve some of the most important problems currently posed by our world. Recently, there is an attempt to build a complex adaptive system, which is rich in autonomy and creativity, with the ideas and methodologies of Artificial Life (A-life). This chapter presents the concepts and methodologies of A-life, and shows some of the typical systems developed based on them. These systems cannot only develop new functionality spontaneously but also grow and evolve its own structure autonomously. They have been applied to categorizing visual patterns, controlling a mobile robot, developing adaptive agents on the WWW, and retrieving media databases based on human preference.

1 Intelligent System and Artificial Life

Intelligent systems can adaptively estimate continuous functions from data without specifying mathematically how outputs depend on inputs [13]. System behavior is called *intelligent* if the system emits appropriate problem-solving responses when faced with problem stimuli. Recently, some researchers have tried to synthesize intelligent systems by using Artificial Life (A-life) technologies.

A-life research aims at studying man-made systems exhibiting behaviors characteristic of natural living systems. It complements the traditional biological sciences concerned with the analysis of living organisms by attempting to synthesize life-like behaviors within computers; extending the empirical foundation upon which biology is based from *life-as-we-know-it* to a larger picture of *life-as-it-could-be* [14]. The essential features of A-life models are as follows:

- They work with populations of simple programs, where no single program directs all of the other programs.
- Each program details the way in which a simple entity reacts to local situations in its environment, including encounters with other entities.
- There are no rules in the system that dictate global behavior, and higher behavior is therefore emergent.

---

* This work was supported in part by a grant from Brain Science Research Center with brain research program of the Ministry of Science and Technology in Korea.

It is the concept of emergent property that highlights the nature of the A-life research. Emergent property is exhibited by a collection of interacting entities whose global behavior cannot be reduced to a simple aggregate of the individual contributions of the entities. Conventional methods in artificial intelligence have difficulty to reveal and explain the emergent properties, because they are generally reductionist that decompose a system into constituent subsystems and then study them in isolation according to top-down approach.

On the other hand, A-life adopts bottom-up approach which starts with a collection of entities exhibiting simple and well-understood behavior and synthesizes more complex systems. There are a lot of technologies in the A-life research such as cellular automata, Lindenmayer system, genetic algorithm, neural networks, and so on, but the key idea is the evolutionary algorithm. In this sense, a practical goal of A-life can be redefined as finding a mechanism for an evolutionary process to be used in the automatic design and creation of artifacts [21].

A genetic algorithm, one of the evolutionary algorithms, is a model of machine learning derived by the procedure of evolution in nature [8]. This is performed by creating the population of individuals that are represented by chromosomes. The chromosome is the string that can be thought of as the human genes. The individuals in the population go through the evolution. This procedure takes the evolutionary procedure in which different individuals compete for resources in the environment. Better individuals are more likely to survive, and propagate their genetic material to offsprings. The procedure of a simple genetic algorithm is as follows.

Table 1. Genetic algorithm for simulated evolution.

```
t = 0;
InitializePopulation P(t);
Evaluate P(t);
while not done {
    t = t + 1;
    P'= SelectParents P(t);
    Recombine P'(t);
    Mutate P'(t);
    Evaluate P'(t);
    P = Survive P, P'(t);
}
```

The algorithm starts with initial population. In the beginning, fitness value of each individual is evaluated to determine how appropriate the individual is for a given problem. In general, individuals in an initial population are randomly generated. Next, two individuals that have relatively high fitness value are selected from the population. Selected individuals can be regarded as parents. New individuals, called children, are created by recombinating the chromosomes of parents. Here, selection and crossover operators are used. Mutation replaces existing gene code with randomly generated code. It performs a function that puts potentially good genes or properties that may be lost during applying genetic operation into the population.

In this chapter, we shall show the potential of the A-life technology in a future direction for intelligent systems and information sciences by introducing four examples of typical applications: categorizing visual patterns; controlling a mobile robot; developing adaptive agents on the WWW; and retrieving media databases based on human preference. Some of the results are not comparable to those of conventional methods yet, but they clearly show the emergent property in which entities interact in non-additive ways and give rise to adaptive behaviors.

2 Categorizing Visual Patterns

The design of neural networks by evolutionary algorithms has attracted great interest in the quest to develop adaptive systems that can change architectures and learning rules according to differing environments. There are more than one hundred publications that discuss evolutionary design methods applied to neural networks. One of the important advantages of evolutionary neural networks is their adaptability to a dynamic environment, and this adaptive process is achieved through the evolution of connection weights, architectures and learning rules [22].

Most of the previous evolutionary neural networks, however, show little structural constraints. Some networks assume total connectivity among all nodes. Others assume a hierarchical, multi-layered structure where each node in a layer is connected to all nodes in neighboring layers. However, there is a large body of neuropsychological evidence showing that the human information processing system consists of modules, which are subdivisions in identifiable parts, each with its own purpose or function.

The question may then be raised on how to design neural networks with various modules. There have also been extensive works to design efficient architectures from an engineering point of view which has produced some success in several problems [1,10]. However, we know of no comprehensive analytical solution to the problem of relating architecture to function.

The architecture of the brain has resulted from a long evolutionary process in which a large set of specialized subsystems emerged interactively carrying out the tasks necessary for survival and reproduction, but it appears that learning a large-scale task from scratch in such networks may take a very long time. From this perspective, we have developed a modular neural network (MNN) [2]. Each module has the ability to autonomously categorize input activation patterns into discrete categories, and representations are distributed over modules rather than over individual nodes. Among the general principles are modularity, locality, self-induced noise, and self-induced learning. We attempt to analyze the cooperative behaviors of the proposed model for showing the usefulness.

2.1 Method

In order to evolve neural processing systems consisting of modules, a population of the individuals having various connectivity and size are maintained. Each individual is a modular neural network represented by a tree-structured chromosome. The module derives its basic internal structure from the neocortical minicolumn: Inhibitory connections are mostly short range, while long-range connections are mostly excitatory.

A single node is either excitatory or inhibitory, but not both (Dale's law). Processes in nodes only rely on information that is locally available through synapses or through locally dispersed neurotransmitters (principle of locality). The internal connections in a module is fixed and the weights of all intramodular connections are non-modifiable during learning process (see Fig. 1).

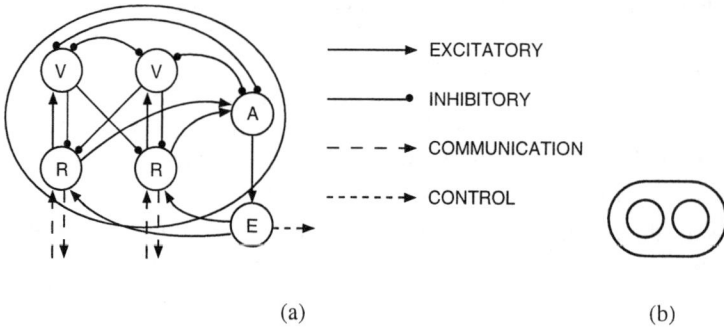

Fig. 1. (a) Schematic diagram of the internal structure of a module. (b) Simplified representation of the module (a).

R-nodes have modifiable connections to the R-nodes in other modules and fixed connections to nodes in the same module. They are the only nodes in a module that may send or receive excitation to or from nodes in other modules. V-nodes only have inhibitory outgoing connections. A V-node inhibits all other nodes in a module, and in particular inhibits each other so strongly that competition arises among the V-nodes. The R and V-nodes form matched pairs as every V-node receives excitatory input from only one R-node. A-nodes are excited by all R-nodes in a module and inhibited by all V-nodes. The activation of the A-node is a positive function of the amount of competition in a module, and E-node activation is a measure of the level of competition going on in a module.

The process goes with the resolution of a winner-take-all competition between all R-nodes activated by input. In the first presentation of a pattern to a module, all R-nodes are activated equally, which results in a state of maximal competition. It is resolved by the inhibitory V-nodes and a state-dependent noise mechanism. The noise is proportional to the amount of competition, as measured through the number of active R-nodes by the A-node and E-node.

The functional characteristics of a module does not depend on the size of the module, i.e. the number of R and V-nodes. This number of the R-V pairs can be determined by the evolutionary process. The most important feature of a module is to autonomously categorize input activation patterns into discrete categories, which is facilitated as the association of an input pattern with a unique R-node.

As a mechanism for generating change and integrating the changes into the whole system, genetic programming produces the networks that have a variety of intermodule connectivity and module sizes, and Hebb learning rule applies to excitatory intermodular connections, whereas all intramodular connections are assumed to remain fixed. Each module serves as a higher order unit with a distinct structure and function.

An initial population is generated with tree-structured chromosomes each of which is developed to a modular neural network of different connectivity and module sizes. After the developmental process, Hebb rule is applied to each network to train the intermodular connections with a training set, yielding a fitness value. If an acceptable solution is found, the algorithm stops. Otherwise, the next population is created from the current one with selection and genetic operations: The selection step accepts the better individuals into the mating pool, where genetic manipulation generates the new population with crossover and mutation operators. We have used an evolutionary algorithm similar to genetic programming with rank-based selection scheme [2].

2.2 Results

For experiments, we have used the handwritten digit database of Concordia University of Canada, which consists of 6000 unconstrained digits originally collected from dead letter envelopes by the U.S. Postal Services at different locations in the U.S. Among the data, 300 digits were used for training, and another ten sets of 300 digits for testing.

The size of a pattern was normalized by fitting a coarse, 10×10 grid over each digit. The proportion of blackness in each square of the grid provided 100 continuous activation values for each pattern. Network architectures generated by the evolutionary mechanism were trained with 300 patterns in two rounds of subsequent presentations. A single presentation lasted for 60 cycles (i.e., iterative updates of all activations and learning weights). A fitness value was assigned to a solution by testing the performance of a trained network with the 300 training digits, and the generalization performance was tested on a set of untrained 300 digits.

Initial population consisted of 50 neural networks having random connections. Each network contains one input module of size 100, one output module of size 10, and different number of hidden modules. The evolution parameters used in this experiment are as follows: crossover is 0.5, mutation is 0.02, and insertion and deletion are 0.001, respectively.

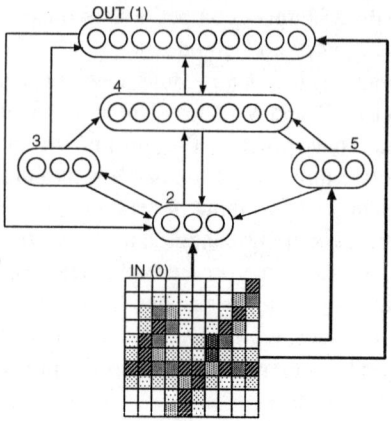

Fig. 2. The best modular neural networks evolved.

Fig. 2 shows the architecture producing the best result. This contains four hidden modules of size 3, 3, 8 and 3, implementing different subsystems that cooperatively process input at different resolutions. The direct connection from the input module to the output module forms the most fine-grained processing stream. It is supplemented by a sophisticated modular structure in which two modules are globally connected with the input. A sort of hierarchical structure (IN → 2 → 4 → OUT) with feedback connections has emerged, and two coarser processing streams as well as local feedback projections support the main processing stream.

In the test of generalization capability, for the patterns that are similar to the trained, the network produced the direct activation through a specific pathway: one of the five basic pathways in Fig. 2 (IN → OUT, IN → 2 → 3 → OUT, IN → 2 → 4 → OUT, IN → 5 → 4 → OUT, IN → 2 → 3 → 4 → OUT). On the contrary, the network oscillated among several pathways to make a consensus for strange patterns. The basic processing pathways in this case complemented each other to result in an improved overall categorization. Furthermore, the recurrent connections utilized bottom-up and top-down information that interactively influences categorization at both directions. The oscillation is stopped when the whole network stabilizes as only one R-node at the output module remains to activate.

It is difficult to fully analyze the neural behaviors because they concern with the oscillatory activation dynamics. To make the analysis simpler, we have presented a sample of the class 1 to the final network and obtained a series of snapshots of the internal activations. This system has turned out to produce the correct result with respect to the input as shown in Table 2. In this network, the coupled oscillatory circuit between Module 4 and OUT module resolves the competition and induces the correct classification.

In order to illustrate the effectiveness of the model proposed, a comparison with modular neural network without evolutionary algorithm and traditional neural network has been conducted. The overall recognition rates for the proposed method

Table 2. A series of snapshots of the internal activations. The numbers in each column represent the activated nodes and * means that there is no node activated in the module.

Step	OUT	Module 2	Module 3	Module 4	Module 5
1	0123456789	012	*	*	*
2	0123456789	012	012	01234567	012
3	0123456789	012	012	0234567	012
4	0123456789	0	012	023456	012
5	0123456789	0	012	03456	012
6	012345789	0	012	3456	012
7	12345789	0	012	456	012
8	178	0	012	5	012
9	1	0	012	5	012

are higher than those for the modular neural network without evolutionary algorithm and the multilayer perceptron. Furthermore, with the paired t-test, for all the cases "no-improvement" hypothesis is rejected at a 0.5% level of significance. This is a strong evidence that the method based on A-life technology is superior to the alternative methods.

3 Controlling a Mobile Robot

There have been several attempts to develop an artificial brain with engineering techniques. Among them CAM-Brain develops neural networks based on cellular automata by evolution. In particular, due to their features Cellular Automata (CA) can be evolved very quickly on parallel hardware such as CAM-8 at MIT, or CBM at ATR [6].

Evolutionary Engineering (EE) is an approach to evolve neural network modules with particular functions to develop artificial brain. It has been extensively exploited to apply each neural network module to a specific problem. We have attempted to evolve a module of CAM-Brain for the problem of robot control, especially Khepera. A simulation indicates that an appropriate neural architecture emerges to make Khepera simulator to navigate the given environment without bumping against walls and obstacles. This section presents the power of the model based on A-life technology by analyzing the robot behaviors and corresponding neural networks evolved.

3.1 Method

CAM-Brain is an evolved neural network based on CA. This section describes one of the CAM-Brain models, CoDi model, and the process of developing neural networks and signaling among neurons. This process consists of two phases. One is growth phase that makes the structure of neural network. The other is signaling phase that sends and receives signals among neurons.

CA in CoDi Cellular automata are composed of state, neighborhood and program [9]. The state of each cell in CA space of CAM-Brain has one of four states: Neuron, Axon, Dendrite and Blank. The cell having the state of blank means empty space. Blank cells do not participate in any cell interaction during the signaling of the neural networks. Neuron cells collect neural signals from the surrounding dendrite cells. If the sum of collected signals is greater than threshold, then neuron cells send them to the surrounding axon cells. Those cells distribute signals originating from neuron cells. Dendrite cells collect signal and eventually pass them to neuron cells [7].

Neighborhood cells are surrounding cells (North, South, West and East) in 2-D CA space (Top and Bottom added in 3-D CA space). A state of each cell and program or rule deciding state of each cell with the state of neighbors are represented at chromosome. One chromosome has the same number segments with cells and can make one neural network. One segment corresponds to each cell and can change blank cell to neuron cell. Also, it decides the directions of sending received signals to neighborhood cells.

Growth phase The growth phase organizes neural structure and makes the signal trails among neurons. Neurons are seeded in CA-space according to chromosome. The neural network structure grows by sending two kinds of growth signals (axon and dendrite) to neighborhood cells. Neuron sends axon growth signal to opposite two directions decided by chromosome and dendrite growth signal to the remaining four directions.

The state of neighborhood cells becomes axon or dendrite depending on the kind of received growth signals. Next, they propagate received growth signal to neighborhood cell. Each axon and dendrite cell belongs to exactly one neuron cell. Once the cell type is decided, it changes no more. Neural network is constructed and encoded to chromosome, and then it is evolved by genetic algorithm [7].

Fig. 3 shows the role of a chromosome in growth phase. A segment of the chromosome corresponding to each cell has information of direction of sending signal when the cell receives signals. In this figure, black arrows represent directions of sending signal, while white arrows represent directions of not sending signal.

Initially, all cells are set to blank type, and some cells are decided as neuron-seed cells by chromosome. Neuron cells send two kinds of growth signals to their neighbors, either "grow a dendrite" or "grow an axon" by the map of Fig. 3. The blank neighborhood cells turn into either an axon cell or a dendrite cell until it is completed to evolve one neural network module.

Signaling phase The signaling phase transmits the signal from input to output cells continuously. The trails of signaling are performed with evolved structure at the growth phase. Each cell plays a different role according to the type of cells. If the cell type is neuron, it gets the signal from neighborhood dendrite cells and gives the signal to neighborhood axon cells when the sum of signals is greater than threshold. If the cell type is dendrite, it collects data from the faced cell and eventually passes

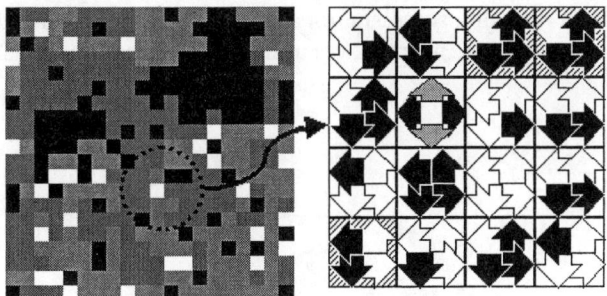

Fig. 3. The directions of sending received signals to neighborhood cells decided by a chromosome.

it to the neuron body. If the cell type is axon, it distributes data originating from the neuron body.

The position of input and output cells in CA-space is decided in advance. At first, if input cells produce the signal, it is sent the signal to the faced axon cells, which distribute that signal. Then, neighborhood dendrite cells belonged to other neurons collect this signal and send it to the connected neurons. The neurons that have received the signal from dendrite cells send it to axon cells.

Finally, dendrite cells of output neuron receive this signal, and send it to the output neurons. Output value can be obtained from output neurons. During signaling phase, the fitness is evaluated. According to given task, various methods can be used. This fitness is used for evolving the chromosome.

Applying CAM-Brain to Robot Control Applying CAM-Brain to a mobile robot control [5,18] requires the following process. Neural network structure is made in growth phase. In signaling phase, sensor values from Khepera simulator [15,18] are used as inputs to CAM-Brain. CAM-Brain transmits signal from input to output cells. As the output values of CAM-Brain are inputed to Khepera simulator, the robot moves. When the robot bumps against the obstacle or reaches the goal, its fitness is computed. Chromosomes are reproduced in proportion to this evaluation.

There are a couple of problems in applying the model to controlling robot. Because CAM-Brain cannot perfectly utilize activation values of robot sensors the activated range of input neuron is made differently according to the value. In addition, because delay time is needed until CAM-Brain makes output value, we execute dummy signaling phase for some duration until the signals started from input cells arrive at output cells. It makes the robot to react appropriately in several situations [3].

3.2 Results

The robot controller evolves in $5 \times 5 \times 5$ CA space to facilitate an easy analysis. After the 21st generation, individuals that have fitness value one keep on appearing. Fig.

4 shows the trajectory of a successful robot. This is less smooth than that obtained in our previous work, but this robot controller has smaller size, which makes the analysis easier.

Fig. 4. The trajectory of a robot.

Fig. 5 shows an architecture of the neural network evolved. Dotted arrow represents inhibitory connection, and lined arrow represents excitatory connection. This has been obtained by tracing the activation values of each neuron. The number of neurons is 12, but neurons 8, 11 and 12 are not functional because these neurons are not in the path of input to output neurons. Neurons 2 and 10 are output neurons to produce the velocity of left and right wheels, and neurons 3, 4, 5 and 6 are input neurons. Neuron 3 is for front sensor of the robot, neurons 5 and 6 are for left sensor of the robot, and neuron 4 is for right sensor of the robot.

The architecture of controller has direct connections from input to output neurons. These connections play a role for turning left and right: If neuron 5 has high activation value (which means no obstacle at the left side of the robot because we scale sensor value of the robot inversely), neuron 10 (right wheel) produces positive signal (because neuron 5 fires excitatory signal to neuron 10). If neuron 4 has low activation value (obstacle at the right side of the robot), neuron 2 (left wheel) produces 0 (because neuron 5 cannot fire neuron 2). By the way, the velocity of each wheel is decided according to the value of the output neurons. It becomes 5 (if the output value is positive), -5 (if negative) or 0 (otherwise). The values of the output neuron make the robot to turn left. Similarly, the robot can turn right with the addition of neuron 7.

4 Developing Adaptive Agents

WWW has a large distributed collection of documents, which can be added, deleted or modified dynamically. Moreover, the document style is various. It takes much time and effort for users to search the Web in this environment. Due to these reasons, several search agents have been developed and investigated.

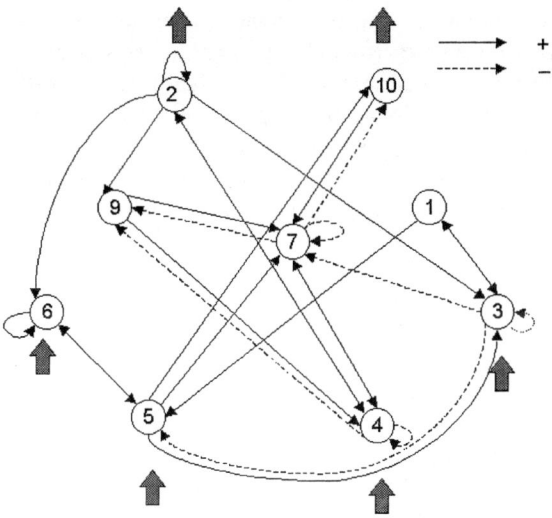

Fig. 5. Schematic diagram of evolved neural network.

Conventional search agents retrieving information on the Web are mainly devised for static and non-distributed environments. With these agents, end user gives queries to the server that maintains index files to get relevant document lists. User's requests are processed through the use of index files, which are made and updated by off-line robot agents that collect and analyze the documents. Because of its fast response time these search agents are prevalent now, but they have several limitations. First, they cannot cope with dynamic change of documents. Second, they can abstract away important data by incorrect indexing, and missing relations between documents. Third, they cannot reflect on the user's preferences or habits. To overcome these limitations, a new method is required to replace the index-based robot agents.

Our A-life agent is very similar to Infospider that was originally proposed by Menczer [17]. It has a population of on-line agents that search documents by deciding own actions locally. Each agent in a population can reproduce the offspring, or disappear, according to the relevancy of the documents retrieved by the agent. The population of agents converges to optimal states through evolution. However, if we incorporate user's preference, we can provide more accurate information quickly and personalize the agents for user. By updating user profile at each queries, we can reflect user's preferences. A-life agents maintain their competence by adapting to the user's preference that may change over time.

4.1 Method

The authors of web document tend to classify the documents according to subjects and connect them in related topic. This tendency results in semantic topology, which defines the correlation of documents. If some documents are relevant to users, the links in current document are also highly relevant to users. Also, the links close to meaningful keywords are more probable than other links. The artificial life agents can reduce the search space by using this property. It has a population of multiple retrieval agents. The energy of each agent in population is increased or decreased by relevance of the document retrieved by agent itself. This method uses genetic algorithm based on local selection. The algorithm is shown in Table 3.

Table 3. Overall algorithm

```
Initialize agents;
Obtain queries from user;
while (there is an alive agent) {
    Get document D_a pointed by current agent;
    Pick an agent a randomly;
    Select a link and fetch selected document D_{a'};
    Compute the relevancy of document D_{a'};
    Update energy (E_a) according to the document relevancy;
    if (E_a > ε)
        Set parent and offspring's genotype appropriately;
        Mutate offspring's genotype;
    else if (E_a < 0)
        Kill agent a;
}
Update user profile;
```

Initialization Each agent's starting point is initialized by user profile. The genotype is composed of confidence and energy. Confidence is the degree to which an agent trusts the descriptions that a document contains about its outgoing links, and energy represents agent's relevancy to the given queries. The energy is initialized to a constant threshold $\varepsilon/2$, and confidence is chosen randomly.

Link Selection Relevancy of each link in current document is estimated by computing physical distances with keywords matched to user queries. This estimation is based on the assumption that any links close to keywords are more relevant to user's interest than other links mostly. For each link l in a document, the relevancy is calculated as follows.

$$\lambda_l = \sum_{k \in tokens} \frac{match(k, Q)}{distance(k, l)} \qquad (1)$$

where k is a token in document D_a, Q is the queries given, and $distance(k, l)$ is the number of links separating k and l in the document. $match(k, Q)$ gets to be 1 if k is in Q. Otherwise, it becomes 0. To select a link to follow, we use a stochastic selector to pick a link with probability distribution which is scaled up and normalized by agent's confidence.

Confidence evolves by selection, reproduction, and mutation. Different confidence values can implement search strategies as different as best-first, random walk, or any middle course. With this distribution, an agent selects a link to follow.

Updating Energy After the agent fetches the document connected by selected link, it estimates the relevancy of a document that is proportional to the hit rate of the number of keywords to whole tokens in the document. Relevancy of the document is represented by

$$r(D_a) = \sqrt[4]{\frac{\sum match(k, Q)}{number(k, D_a)}} \tag{2}$$

where $number(k, D_a)$ is the number of keywords in D_a. An agent's energy is updated according to the relevancy of document. The use of network resource means the loss of the energy. If the document is already visited, the increase of energy is not expected.

$$E_A = E_A - expense + \begin{cases} r(D_a) & \text{if } D_a \text{ is new,} \\ 0 & \text{otherwise.} \end{cases} \tag{3}$$

where $r(D_a)$ is the relevancy of document, and $expense$ is the loss of energy.

Reproduction Each agent can reproduce offspring or be killed by comparing the agent's energy with constant threshold ε. If the agent's energy exceeds the threshold, it reproduces offspring. The offspring's energy is fed by splitting the parent's energy, and the offspring is mutated to provide evolution with the necessary variation. The bound of confidence is determined by the current document relevancy. This mechanism can cause the population of agents to be biased to the regions where the relevant documents exist.

Updating User Profile User profile should respond to user's interests. Since the agents learn about the user's interests by getting user's queries and feedback, it is important to update user profile after searching. The updated user profile is composed of relevant document URLs and other interesting subjects. With this property, user can personalize the agents as he gives queries repeatedly.

4.2 Results

In order to provide a fair and consistent evaluation of the system's performance, we have restricted search space to the local machine, instead of real Web. We have collected a number of HTML pages with various topics, classified the pages according to the subjects, and put them in different directories. The initial user profile is composed of the top directories of local machine. The initial population size of agents depends on the number of URLs in user profile. We have compared the A-life agents with Breadth First Search (BFS) and Random Search agents. BFS searches every documents exhaustively, while the A-life agents can search documents selectively.

The initial population is composed of 10 agents. The population size has no limitation in run time. The constant threshold ε is set to 0.4. Agent whose energy is greater than ε can reproduce offspring. Initial agent's energy is set to $\varepsilon/2$. The agent uses the network resource, which means loss of energy. This loss of energy is called *expense*, and set to 0.1. The process is influenced by the *expense* value. If we increase the *expense* value, the agents have less chance to search further. Irrelevant agents may disappear quickly, and there is some possibility that even some relevant agents can disappear without searching the regions sufficiently. We select the *expense* value with trial and errors.

The most important property of the A-life agents can discard the useless agents irrelevant to the user's preference effectively. Fig. 6 shows the hit rate of relevant documents. In the beginning, the performance is not better than other search methods, but the performance of the proposed agents is improved rapidly. This result implies that each agent can cut irrelevant paths of documents effectively. The action of each agent goes toward relevant document paths gradually. By using this property, A-life agents can reduce the access of irrelevant documents.

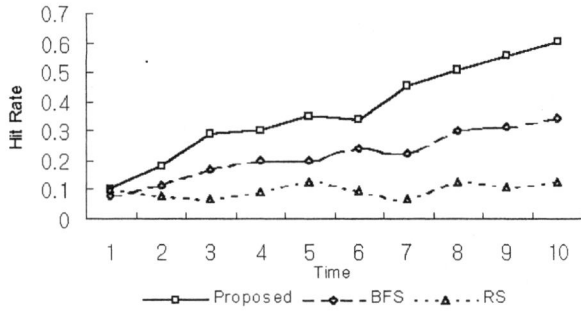

Fig. 6. Hit rate.

We have tested the performance improvement in case the user gives queries in the same category. For each query, the user profile is updated according to document relevancy, and the agents reflect the user's preference adaptively. As the user gives the queries in the same category, the proposed agents improve the response time according to the queries. Fig. 7 shows the results on the two tasks. In task 1, a sequence of queries is provided as computer, artificial intelligence, neural network, agent, evolution, user feedback, retrieval, and search. In task 2, the sequence is computer, DSSSL, SGML, grove, property set, repository, and database. Initial response time is not good, but as the queries are given repeatedly, we can see the improvement of the response time for the two tasks.

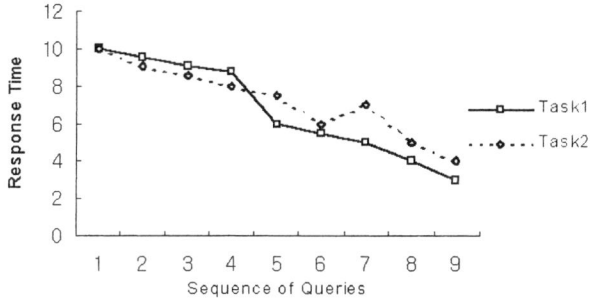

Fig. 7. Response time for the queries in the same category.

5 Retrieving Media Databases

As technology in computer hardware and software advances, efficient information retrieval from multimedia databases gets highly demanded. Recently, content based image retrieval method has been studied as a core technique. It can be applied to digital library, medical management systems, home shopping, and so on. Several working systems have already been developed: QBIC system of IBM [19], QVE of Hirata and Kato [11], chatbot of UC Berkeley [20], photobook of MIT, and Image Surfer of Interpix software.

In our previous work, we have proposed an image retrieval method based on human intuition and emotion [16]. It utilizes the interactive genetic algorithm for the retrieval of images based on user's preference and emotion. In this method, wavelet transform is used to extract features from images. This paper attempts to analyze the effectiveness of the wavelet features in the content-based image retrieval with interactive genetic algorithm.

5.1 Method

Discrete Wavelet Transform The wavelet transform is a mathematical function that decomposes data into various kinds of frequency components. It is better than

Fourier transform in the case of analyzing physical situation that contains sharp area, and it is applied to many image processing areas, such as edge preserving smoothing, de-noising, compression, and so on. These processes are based on the fact that by smart selection and modification of the wavelet coefficients the content of the image is preserved. We use two-dimensional wavelet transform to extract features from images and use the result of the transform for a search. Wavelet transform allows for very good image approximation with just a few coefficients. This property has been exploited for lossy image compression. In particular, we use Haar wavelet transform because it can be easily implemented and is fast. The following equations are basis functions of Haar wavelet transform.

$$average = (a+b)/\sqrt{2}$$
$$difference = (a-b)/\sqrt{2}$$
$$a = (average + difference)/\sqrt{2}$$
$$b = (average - difference)/\sqrt{2}$$

(4)

A standard two-dimensional Haar wavelet decomposition of an image is implemented by pyramid algorithm [4]. This algorithm operates on a finite set of N input data. These data are passed through two convolution function, each of which creates output stream that is half the length of the original input. One half of the output is produced by the low pass filter, and the other half is produced by the high pass filter. Low pass output contains information content and is used as an input of filtering in the next step. High pass output contains the difference between the true input and reconstructed input. In general, high-order wavelets tend to put more information into the low-pass output. Wavelet based compression is caused by making the high-pass output to be nearly zero. Pyramid algorithm involves one-dimensional decomposition on each row of the image followed by one-dimensional decomposition on the column of the result.

Interactive Genetic Algorithm It is a genetic algorithm that adopts user's choice as fitness, when fitness function cannot be explicitly defined. This allows to the development of systems based on human intuition or emotion. This paper obtains fitness value for an image from user, and it is used to select better images for the next generation.

The strategy of selection is governed by expected frequency of each individual, and one point horizontal and vertical crossover is used [16]. Mutation is not adopted. Here, a chromosome is represented by an array that consists of index of wavelet coefficients.

Wavelet coefficients are obtained by decomposing an image using wavelet transform. The $r \times r$ matrix, T, through the above procedure, has the average color of the image in entry $T[0,0]$ and wavelet coefficients in the other entries of T. We can reconstruct original image without loss using this information, but because there is no need to maintain information to search, we extract the largest 50 coefficients in

RGB channels and use for constructing a chromosome in 3×50 array. Jacobs' work shows that storing 40 to 50 largest-magnitude coefficients in each color works best and truncating the coefficients appears to improve the discrimination power of the metric [12]. We only store the sign information of coefficient values. The entire system is constructed as shown in Fig. 8. In preprocessing step, we perform wavelet transform for every image in the database and store the overall average color, and the indices and signs of the m magnitude wavelet coefficients in search table.

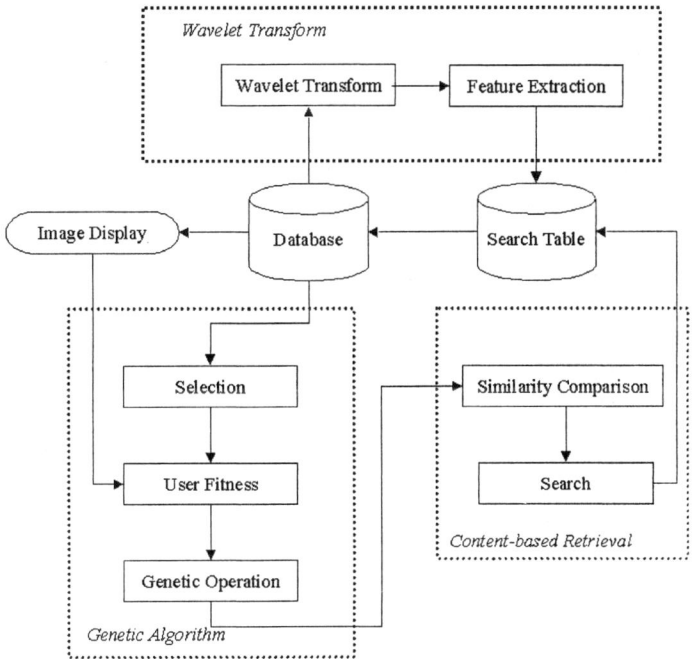

Fig. 8. System structure.

To search images, we use interactive genetic algorithm. A system displays twelve images, obtains the fitness value from user, and selects candidates based on the fitness. A genetic operation, vertical or horizontal crossover, is applied to the selected candidates. To find the best twelve images, the stored image information is evaluated by each criterion. Twelve images of the higher magnitude value are provided as a result of the search. At this time, the similarity between potential target image and candidate image is calculated by the following equation.

$$\|Q,T\| = w_{0,0}|Q[0,0] - T[0,0]| + \sum_{i,j}|Q[i,j] - T[i,j]| \tag{5}$$

$Q[0,0]$: overall average color of a candidate image
$T[0,0]$: overall average color of a target image
$Q[i,j]$: wavelet coefficients of a candidate image
$T[i,j]$: wavelet coefficients of a target image $(i,j \neq 0)$

$Q[i,j]$ and $T[i,j]$ represent single color channels of wavelet decomposition of the query and target images, and $Q[0,0]$ and $T[0,0]$ mean overall average intensities of those color channels. The system repeats this process to search new candidates until user finds the image that he wants.

5.2 Results

The system is written in Microsoft Visual C++, and runs on Pentium PC. A searching table is constructed by a batch job over 256 × 256 JPEG images. The size of the image database is 2000.

Initial population consists of randomly selected images. A user gives the fitness of images based on the similarity of what he wants, and the system creates and displays the next generation using the genetic algorithm.

This procedure is repeated until the user obtains an image that is most similar to what he has in mind. In case that the results of the next generation are worse, the system allows to go back to the previous generation. Moreover, the user can increase or decrease the effectiveness of the color to search images.

To evaluate the performance of this system, we requested 10 graduate students to search cheerful, gloomy, and cool images and answer the questions: How similar the result image is to what you want and how long it has taken to find it. We tested the satisfiability with the images and search time over cheerful, gloomy, and cool emotions. The test result is shown in Table 4.

Table 4. Response statistics from 10 users for the three queries.

	Cheerful	Gloomy	Cool
Very good	2	1	3
Good	3	0	4
Normal	2	3	0
Bad	0	2	2
Very bad	3	4	1

In this test, 70 percents of the subjects are satisfied with the tasks for finding cheerful and cool images, but in the case of gloomy image most of the subjects are unsatisfied. A major reason might be in the small size of database, which does not contain as many images as user wants. The retrieval of the cool image shows that a content-based image retrieval method by human intuition or emotion can have good performance if large database is given. When we limit the number of trials to 5, a half of subjects can find a satisfied image within the limitation, and the remainings take five trials to find it.

Next, we have tested whether wavelet coefficients are appropriate for genetic representation or not. To do this, we have grouped images into two classes: one is gloomy and the other is not. For each image, we calculate similarity between images contained in the same group (within-class similarity) and the similarity between images contained in different groups (inter-class similarity), respectively. The similarity is evaluated through the same method that is used to compare the target image with the candidate image.

The higher the value is, the closer the images are. High similarity would mean that the two images are in near place on the wavelet space. Most of the gloomy images produce higher similarity value in the same class. The averages of within-class similarity and inter-class similarity are 1154.692 and 1114.939, respectively. It is shown in Table 5. Though the mean of within-class similarity is larger than that of inter-class, we want to determine whether this difference is statistically significant or not.

We have performed a paired t-test. This test is useful for testing whether the difference of mean values is significant or not. We hypothesize that the mean values of within-class similarity and inter-class similarity are not different. The result of the t-test is 5.508 and our hypothesis is rejected at 99.995% level of significance. This means that images in the same group are closer on the wavelet space and we can assert that wavelet coefficients represent the mood of images well.

Table 5. Within-class and inter-class similarities for gloomy images.

	Within-class	Inter-class	Difference
Mean	1154.692	1114.939	39.752
Variance	27368.739	24196.524	3229.317
SD	165.434	155.552	56.827
t-test			5.508

6 Concluding Remarks

This chapter introduces the key concept of A-life technology and shows the potential with the applications such as categorizing visual patterns, controlling a mobile robot, developing adaptive agents on the WWW, and retrieving media database based on human preference. While artificial intelligence uses the technology of computation as a model of intelligence, A-life is attempting to develop a new computational paradigm based on biological processes. This technology surely opens a new era for intelligent systems and information sciences.

Acknowledgements

The author thanks Prof. N. Kasabov for his comments on the earlier drafts of this article.

References

1. S.-B. Cho, J.H. Kim (1995) Combining multiple neural networks by fuzzy integral for robust classification, IEEE Trans. Systems, Man, and Cybernetics, 25(2): 380–384
2. S.-B. Cho, K. Shimohara (1996) Modular neural networks evolved by genetic programming, Proc. IEEE Conf. Evolutionary Computation, Nagoya, 681–684
3. S.-B. Cho, G.-B. Song, J.-H. Lee, S.-I. Lee (1998) Evolving CAM-Brain to control a mobile robot, Proc. Int. Conf. on Artificial Life and Robotics, 271–274
4. T. Edwards (1991) Discrete wavelet transform: Theory and implementation, Technical Report, Stanford University
5. D. Floreano, F. Mondada (1996) Evolution of homing navigation in a real mobile robot, IEEE Trans. Systems, Man, and Cybernetics, 26(3):396–407
6. H. de Garis (1996) CAM-Brain : ATR's billion neuron artificial brain project : A three year progress report, Proc. Int. Conf. on Evolutionary Computation, 886–891
7. F. Gers, H. de Garis, M. Korkin (1997) CoDi-1Bit : A simplified cellular automata based neuron model, Proc. Artificial Evolution Conf.
8. D.E. Goldberg (1989) Genetic Algorithms in Search, Optimization, and Machine Learning, Addison-Wesley Publishing Company Inc.
9. T. Toffoli, N. Margolus (1987) Cellular Automata Machines, The MIT Press
10. J.B. Hampshire II, A. Waibel (1992) The meta-pi network: Building distributed knowledge representations for robust multisource pattern recognition, IEEE Trans. Pattern Analysis and Machine Intelligence, 14(7): 751–769
11. K. Hirata, T. Kato (1992) Query by visual example: Content based image retrieval, Advances in Database Technology, 56–61
12. C.E. Jacobs, A. Findkelstein, D.H. Salesin (1995) Fast multiresolution image querying, Proc. of SIGGRAPH 95, 277–286
13. B. Kosko (1992) Neural Networks and Fuzzy Systems, Prentice-Hall International, Inc.
14. C.G. Langton (1989) Artificial life, Artificial Life, Addison-Wesley Publishing Company, 1–48
15. S.-I. Lee, S.-B. Cho (1996) Evolutionary learning of fuzzy controller for a mobile robot, Proc. Int. Conf. Soft Computing, 745–748
16. J.-Y. Lee, S.-B. Cho (1998) Interactive genetic algorithm for content-based image retrieval, Proc. Asian Fuzzy Systems Symposium, 470–484
17. F. Menczer (1997) ARACHNID: Adaptive retrieval agents choosing heuristic neighborhoods for information discovery, Proc. 14th Int. Conf. on Machine Learning
18. O. Michel (1995) Khepera Simulator Version 1.0 User Manual
19. W. Niblack, R. Barber, W. Equitz, M. Flickner, E. Glasman, D.Pekovic, P. Yanker, C. Faloutsos, G. Taubin (1993) The QBIC project : Querying images by content using color, texture, and shape, Storage and Retrieval for Image and Video Databases, 173–187
20. V.E. Ogel, M. Stonebraker (1995) Chatbot: Retrieval from a relational database of images, IEEE Computer, 28, 109-112
21. J. Vaario (1993) Artificial Life Primer, Technical Report TR-H-033, ATR Human Information Processing Research Labs

22. X. Yao (1993) Evolutionary artificial neural networks, Int. Journal of Neural Systems, 4(3): 203–222

Chapter 3. Evolving ANN Controllers for Smart Mobile Robots

R. J. Duro, J. Santos and J. A. Becerra

Grupo de Sistemas Autónomos, Universidade da Coruña, Spain
richard@udc.es, santos@udc.es, ronin@cdf.udc.es

Abstract. *In this work we present an overview of the application of evolution for obtaining autonomous robot controllers. It concentrates on controllers for behavior based robots implemented through Artificial Neural Networks. Specific approaches for taking into account temporal relationships are presented as well as a methodology for the progressive implementation of controllers comprising multiple behaviors. In addition we will consider the problem of transferring simulation results to real robots and the conditions that must be met for this process to be effective. Some examples of the application of these techniques to simple problems are included.*

Keywords. *Behavior based robotics, evolutionary algorithms, artificial neural networks, temporal processing, hierarchical behavior controllers*

1 Introduction

If you want a robot to operate in an unstructured, partially unknown dynamic environment, it cannot be fully pre-programmed. It must be provided with some type of cognitive architecture that allows it to establish relationships between its sensorial inputs and its actions on the world. In other words, an architecture which allows it to generate its sensori-motor mappings autonomously, so that it can survive and achieve its goals.

Many authors have resorted to traditional symbolic AI as a first attempt to achieve these goals. The main assumption made by traditional AI is that intelligence is intrinsically a computational phenomenon and can be studied in disembodied systems. Architectures following this traditional approach are usually based on decomposing the processes the robot has to perform into independent tasks and arranging them in some type of hierarchical fashion. These tasks, such as planning, sensing, etc., are carried out by modules developed for the particular task, generally without taking into account the rest of the modules participating in the architecture or even within what architecture they are going to be included.

This divide and conquer approach to the problem produced some interesting results for individual modules, such as vision or motion detectors, but in general proved to be very brittle when the robot had to operate in dynamic environments. The reason for this is that to apply divide and conquer, one is ignoring the interactions between modules [36]. Some authors, such as Brooks [8][9][10] with his behavior based robotics, Maes [52] with her competence agents or Mataric [54] with her bottom-up approach without symbolic representations or manipulable data structures, began to lay down the foundations of what Sharkey [74] called "a new wave in robotics". This wave, is situated between the high level planning robotics of deliberative artificial intelligence and low level control theory. It drew inspiration from natural phenomena and the emphasis was on behavior and fast reaction and not on knowledge and planning. As Brooks [7] stated "planning is just a way of avoiding figuring out what to do next".

The main idea behind these proposals was that representations are not necessary for intelligence - except in a very limited sense [9] - as the world is a good enough model of itself. Moreover, there must be a direct interaction between the robot and the world and the robot must be taken as a whole and constructed from the bottom up, and not as in traditional AI as a sum of independent knowledge modules. The focus of these reactive systems was on behaviors, which can be taken as "stimulus response pairs modulated by attention and determined by intention" [1]. Attention prioritizes tasks and provides some organization in the use of sensorial resources, while intention determines the behaviors to be activated, depending on the objectives or tasks the robot must achieve.

Some precursors of these ideas were to be found in Walter´s [76] turtles and in Braitenberg´s vehicles [6]. They showed that what seemed to be complicated intelligent behavior could be derived from direct physical connections between sensors and actuators.

For the systems to produce truly robust and useful behavior, Brooks [9] and Arkin [1] enumerated some aspects that should be taken into account such as:

- *Situatedness* - the robot is immersed in a real world and does not operate with abstract representations of it, but with the world itself.
- *Embodiment* - a robot presents some physical characteristics that have a bearing on its actions in the world and which must be taken into account.
- *Emergence* - intelligence arises from the interaction with the environment, it is not a property of the agent or of the environment alone.
- *Grounding* - the information the robot works with comes directly from the world and not through symbols as in traditional AI.
- *Ecological dynamics* - the world is continuously changing and this variability must be addressed.
- *Scalability* - the systems must scale up to larger problems.

To summarise, in the behavior based approach to robotics, behaviors are the building blocks of the robotic cognitive system. Explicit abstract representations of knowledge are avoided. Animal behavior is taken as a model and the systems are inherently modular from a software point of view.

In this approach to robotics, two basic decisions must be made. On one hand one has to define the paradigm to be used in the implemetation of the individual controllers or behavior modules and, on the other, how these are going to be coordinated.

It is not easy to generate mechanisms for relating sensors and actuators that lead to a level of coordination capable of sustaining adaptive behavior in a real environment. Thus, when more than a small number of behaviors are required, the process of connecting the different modules becomes very difficult to perform due to the fact that the interactions are very hard to follow. As Cliff et al. [14] point out, the complexity in a design scales faster than the number of parts or modules that make it up. In fact, it scales with the number of possible interactions among them. In addition, the hand-designed behaviors are not necessarily the best, or, in some cases, adequate for the task. A human designer is interpreting what the robot perceives and how it should act, and thus often does not take into account elements of the robot, such as its body or other resources that could make the task easier and behavior more efficient.

Another problem is that if so much emphasis is placed on non-deliberative elements, it may actually hinder the scalability of the systems. When one of these cognitive architectures is operating in an environment in which there is a certain amount of regularity, one may want to make use of it, epecially in terms of prediction. We believe that most unstructured dynamic environments that any robot is going to be faced with will present regularities in some of their aspects, otherwise no operation would be possible. Not even a human can operate successfully in a completely random environment. Therefore, we believe that there is always some room for deliberative processes, although they may have to be introduced in the architecture in more flexible ways than has been traditionally done.

As a consequence of all this, a new procedure is required to obtain behaviors that make better use of the body and resources of the robot, along with arbitrating mechanisms that are too complex to be designed by man. One must shy away from schemes that rely on the intuition or competence of the designers seeking solutions that permit an automatic design of autonomous systems that can survive in the environments they encounter with as little human intervention as possible.

As Kodjabachian and Mayer indicate in [43], nature has already invented and used three automatic design procedures: evolution, development and learning. The first two permit the creation of the organism and its internal organization, and the third is employed for the purpose of short term adaptation.

In the late Eighties and early Nineties some authors, such as Harvey et al. [37], Cliff et al. [15], and Beer and Gallagher [4] proposed artificial evolution as a means to automate the design procedure of these types of systems. Many authors have taken up this issue and have developed different evolutionary mechanisms and strategies to obtain robotic controllers for the autonomous operation of robots in structured and unstructured environments. Using this approach, the focus is changed from deciding how the adaptive behaviors are to be generated to deciding what behaviors are generated [14].

In this chapter we have attempted to summarize the work carried out during the last few years on the application of evolution to ANN based behavior based robotics, through a vision of what other authors have done, together with a few illustrative practical examples obtained by our group. These examples consider the most relevant problems encountered in the application of these techniques, such as: which evolutionary methodology should be employed, how we can represent the controllers in the chromosomes, inclusion of temporal processing (as a first step to more deliberative behaviors), approaches to bridging the reality gap through the inclusion of different types of noise and the generation of global controllers containing several behaviors and their coherent coordination. At the end, we include an overview of what we consider to be the open issues in this field and what we feel should be studied in the near future to produce applicable results.

2 Evolutionary Paradigm

Several evolutionary methods have been considered by the robotics community to produce the controllers or behaviors required for the robots to perform their assigned tasks and survive in their environments. All of them may be taken as variations of a general process whereby a population of individuals is evaluated each generation. These individuals procreate according to their fitness. The population undergoes mutation processes and a new population is in some way selected from the old, to continue with the process. These methods include Genetic Algorithms, Evolutionary Strategies, Genetic and Evolutionary Programming and Coevolution. In this section we are going to provide a brief characterization of these methods in general and their bearing on robotics, assuming a general understanding of evolutionary processes. In addition, we will consider the relevance of learning in the evolution of behavior.

2.1 Genetic Algorithms

The bases of genetic algorithms (GA) were proposed by John Holland in the Sixties and the idea of genetic algorithms followed in 1975 [38]. These types of algorithms are characterized by the fact that the dominant factor for the purpose of evolution is crossover. Mutation is just a way of generating variety in the populations and always presents a very low value. Many variants of genetic algorithms have been applied to the generation of robot controllers, with higher or lower crossover and mutation probabilities and different population updating and selection strategies, etc. The use of any one of them will depend on the problem domain, as no strategy seems to be better in every case.

Some authors, such as Harvey [35], claim that genetic algorithms are not very adequate or applicable to robotics in their standard form, and propose some adaptations. An important problem, which is very relevant to the evolution of robot controllers when using GAs, is that of deceptivity [29]. Often, an individual coming from the crossover of two high fitness parents and who preserves the genes that determine this high fitness displays a very low fitness value. When a problem is deceptive, it is very difficult for the GA to obtain an optimal solution and its performance may degrade down to a random search procedure.

One of the causes of deceptivity is epistasis, i.e. the fitness of a gene within the chromosome is strongly dependent on the value of other genes. Epistasis very often arises when the controllers that a robot must implement are encoded through Artificial Neural Networks (ANN) or any other type of distributed architecture. Unfortunately, when the chromosomes represent the weights or other distributed parameters of ANNs, there is usually a high correlation between the genes in terms of fitness. In the case of behavior controllers based on ANNs, which is the dominant paradigm in the field, this problem must be taken into account and a solution sought. An approximation that is frequently employed is that of not making use of the crossover operator and, instead, just copying the individual several times in the new population after the pertinent mutations, or making use of evolution strategies instead of genetic algorithms [49][58][59][61][63][65].

2.2 Evolution Strategies

Evolution strategies (ES) were proposed by Rechenberg [68] and Schwefel [73], and their main difference with respect to genetic algorithms is that the operator guiding the evolutionary process is mutation and not crossover.

There are basically two types of evolutionary strategies:

- $(\mu+\lambda)$ ES, where μ individuals generate λ offspring and select for the next generation the best μ individuals out of the total $\mu+\lambda$.

- (μ,λ) ES, where μ individuals generate λ offspring and select for the next generation the best μ individuals out of the λ offspring.

In both cases, the λ offspring are generated by mutating all the genes of the chromosomes. This mutation typically implies the addition of a random value from a zero mean distribution to each gene.

The use of evolution strategies is still not the most extended way of generating robot controllers evolutionarily, although some authors such as Salomon [71] and Becerra [3] have employed them and demonstrated, at least in the particular cases they examined, that they were faster than genetic algorithms.

Several variants of evolutionary strategies have been contemplated by the robotics community, such as applying the crossover operation between two individuals before applying mutation. Many of these variants blur the difference between genetic algorithms and evolution strategies, highlighting the fact that in this field there is no consensus on the elimination of crossover. Even though crossover seems to be of little use when epistasis is present, it is an operator that permits the traversing of valleys in the search space in a fast and efficient manner and thus helps to prevent the process from becoming trapped in local minima.

It seems reasonable to assume that the higher the level of epistasis present in a problem, the more need for the mutation operator and less need for the crossover operator, usually without either of them disappearing completely.

2.3 Genetic and Evolutionary Programming

Evolutionary programming (EP), originally defined by Fogel [25], is similar in conception to evolution strategies, but the objects of evolution are programs in a high level language. They may describe a finite state automaton (as in its original version), a set of rewriting rules, or any other type of structure defined by means of a program. Obviously, as the operations are carried out on programs, mutation is more complex than in the previous cases. It implies changing a command or group of commands in the high level language program, with the added constraint that the new individual has to be syntactically and semantically valid.

There is a variant, between evolutionary programming and genetic algorithms, called genetic programming (GP) [46][47] in which the object of evolution are still programs but where a crossover operator is used. This crossover operator usually acts over tree representations of the program. Examples of the use of GP are found in [16][48][66][69].

The main drawback of these approaches is that the operations allowed over the chromosomes must be established for each problem. In addition, due to complexity considerations, the language employed - usually a subset of another language - is particular to the problem in hand and may not be extrapolated to other problems. From the viewpoint of behavioral robotics, the results present the same problems of adaptivity and robustness as the traditional explicit control programs: no learning, a set of functions fixed beforehand, low scalability, etc. Notwithstanding their drawbacks, evolutionary and genetic programming are finding a niche in behavior based robotics, in the area of evolving developmental programs for other structures.

2.4 Coevolution

Coevolving implies evolving more than one population in a coordinated fashion, either by obtaining a global fitness value from the use of a combination of elements from the different populations (cooperative coevolution), or by obtaining the fitness of a population through an evaluation of its interaction with another population and viceversa (competitive coevolution).

Much of the coevolution work in robotics has been in the direction of cooperative coevolution in the context of hardware or morphology being coevolved with the controllers. Notwithstanding this fact, there is a special case when coevolution involves the competition between populations specialized in different and competing behaviors. Nature presents examples of this in cases such as host and parasites or predators and preys. When these types of processes are considered, "competitive coevolution fitness progress is achieved to the disadvantage of the fitness of the other population" [23][24]. Because of this, on the evolutionary time scale, the coupled dynamics of the system give rise to the "Red Queen Effect", whereby the fitness landscape of each population is continuously modified by the competing population. In [23][24] Floreano and Nolfi study competitive coevolution for a population of predators and preys implemented with Khepera robots and in [13] Cliff and Miller study co-evolution for pursuit.

These types of coevolutionary processes are very promising, in the sense that they allow for a certain automatic definition of the fitness criteria. There is no single hand designed fitness function that leads to the desired solution. The fitness function arises from the competition of the different populations in the artificial (or even natural) environment. In a certain sense, it is an emergent function. In the field of robotics this is very interesting, due to the difficulties found in defining fitness functions that lead to complex behaviors and permit an appropriate apportionment of credit. Coevolution allows for an automatic incremental complication of the environment and thus facilitates this task. Initially all populations are not good at their tasks, so they are playing in a leveled field. As evolution progresses, the different populations become better adapted, but, hopefully, at a similar rate.

2.5 Learning

There are two ways in which evolutionary algorithms and learning can interact. On one hand, learning may be employed to fine tune individuals obtained, i.e. move around the point obtained by the evolutionary algorithm in the search space. On the other hand, learning may be taken as an intrinsic capacity to be evolved.

The main advantages of making use of an evolutionary algorithm for developing robot controllers is that a search for the optimal controller or controllers is carried out from multiple points in the search space in parallel. As opposed to the case of, for example, ANN learning, which in general involves the search of the problem space using a single point. With the aim of making the best use of both worlds, some authors have employed a combination of both strategies to refine the controllers obtained by an evolutionary algorithm. So, in these cases, an exploration of the surroundings of the points in the search space selected by the evolutionary algorithm is carried out using a traditional learning procedure, such as a backpropagation algorithm.

In the particular case of behavioral controllers, some authors have attempted to provide an edge to the individuals evolving in the population by endowing them with mechanisms that allow them to learn during their lifetime, in the hope that this would provide an evolutionary advantage to the individuals that have the best aptitude for learning [22][57][63][64]. This second approach is also called the Baldwin effect [2] and is very useful for obtaining individuals that adapt well to changing environments in ontogenetic time.

3 Evolution of Behavioral Controllers

According to [37] there are three types of structures that may be employed and thus evolved to obtain a behavioral controller: explicit control programs, mathematical expressions, or a blueprint for a processing structure. The first two types of structures are seldom used in behavior based robotics and we are therefore going to concentrate on the third option.

The processing structures most often found in the literature are Fuzzy Rule Systems [55], Classifier Systems [18] and Artificial Neural Networks, with a clear preeminence of the latter [4][14][22][43][49][64][72]. ANNs form a powerful distributed processing model with very interesting characteristics for use in autonomous robotics [61], such as their fault tolerance, noise tolerance (a crucial property because of the presence of noise in real environments) and the possibility of using the traditional connectionist learning algorithms for obtaining the ANNs in ontogenetic time, or as we have already commented, for the combination of evolution with connectionist learning. In fact, it has been shown that introducing noise when training neural networks may be beneficial [19]. Another advantage is derived from the fact that the weights and nodes are low level syntactic primitives, well below the semantic level [12][34], and, as [65] argues, the primitives considered by the evolutionary process must be of the lowest level possible so that undesirable selections produced by a human designer are avoided.

For the sake of simplicity in their structure and ease of training, in most practical tasks where ANNs have been applied, these networks consisted of simple nodes and scalar connections. One of the most frequently used schemes, because of the existence of an easily implemented training algorithm, has been the Multilayer Perceptron (MLP). These types of networks have proven to be capable of great plasticity and adaptation to different types of problems. However, due to the low processing capacity of their nodes and connections, their scalability to more complex problems has usually implied a significant increase in their sizes, and thus has made their training and/or evolution more time consuming and liable to local minima.

To prevent these drawbacks, one approach of great interest is to seek structures that correspond in a more direct manner to the context in which they are employed. For instance, if a problem based on temporal relations is considered, it will be of great help to employ networks that explicitly take into account time. Obviously, if we increase the processing order of nodes and/or connections appropriately, we may obtain a larger processing capacity with an architecturally simpler structure. This simplicity will make the structure easier to evolve. If this structure is clearly adapted to the problem, the generalization and learning capacities of the system will not be affected, and may even improve. In fact, what we are discussing is going from general purpose networks to networks that allow for an easy procedure to concentrate their attention on whatever is relevant to the problem in hand and ignore whatever is not, through an appropriate architecture and training process.

The power of an MLP network lies, almost exclusively, in its high connectivity. The components of the network do not present sufficient complexity for them to individually handle an important part of the information processing. Thus, one possible strategy is to employ activation functions that differ from the usual step or sigmoid: for example, using higher order neurons [28], or sigma-pi neurons [70]. Another option is to increase the processing power of the synapses. In typical MLP architectures, the weight is a numerical value that, once the network has been trained, will have the same effect on any value that circulates through this synapse. On the other hand, the use of functions converts the synapses into active elements, whose outputs depend not only on the higher or lower intrinsic importance of the connection, but also on the value that reaches it in each moment of time.

We have made use of the idea through the introduction of different time and space processing networks, where some of the processing load has been moved to the synapses. An example of space processing networks has been achieved by adding a gaussian function to the synapses between the nodes. In this way, the output of each connection depends not only on the traditional weight, but also on how the corresponding gaussian function is centered with respect to the current input.

Consequently, these higher order synapses lead to a simplification of the inputs to the behavior based controllers. In fact, the function of these networks is to reduce the dimensionality of the input space by making use of the "need to know" principle. For example, in [20] we make use of gaussian synapse based networks to allow a robot to detect another robot in its visual field independently of the background, and for tracking a mobile light. In both cases, the networks require a very small number of nodes to carry out the required task.

Another type of network where processing has been transferred to the synapses are synaptic delay based networks. These networks can explicitly process temporally dependent events, without having to resort to any type of predefined temporal window.

We call the combination of sensors and dimensionality reduction networks *Virtual Sensors*. In most of the work carried out in the field of behavior based robotics until now, the sensors in the robots employed have been very simple. This was due to the fact that when this strategy was employed, the values of the sensors were taken as direct input to the controllers, and thus, when anything other than a very small set of values were taken into account, the networks became too complex to be easily obtained. Virtual sensors using the need-to-know principle are one way of overcoming this problem.

The use of virtual sensors incorporates the possibility of any type of hybridization. This entails the introduction, over behavior based models, of preestablished modules of any other paradigm (cognitive, programmed, etc), which allows different types of preprocessing for the reduction of the dimensionality of the information that the behavior based modules must manage directly. In this manner, the search space is drastically reduced through the design or evolution of some *ad hoc* components. This introduces disadvantages, but provides a way to dramatically reduce the complexity of the autonomous design of the behavior based modules and their interrelation.

In this chapter we include some simple examples, which make use of higher order structures applied to time dependent phenomena, to illustrate this fact. These structures can be employed to allow the robot to perform tasks it could not perform before, to make more efficient use of its sensing apparatus, or to improve the strategies it uses for certain tasks.

3.1 Representation of the Controllers

As we have already pointed out in the introduction, "Nature has invented three automatic design procedures, namely those of evolution, development and learning" [43]. The first two are related to the organization of an organism, while the third is used to tune that organization to its specific environment throughout its lifetime. Most of the work in evolutionary robotics has either made use of the evolutionary procedure - through a direct genotypical representation - for the design of controllers in robots [53][65], or employed a combination of evolution with learning, especially in structures, such as certain topologies of ANN, where learning algorithms are well established. Some authors make use of development [5][11][17][31][33][43][44][45] This alternative implies that the genotype now encodes the development of the phenotype, as opposed to the most usual case in ANN controller evolution, where a direct genotype-phenotype translation is applied.

Whatever the chosen representation, there must be some type of encoding of the ANN architecture and, as Husbands et al. [39] have already stated, this issue of encoding is central to the evolutionary development of control systems. Gruau [32] indicated that a scheme for encoding a neural network must be able to encode any ANN (completeness), should be compact. Every genotype should encode some architecture (closure). It should also be modular and present smooth interaction with genetic operators. It should not presuppose the dimensionality of the search space and should allow the specification of sensor and actuator properties as well as the control network.

Direct genotypical representations present the problem, among others, that the designer must specify a priori some set of parameters concerning the final controller. For example, in a feedforward ANN, it is necessary to specify the number of layers and the number of nodes in each layer. In addition, in Kodjabachian and Meyer's words "numerous encoding schemes implement a direct genotype-to-phenotype mapping and are hampered by a lack of scalability, according to which the genetic description of a neural network grows as the square of the size of the network. As a consequence, the evolutionary algorithm explores a genotypical space that grows larger and larger as the phenotypic solutions that are sought become more and more complex"[45].

With regard to the developmental approach, it is very difficult to obtain a developmental program, due to all the problems with respect to genetic or evolutionary programming that we have already commented. The problem of epistasis becomes even more marked in these cases and, thus, evolution is much slower. In addition, most developmental schemes employed up until now have problems meeting the above stated conditions.

In the examples we present in this chapter, we have chosen the simplest encoding scheme to make these illustrations as clear as possible. This encoding scheme is simply a direct genotypic representation of the phenotype. The phenotype is a neural network that will implement a behavior. It is represented by a simple floating point number string that contains the weights or other parameters of the neural network we represent and want to evolve. Among the different parameters we can include in the string, we obviously find the synaptic weights. Delays in synaptic connections, slopes of habituation neurons, etc., are also of interest sometimes.

Notwithstanding its simplicity, this representation can accommodate Gruau's [32] propertie, except perhaps the modularity constraint. This is because, in the examples provided, this modularity is achieved through the concurrent operation of several evolved or coevolved neural networks and not by subparts of a given network.

3.2 Fitness Evaluation

Whatever the evolution mechanism and representation employed, the controllers corresponding to the different individuals must be evaluated and their fitness obtained. This fitness corresponds to how well the robot performs in an environment during its lifetime. Two different perspectives are possible from the point of view of evolutionary robotics: a local perspective or a global one [60]. With the first it is necessary to establish, for each step of robot life, a degree of the goodness of its actions in relation to its goal. The final fitness will be the sum of the fitness values corresponding to each step. For example, in a wall following behavior, fitness increases for actions of the robot that take it closer to the wall (without contact) and make it move in a straight line and at high speed. This strategy has its drawbacks, as, except in toy problems, it is very difficult to decide beforehand the goodness of each action that is taken towards a final objective. Sometimes the same action may be good or bad, depending on what has happened before or the context in which it has occurred. Obviously, this approach implies an external determination of goodness. It is the designer who is imposing, through these action-goodness pairs, how the robot must act. It is the most common solution employed in the field, and always implies tuning (by trial and error) the terms and variables of the fitness function to obtain an adequate controller.

The global approach, on the other hand, implies defining fitness criteria that is not based on the goodness of each particular action in the life of the robot, but on how good the robot was at achieving its final goal. The designer does not specify how good each action is in order to maximize fitness and the evolutionary algorithm has a lot more freedom to discover the best final controller. This approach presents several problems, but the most important is the credit-apportionment problem. The final fitness value does not directly reflect the goodness of each action taken by the robot to achieve this final fitness. Therefore, a lot of information that could accelerate the evolution process is lost. In addition, as much less knowledge is injected into the evolutionary process, it may not work at all. In fact, when using global strategies, what we often encounter is that the resulting controllers find loopholes in the problem specification. They thus maximize their fitness without really achieving the function that we want them to achieve, or in what we, as humans, consider to be an efficient manner.

There are two principal ways for obtaining a global fitness level: external and internal. By external we mean fitness assigned by someone or some process outside the robot. This evaluator has much more information than the robot and can thus produce all kinds of fitness values. For instance, an external evaluator can judge the fitness of a robot that is supposed to catch a mobile target as a function of the distance it was from the target at the end of its life. This obviously implies some "God-like" information, i.e. information from an external observer that can perceive data such as exact positions and distances, which the robot is incapable of doing. This approach is very efficient in some cases, where the designer wants a particular task performed, in a very well defined environment. It is not so satisfactory when these conditions are not met, i.e. when the environment is partially unknown or when there is noise in the perception of it, or when the robot controller is being evolved in one environment (simulated or not) and employed in another. An extreme example of external global evaluation of robot behavior is presented by Lund et al. [50][51]. In this work the authors apply an interactive genetic algorithm for the development of Lego robot controllers and morphology, and then let a group of children evaluate each robot while it was operating in a real environment.

The other possible approach is to employ an internal representation of fitness. This means resorting to clues in the environment that the robot is conscious of and can use to judge its level of fitness without help from any external type of evaluator (apart from the environment itself). A concept often used in to implement this approach is that of internal energy. The robot has an internal energy level and this energy level increases or decreases according to a set of rules or functions related to robot perceptions. In a sense it is very similar to the concept of eating or being hungry in human terms. The final fitness of the robot is given by the level of energy at the end of its life. The only intervention of the human designer here is in assigning the relationship between energy increases or decreases and robot perceptions.

In the examples presented here, we assumed that in allowing the robot to freely obtain the best behavior possible to achieve its objectives (related to some task we wanted it to perform), we had to let it judge its environment from the data it had available, with as few preconditions as possible. The strategy consisted of sticking food items on objects in the environment that could lead the robot to performing its task, and then let the robot work only on a global energy criterion. That is, robot fitness was determined by the energy that it had available at the end of its life period. In the case of our robot (in the examples we will use a Rug Warrior), we defined an internal energy variable whereby the robot loved fungi and that eating fungi would raise its energy level. On the other hand, other actions can cause a decrease in its energy level. It is clear that depending on how we distribute the fungi throughout the environment (in terms of amount, spacing, and position) and depending on the relationships between energy gains and loses, significantly different behaviors will result.

Obviously, two different lives of a given robot in the same environment can produce wildly different values for final energy, due to noise in the environment, its starting position, or orientation, etc. To prevent these misleading fitness values, the fitness of an individual is calculated as the average fitness of this individual after several lives in the same or slightly changed environments.

3.3 Behavior Evaluation

It would be ideal to evaluate a behavior in the real robot and in a variety of representative environments in which the robot will operate. Some authors [21][22][50][60][71] have carried out experiments with this approach, but the slowness of the process makes alternatives based on the simulation of the characteristics of the robot and the environment necessary. Either one of the methods presents its own advantages and disadvantages, but in any case, for the validation of the results, it is necessary to test them on a real robot and in a real environment.

When the evolution is carried out in a simulator of the robot and environment [4][56][58][59] the process is greatly accelerated with respect to evolution in real robots because the actions of the robot are faster in the simulation and the sensor stabilization times are now unnecessary. In addition, simulations may be run in powerful machines or even in parallel or distributed computer architectures. Also, in the simulation it is possible to take into account information that would not be available if the evolution took place on the real robot. For example, the distance between the robot and a possible goal may be included in the fitness function of the controller.

However, when controllers evolved in simulation are tested on real robots, they usually suffer a considerable loss of performance. This may be a consequence of a poor simulation of some aspect of the behavior of the robot (sensors, actuators,...) or the environment, or due to the presence of noise in the real environment. Therefore, mechanisms must be obtained for the simulation to minimize the "reality gap" [40]. A possibility for this minimization is to obtain the response of the sensors and actuators from their real behavior in the real robot, and not from their theoretical specifications. Along these lines, some authors prefer to capture the real response of sensors and actuators using the real objects the robot will encounter in reality, and not to make use of a mathematical model of these sensors and actuators, as in [42]. Two sensors or actuators, even though they may be of an identical model, may produce significantly different responses [58][65]. This alternative works well when the variety of objects and surfaces is not large, but it is difficult to model sensors and actuators in more dynamic environments in this way.

Another alternative is a mixed evaluation, where the majority of the evolution can be carried out in simulation and the last few generations with the real robot in the real environment, until a comparable fitness to that of the simulation is obtained. This is the solution adopted by Miglino et al. [58].

Most controllers are evolved in simulation for the reasons mentioned at the beginning of this section. The problem here is, how to evolve these controllers so that they scale up to reality or even to a different reality. If we take into account that the controllers only work in the real world when the simulations are perfect replicas of this world, as the world becomes more complex, the simulators become unattainable both in terms of complexity and of execution time. As a result, some authors have addressed the problem of deciding whether or not it is possible to evolve controllers for real robots through simulations that are not as complex as the real world. Jakobi [41] calls this *behavior transference* and obtains a minimal set of conditions for the successful transference from a controller evolved in a simulated setting of one in the real world or to another system. He indicates that in order for this to happen certain conditions must be met. Both systems must be able to support the behavior. Furthermore, there must be a mapping from the base set of external variables of the real world onto the base set of external variables of the simulation and two sets of functions must exist so that the controller is base set exclusive and base set robust.

He contends that base set exclusive controllers are those whose internal state depends exclusively on those features of the environment that are described by the base set of external variables. Base set robust implies that the controller must operate reliably, not only in the agent environment system, but also in a whole set of associated agent environment systems.

In order to use these conditions, Jakobi [41] starts by identifying and expliciting a base set of robot environment interactions to be included in the simulation and randomly varies all of the implementation aspects of the simulation from trial to trial to ensure that no controller will be reliably fit unless it is base set exclusive and varies every base set aspect of the simulation to make the fit controllers base set robust.

Summing up, to ensure that a behavior controller developed in a relatively simple simulation operates adequately in a real environment or in a different system from that for which it was designed, the simulation must be carried out with random noise in the characteristics of the simulation, including the environment employed, sensors, actuators and readings obtained by the robot. It must also include sufficient cases for the robot to evolve behaviors based only on those clues from the environment that it really needs (base set).

3.4 Noise

To minimize the reality gap, several authors have proposed different types of noise that should be applied to the simulated sensors, actuators or environment. Traditionally, the noise applied to the simulated sensors and actuators consisted in small random variations [56][58], or variations with a gaussian distribution [27] in the values sensed or applied. Another type of noise that seems to produce good results is "conservative noise" [58], which consists of simulating small alterations in the positions of the objects perceived by the sensors that can be translated into large variations in the sensed values depending on the sensor model. This type of noise simulates the different behaviors exhibited by the sensors depending on "differences in the illumination of the objects, shadows, or because of slight physical differences between objects of the same type" [58]. It is specially useful with some types of sensors such as infrared ones.

An extension of Jakobi´s [41] concepts have been applied to the simulations employed in work carried out by our group and we have seen that, to obtain robust behaviors, different types of noise were necessary. On one hand, we considered the most obvious types of noise, i.e. noise in the values sensed and actuator noise. The same object in the same position does not necessarily mean the same value returned by a given sensor. Different causes beyond our control, such as changes in ambient light, infrared reflections, etc., are impossible to predict, and the only way to take them into account is through the addition or subtraction of small (sometimes not so small) amounts to the values returned by the simulated sensors. The same applies to the operation of the actuators.

We also included generalization noise. This type of noise is part of the group of artifacts for making the behaviors more robust and thus allowing them to compensate different environmental conditions and changes in the operation of the robot. It consists of randomly changing some characteristic of the robot or the environment (such as the orientation of a sensor) and maintaining this change for one whole life of the robot (one evaluation of the robot). This is repeated for every life of the simulated robot. There are two instances were this is absolutely necessary. One is when the robot is working on battery power. Different levels of battery charge imply different speeds for the wheels of the robot when presented with the same command. The other is when there are changes in ambient light, which causes the same object to be perceived differently.

Another especially important type of noise we take into consideration is noise that has to do with defects or particular traits in the operation of the real robot (Systemic noise). For instance, the Rug Warrior we employ has a defect in the left motor-wheel system, which causes the left wheel to be much slower than the right one for the same command. However, this difference is not constant.

Finally, temporal noise is used to obtain behaviors that are tolerant to variations in the time that elapses between events. This is especially relevant if one takes into account that simple robots have very limited processing capabilities and therefore, depending on what networks must be run before a given action is taken or perception measured, the time elapsed will be very different.

Obviously, any type of noise makes the process for obtaining a given behavior slower and more difficult, especially with regard to the last three types of noise. These may cause the environment to be perceived very differently in each evaluation of the robot, and thus implies that the optimal strategy is also different, forcing the robot to obtain compromise solutions.

4 Evolution of Temporally Dependent Behaviors

When acting on the real world, there are behaviors that require the handling of temporal information, in the sense that the robot must perform actions depending on previously sensed values or previous actions. Reasoning with the temporal values of perceived information can also improve tasks, such as capturing and escaping, by means of predictions of the motions of the objects in the environment, or such as wall following with infra-sensorized robots. The inclusion of time in the behavior controllers represents a step, albeit small, in the direction of deliberative systems, as opposed to purely reactive systems.

In the case of ANNs, there are several possible solutions that take into account past information. The first possibility is to use recurrent artificial neural networks that encode a summary of past information received by the inputs, i.e. the "state", encoded in the nodes with recurrences as in [30][37][40][44][50][59]. These networks are useful for summarizing the history of previous activation states, but do not permit a simple storage of fixed temporal patterns, which are necessary for a large number of applications.

On the other hand, temporal windows, consisting of several inputs (each one usually with a different weight) corresponding to consecutive temporal values of the same sensor, to a given neuron permit the storage of these temporal patterns. However, they present the drawback that a window must be defined and that connections for all the temporal instants within the window must exist, even when they are not necessary. This leads to a large number of connections, lengthens processing time and obscures the processing that the network must perform. In some applications, such as mobile robotics, the processing time is very important, as it determines the reaction speed of the robot. The importance of this must be stressed, as the robot may be faced with dangerous situations and reaction speed is crucial, especially when noise may cause a delay in the perception of the dangers, or in the case of simple robots whose processing capacity is very small.

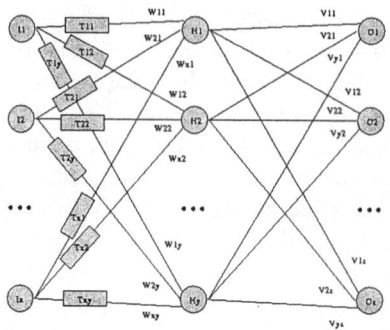

Fig. 1: Neural network with synaptic delays.

To prevent these drawbacks, the architecture for the ANN we employ (Fig.1) consists of several layers of neurons interconnected as a multiple layer perceptron, where the synapses, in addition to the weights, include a delay term that indicates the time an event takes to traverse it. These delays, as well as the weights, are trainable and/or evolvable, allowing the network to obtain a model of the temporal processes required from its interaction with the world. A fact that must be taken into account is that, in general, to have delays only in the first layer is equivalent to having them in all the layers for the processes we are going to consider.

5 Monolithic Behavior Controller

When designing a global behavior controller, two options are possible. On one hand, a single module that includes all the necessary mappings between the sensors and actuators, i.e. a monolithic approach, may be used. On the other hand, the global behavior may be decomposed into a set of simpler ones, each implemented in its own controller, and a method obtained to interconnect them in such a way as to present a final behavior to the actuators.

In this section and to illustrate the concepts and problems found when using these approaches, we will begin to work and really start to obtain behaviors. Initially, these behaviors will be implemented in a monolithic manner; that is, a single controller will produce the desired global behavior. In section 6, we will present a methodology for automatically obtaining behaviors implemented through the coordination of different modules and, in section 7, we will provide some examples of them. All of the behaviors considered are tested on a real Rug Warrior robot in to ascertain their validity or decide what type of modifications the simulation/evolution process requires.

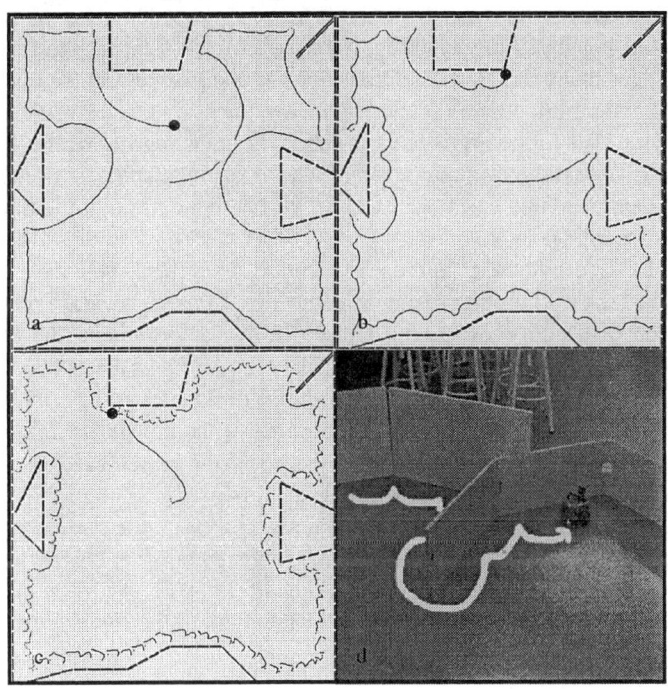

Fig. 2: Robot controllers evolved a) without synaptic delays, b) with synaptic delays but no simulation noise, c) same as b with temporal noise, and d) operation of the real robot using the controller shown in c).

Wall following is one of the most usual behaviors found in autonomous robotics literature. Typically, this behavior consists of the robot following the walls of an enclosure at the highest possible speed, minimizing the distance to the wall it is following at each instant of time while avoiding collisions. It is usually implemented in robots where the sensors employed in this task provide values for the robot to be able to distinguish between when it is approaching a wall or going away from it. The biggest problem found when obtaining these behaviors is caused by the presence of noise in the sensors. The infrared sensors of the Rug Warrior, which are the only ones we can use for this task, are binary. This fact makes it impossible to decide when we are approaching or going away from an object without taking into account the readings corresponding to previous instants of time.

Some authors, as in our case, include in this behavior the need of finding a wall to follow in order to make it more interesting. Obviously, this makes it much more difficult to obtain the behavior. Thus, when the robot is not sensing anything, it must select a trajectory that leads it to a wall. This trajectory cannot be a tight circle, or it would never find the wall. When the robot leaves a wall that it is following and is not sensing anything, it must perform a tight turn to return to the wall. It is clear that these two conditions are not compatible and the consequence is that a robot controlled by a regular feedforward neural network usually presents very deficient wall following behavior, generally bouncing off walls. To fix this problem it is necessary to include some way of distinguishing between the two situations mentioned above. In our case this will be done through the introduction of variable synaptic delays.

Using the energy-based approach mentioned earlier, we have built an environment for the evolution of the behavior that contains several types of turns with a broad distribution of turning angles. In this environment, the bricks that make up the walls have fungi attached to them. These fungi are taken as food by the robot. When the robot senses a brick with one of its infrared sensors, its energy level is raised in a quantity directly proportional to the amount of fungi stuck to the brick and the fungi is eliminated from the brick (the robot "eats" it). The fungi begin to grow slowly until they reach their maximum size or the robot eats them again. The robot fitness is its energy level at the end of its life, which is given by how much it has eaten. It is obvious that the robot, to raise its energy level, must sense as many different bricks as possible in its lifetime, i.e. the robot must follow walls. There is no designer-led indication of the method the robot must follow to accomplish its mission, thereby minimizing problems due to a misinterpretation of the environmental conditions.

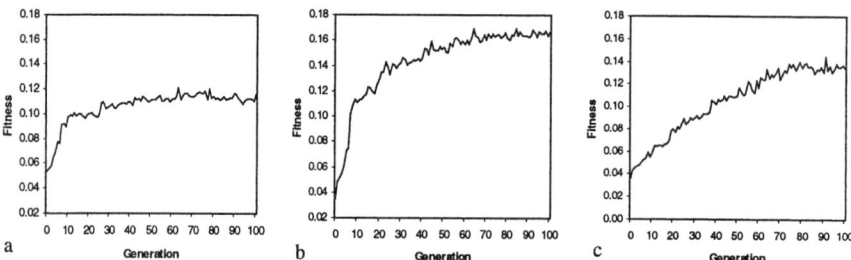

Fig. 3: Fitness of the best individual throughout the evolution process for the robot controllers evolved: a) without synaptic delays, b) with synaptic delays but no simulation temporal noise, and c) same as before with temporal noise.

It is obvious from Fig. 2a and Fig. 3a that without the use of time delays, the behavior obtained by the evolutionary procedure is very poor, and the robot, as we mentioned before, can never really hug the wall. Adding delays in the network structure, as shown in Fig. 2b and Fig. 3b, really improves the fitness of the best individual, and the behavior obtained is the one desired. These results, as we have already indicated, are misleading, as the world is not modeled very accurately. The real world involves noise and if one attempts to transfer this behavioral module to the real robot, which suffers even more noise types than included in the simulations, it will not work at all. There is a reality gap between the simulation and the real thing, and to overcome this, noise must be considered in the simulations employed for evolution.

By introducing different levels of noise in the simulations during the evolution of the robot, we obtain the behavior of Fig. 2c and Fig. 2d. In these, we display the behavior of the robot when it is evolved with an environment that considers the different types of noise. In addition to 15% random noise in sensors and actuators, we have included generalization noise with variations in the orientations of the sensors and on the charge of the battery. We have also included systemic noise, by which the left wheel of the robot operates at random speeds of down to 10% below its specified value, and temporal noise. The evolution and ANN parameters are the same as before. This behavior obtains a result in simulation that is very similar to case b, but when it is implemented in the real robot, it works as shown in Fig. 2d.

From an evolutionary point of view, as we have already mentioned, it becomes more difficult to evolve a controller when the environment is so noisy. This is evident in the evolution diagrams of Fig. 3. In Fig. 3.a, the fitness of the best individual never reaches very high values, because of the lack of an internal structure of the network that would allow it to perform the task appropriately. In the central diagram of this figure (3.b), we display the fact that the best individual obtained when using delays presents higher fitness than in the previous case, but this controller does not perform appropriately in the real robot. Finally, in Fig. 3c we display an evolution process where the different types of noise have been considered and it can be seen how evolution is slower than in Fig. 3b. This is because, depending on the particular noise parameters of each life a robot lives, this robot can be deemed better or worse. Therefore, more robust compromise solutions to the problems it finds must be obtained. The fact is that, in the end, a fitness level can be achieved that is almost as good as that obtained without noises in the simulations and, more importantly, the behavior works perfectly on the real robot.

6 Generating Coherent Global Behavior

One of the most important problems found when constructing behavior based control structures for autonomous robots with a certain complexity, is how to choose among the different activities that must be considered. This is what

Hendriks calls the high level of design, as opposed to the micro level of design that considers the implementation of particular behaviors.

In monolithic architectures all the behaviors are implemented in the same controller, whether or not it is an ANN, a LCS, or an AFSM, etc. The advantage of the alternative is that it is not necessary to have prior knowledge about possible sub-behaviors and the interrelations between them. The disadvantage is that it is not possible to reuse the individual behaviors. If a new behavior is required it is necessary to evolve the complete module again, even if it could have reused previously learnt or evolved behaviors.

When considering a modular approach, various options are possible. On one hand, we have the possibility of hierarchical architectures, where higher level controllers decide which lower level controllers are activated in each situation. On the other, hand we have distributed architectures, where there are no priorities and all the controllers compete for control of the actuators at each instant of time.

In hierarchical modular architectures, the global behavior is decomposed, as necessary, into lower level behaviors that will be implemented in particular controllers. The higher level controllers can take information from the sensors or from low level controllers, and depending on the architecture, act over the actuators or select a lower level controller for activation. The advantage of these methods is that the behaviors can be obtained individually and then the interconnection between them can be established. Also, it is possible to reuse the behaviors obtained when implementing higher level behaviors. The problem that arises is that the decomposition is not clear in every case, as it implies a specific knowledge of what sub-behaviors must be employed.

In distributed modular architectures there is no hierarchy of sub-behaviors and all of them compete for the use of the actuators by means of some type of arbitration or inhibitions system. This alternative has seldom been used because it is not clear how to generate or expand these representations in a structured manner. Intermediate solutions between this last alternative and the monolithic one are found in the work by Nolfi 6162 and that of Gomez and Miikkulainen 30.

In the examples presented here, multiple behaviors are generated through a hierarchical structure where some high level and low level behaviors may be coevolved, so that they fit snugly, and the flexibility of not having to define all the low level behaviors is preserved.

In a first step, a designer must identify sub-behaviors that may be useful to generate the global behavior required. The designer need not be exhaustive nor does he need to be concise, i.e. there is no problem if useless behaviors are included in this preliminary set. The selection and interconnection of these behaviors are carried out by means of an additional ANN that is evolved. In this way, the designer does not have to specify which low-level module must be executed at each moment in time. The evolutionary algorithm will evolve the ANN so that it makes this selection appropriately, based on data from sensors or from the outputs of other behavior controllers. This higher level ANN differs from the lower level ones in that its output does not always act over the actuators; it mostly selects the adequate lower level behavior. There is nothing in this architecture to limit the number of levels to two. The lower level behavior selected by a given higher level behavior may itself select a lower level behavior that acts over the actuators.

Using this structure, the controllers may be used in multiple composite behaviors without duplicities, and, as new behaviors are evolved, this evolution becomes faster and simpler due to the fact that a lot of experience in the form of previously evolved behaviors is available.

To prevent the problem of the designers having to be exhaustive in their determination of all the necessary lower level behaviors, we have included the possibility of cooperatively coevolving lower and higher level behaviors. That is, a higher level behavior may be evolved by itself using previously evolved lower level behaviors, or it may be coevolved with part of the lower level behaviors and use the previously evolved ones. This way, when the designer is faced with a problem where he is only able to identify part of the behaviors that may be involved, the unidentified ones will be evolved at the same time as the higher level controller. This option makes the evolution of complex behaviors a lot easier for the designer, as it works even when the designer is not capable of identifying any of the behaviors involved. The designer can provide a set of previously evolved behaviors as a guess and let the evolver do the rest.

7 Compound Behavior Controller

To illustrate the use and results of the methodology presented above, we consider here a compound controller that implements another typical behavior in the evolutionary robotics literature: homing. It consists of finding a given object in the environment. The robot will interpret this object as its home and will go towards it to feed. In this case we have prepared the following scenario. The environment has some walls which block direct sight of home, thus forcing the robot to search for it. A flashing light represents home, but there is an object that is very similar to home and which acts like a trap, represented by a static light. The robot will necessarily require temporal information to distinguish one object from the other.

In the scenario we have prepared, the objective is for the robot to seek home, but the flashing light is hidden. To obtain a monolithic controller for this task is very difficult and time consuming, and the network that implements it is necessarily quite complex. An easier way of producing the global controller is simply to obtain a higher level controller that coordinates lower level controllers. Then, if home is seen, a lower level homing behavior evolved for previous experiments in environments that were free from obstacles will be active, and in any other case, wall following or escape from collisions, or any other behavior that makes the search procedure efficient will be the active one.

This is the approach we followed and the results are shown in Fig. 4. The robot, depending on the levels of noise present and on the complexity of the environment, makes use of different strategies. If the environment is relatively free from obstacles, the robot uses an escape from collisions behavior that allows it to explore its world relatively quickly. When the environment is more complex, such as in the case presented in the figure, this type of behavior is not efficient, as the probability of the robot going into one of the boxes containing home is quite low. In this case, the robot makes use of either a wall following behavior or the lower level monolithic homing behavior, taking into account that this behavior when it does not

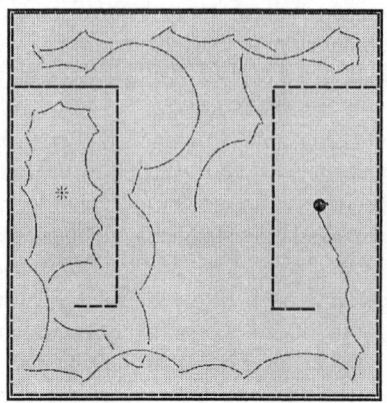

Fig. 4. Compound homing behavior. Home is in the right box and the trap in the left one.

detect home produces a motion with a slight turn. Thus, the robot uses it until it collides with or detects the wall, and then the escape from collision controller kicks in, allowing for a type of bad wall following behavior that permits a faster exploration of the wall contours. When the robot enters the left box, it must go around the fixed light and look behind it, as it may be masking its home. Fig. 5 shows the three level architecture evolved for this behavior, consisting of the monolithic homing behavior, an escape from collisions behavior, a two level wall following behavior that shares the escape from collisions controller, and a high level module that coordinates them.

One of the controllers obtained was implemented on a real robot and in Fig. 6 we present four pictures of the operation of a real Rug Warrior, in a real environment in which home was hidden. This example displays the ability of the robot to search for its home and finally find it through the use of a modular architecture containing wall following, escape from collisions and monolithic homing lower level modules.

As we will comment on the next section, one of the main lines of work in the context of evolutionary robotics is the problem of establishing global architectures that permit the automatic implementation of complex behaviors efficiently and that do not present duplications of task solvers. At the same time, these should allow for the consideration of previous experience through the reuse of previously evolved controllers. In this section we have briefly summarized one approach to the problem as an illustration so that the reader obtains an idea of what difficulties are encountered.

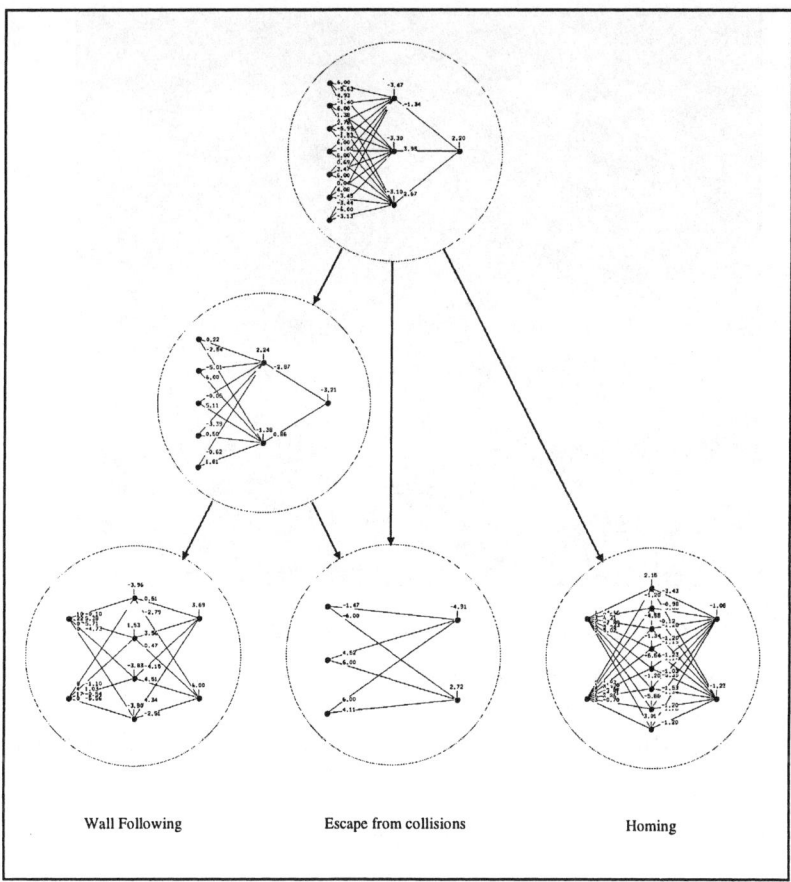

Fig. 5. Three level compound homing architecture.

8 Open Issues

The use of evolutionary techniques in the design and implementation of robots is an exciting and growing field of research. There are still many problems to be solved to bring the technology out of petit problems and into the real world of performing useful tasks. Stone [75] argues that we are still far from the construction of even the simplest artificial animals for three reasons. There is no "universally accepted paradigmatic approach" to the study of nervous systems. Moreover, animats still do not possess some types of learning or, as he calls them, "functional primitives" in their nervous system that are necessary for perceptual

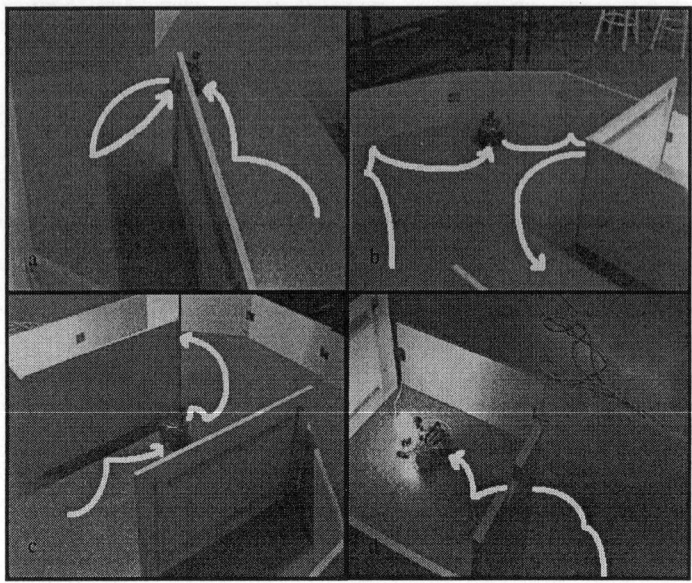

Fig. 6. Compound homing behavior implemented on a real robot.

processes. Finally, currently employed evolutionary techniques only model a small, finite set of the properties found in natural evolutionary systems.

These basic points lead to the consideration of several aspects of evolutionary robotics that will have to be studied in more detail. For instance, as shown throughout the chapter, most authors employ just a small number of standard, general purpose, neural network topologies. In contrast, nature has followed a different path and designed distinct types of neural circuits, or even neurons for different tasks. Thus, specific time processing, spatial processing or other types of structures should be made available to the evolutionary processes in order to obtain behavior controllers that are better adapted through the combination of different structures. An additional architectural aspect that has seldom been taken into account is that of modulation, or modulating behaviors. It is very interesting to study possible structures that may lead to different types of networks modulating each other, so that behaviors interact seamlessly and a higher degree of behavioral continuity and coherence may be achieved.

We believe that, to obtain complex heterogeneous cognitive architectures, it is necessary to take into account that evolution cannot start from the beginning. We are not trying to replicate many million years of evolution in a fast computer program. Instead, we must provide the evolutionary process with a set of basic structures that will be adapted to the environment or problem in hand. To do this without creating an immense search space, one promising approach is that of evolving the development programs or even some type of development environments guided by evolved rules, which, starting from basic structures, generate the desired controllers.

As for not starting evolution from the absolute beginning, very little work has been carried out on the reusability of pieces or behaviors by other systems [22]. It is necessary to find new mechanisms that permit the use of the experience gained through previous evolutions: mechanisms for transporting behaviors learnt in one platform for some type of environment to other platforms and other environments.

From the point of view of the hardware employed, there is a need for more complex sensors to involve more complete knowledge of the environment in the cognitive process. This has been a problem for behavior based evolutionary robotics because of the dimensionality of the sensor space this would imply due to the emphasis on direct connection between the behavior controllers and the world. One possible path to solving this problem is through hybridization with other paradigms and the inclusion of virtual sensors. Obviously, incorporating other paradigms also entails their drawbacks, so one must be very careful not to distort the behavior based controller to such an extent that it loses its advantages. In addition, very few of the architectures support active or goal-driven perception, i.e. taking action to improve the data scenario being handled. This leads to poor or ambiguous perception in situations where very simple active sensing mechanisms would greatly improve the controller's image of the world.

Attention is another aspect that is complementary to active sensing and which has not been sufficiently considered in the evolutionary robotics community. It prevents the system from becoming overwhelmed by the amount of information it receives from the world. In other words, it allows it to select what is relevant to its goals. Some attention mechanisms have been devised, such as those presented by Foner and Maes [26], but much more work along these lines is still required.

From the point of view of information processing and interaction with the world, most behavior based robotic systems have been implemented as purely reactive systems. We believe that this is a very limited approach. There are many instances where some type of deliberative mechanism would be useful for the integration of these types of architectures in real world problems. Examples of this have already been presented here.

Summing up, the behavior based approach to robotics has been shown to be very promising for the development of really autonomous systems that can operate in partially unknown dynamic environments. This approach still has many unresolved problems, especially those caused by the complexity arising from the interactions between the robots and the environment and their behaviors among themselves. One path to solving these complexity problems is through the use of evolutionary techniques, and many authors have resorted to this in the last few years. Currently, there is a very intense research effort in this direction. New systems are being developed every day and new and exciting sub-fields are constantly being generated. Predictably, within a few years, this chapter will be out of date and solutions will have been found for many of the problems presented here. We sincerely hope so.

Acknowledgements

This work was funded by the CICYT of Spain under project TAP98-0294-C02-0 and Xunta de Galicia under project PGIDT99PXI10503A.

References

1. Arkin, R.C. (1998), Behavior Based Robotics, MIT Press, Cambridge, MA.
2. Baldwin, J.M. (1896), "A New Factor in Evolution", American Naturalist, Vol. 30, pp.441-451.
3. Becerra, J.A. (1999), Diseño de Comportamientos de un Robot Autónomo Considerando Información Temporal, Master's Thesis, Universidade da Coruña.
4. Beer, R.D. and Gallagher, J.C. (1992), "Evolving Dynamical Neural Networks for Adaptive Behavior", Adaptive Behavior, Vol. 1, No. 1, pp. 91-122.
5. Boers, E. and Kuiper, H. (1992), Biological Metaphors and the Design of Modular Artificial Neural Networks, Master's Thesis, Leiden University.
6. Braitenberg, V. (1984), Vehicles: Experiments in Synthetic Psychology, MIT Press, Cambridge, MA.
7. Brooks, R. (1987), Planning is Just a Way of Avoiding Figuring Out What to Do Next, Working Paper 303, MIT AI Laboratory.
8. Brooks, R.A. (1990), "Elephants don´t Play Chess", In Designing Autonomous Agents: Theory and Practice from Biology to Engineering and Back, Patti Maes (Ed.), MIT Press, pp. 3-15.
9. Brooks, R.A. (1991), "Intelligence without Representation", Artificial Intelligence, Vol. 47, pp. 139-159.
10. Brooks, R.A. (1991), "New Approaches to Robotics", Science, Vol.253, pp. 1227-1231.
11. Cangelosi, A., Parisi, D., and Nolfi, S. (1994), "Cell Division and Migration in a Genotype for Neural Networks", Network, Vol. 5, pp. 497-515.
12. Chalmers, D.J. (1992), "Subsymbolic Computation and the Chinese Room", Symbolic and Connectionist Paradigms: Closing the Gap, Dinsmore, J., (Ed.).
13. Cliff, D. and Miller, G. F. (1996), "Co-evolution of Pursuit and Evasion II: Simulation Methods and Results", From Animals to Animats 4, MIT Press, pp. 506-515.
14. Cliff, D., Harvey, I., and Husbands, P. (1992), Incremental Evolution of Neural Network Architectures for Adaptive Behaviour, Tech. Rep. No. CSRP256, Brighton, School of Cognitive and Computing Sciences, University of Sussex, UK.
15. Cliff, D.T., Husbands, P., and Harvey, I. (1993), "Evolving Visually Guided Robots", From Animals to Animats 2. Proceedings of the Second International Conference on Simulation of Adaptive Behaviour (SAP 92), J-A. Meyer, H. Roitblat and S. Wilson (Eds.), MIT Press Bradford Books, Cambridge, MA, pp. 374-383.
16. Dain, R.A. (1998), "Developing Mobile Robot Wall-Following Algorithms Using Genetic Programming", Applied Intelligence, Vol. 8, pp. 33-41.
17. Dellaert, F. and Beer, R.D. (1996), "A Developmental Model for the Evolution of Complete Autonomous Agents", From Animals to Animats 4, MIT Press, pp. 393-401.

18. Dorigo, M. and Colombetti, M. (1998), Robot Shaping: An Experiment in Behavior Engineering, MIT Press.
19. Duro, R.J. and Santos, J. (1999), "Bearance of Noise on Performance and Distribution of Computations Using DTB", Ninth International Conference on Artificial Neural Networks, Edinburgh, UK.
20. Duro, R.J., Crespo, J. L., Santos, J. (1999), "Training Higher Order Gaussian Synapses", Foundations and Tools for Neural Modeling, J. Mira & J.V. Sánchez-Andrés (Eds.), Lecture Notes in Computer Science, Vol. 1606, Springer-Verlag, Berlín, pp. 537-545.
21. Floreano, D. and Mondada, F. (1996), "Evolution of Homing Navigation in a Real Mobile Robot", IEEE Transactions on Systems, Man, and Cybernetics, Part-B, Vol. 26, pp. 396-407.
22. Floreano, D. and Mondada, F. (1998), "Evolutionary Neurocontrollers for Autonomous Mobile Robots", Neural Networks, Vol. 11, pp. 1461-1478.
23. Floreano, D. and Nolfi, S. (1997), "Adaptive Behavior in Competing Co-evolving Species", P. Husbands and I. Harvey (Eds.), Proceedings of Fourth European Conference on Artificial Life (ECAL 97), Complex Adaptive Systems Series, MIT Press, pp. 378-387.
24. Floreano, D. and Nolfi, S. (1997), "God Save the Red Queen!. Competition in Co-evolutionary Robotics", J. Koza, K. Deb, M. Dorigo, D. Fogel, M. Garzon, H. Iba, and R.L. Riolo (Eds.), Proceedings of the 2nd International Conference on Genetic Programming, Stanford University, pp. 398-406.
25. Fogel, L.J. (1964), On the Organization of Intellect, PhD Dissertation, University of California.
26. Foner, L.N. and Maes, P. (1994), "Paying Attention to What's Important: Using Focus of Attention to Improve Unsupervised Learning", Proceedings of the third International Conference on Simulation of Adaptive Behavior (SAB94), pp. 256.
27. Gallagher, J.C., Beer, R.D., Espenschied, K.S., and Quinn, R.D. (1994), Application of Evolved Locomotion Controllers to a Hexapod Robot, Technical Report CES-94-7, Dept. of Computer Engineering and Science, Case Western Reserve University.
28. Giles, L. and Maxwell, T. (1987), "Learning Invariance and Generalization in High Order Neural Networks", Applied Optics, Vol. 26, No. 23, pp. 4972-4978.
29. Goldberg, D.E. (1989), Genetic Algorithms in Search, Optimization and Machine Learning, New-York, Addison-Wesley.
30. Gomez, F. and Miikkulainen, R. (1997), "Incremental Evolution of Complex General Behavior", Adaptive Behavior, Vol. 5, No. 3/4, pp. 317-342.
31. Gruau, F. (1992), "Genetic Systems of Boolean Neural Networks with a Cell Rewriting Developmental Process", Combination of Genetic Algorithms and Neural Networks, IEEE Computer Society Press.
32. Gruau, F. (1992), Cellular Encoding of Genetic Neural Networks, Technical Report 92-21, Laboratoire de L'Informatique du Parallelisme, Ecole Normale Superieure de Lyon.
33. Guillot, A. and Meyer, J.A. (1997), "Synthetic Animals in Synthetic Worlds", In Kunii et Luciani (Eds), Synthetic Worlds, John Wiley and Sons.
34. Harnad, S. (1990), "The Symbol Grounding Problem", Physica D, Vol. 42, pp. 335-346.
35. Harvey, I. (1992), "Species Adaptation Genetic Algorithms: A Basis for a Continuing SAGA", F. Varela and P. Bourgine (Eds.), Towards a Practice of Autonomous

Systems: Proceedings of the First European Conference on Artificial Life (ECAL91), MIT Press, Cambridge, MA, pp. 346-354.

36. Harvey, I. (1996), "Artificial Evolution and Real Robots", Proceedings of International Symposium on Artificial Life and Robotics (AROB), Masanori Sugisaka (Ed.), Beppu, Japan, pp. 138-141.
37. Harvey, I., Husbands, P., and Cliff, D. (1993), "Issues in Evolutionary Robotics", J-A. Meyer, H. Roitblat, and S. Wilson (Eds.), From Animals to Animats 2. Proceedings of the Second International Conference on Simulation of Adaptive Behavior (SAB92), MIT Press, Cambridge, MA, pp. 364-373.
38. Holland, J. (1975), Adaptation in Natural and Artificial Systems, University of Michingan Press, Ann Arbor.
39. Husbands, P., Harvey, I., Cliff, D., and Miller, G. (1994), "The Use of Genetic Algorithms for the Development of Sensorimotor Control Systems", P. Gaussier and J.-D. Nicoud (Eds.), From Perception to Action, IEEE Computer Society Press, Los Alamitos CA, pp. 110-121.
40. Jakobi, N. (1997), "Evolutionary Robotics and the Radical Envelope of Noise Hypothesis", Adaptive Behavior, Vol. 6, No. 2, pp. 325-368.
41. Jakobi, N., (1997), "Half-Baked, Ad-Hoc and Noisy Minimal Simulations for Evolutionary Robotics", Fourth European Conference on Artificial Life, P. Husbands and I. Harvey (Eds.), MIT Press.
42. Jakobi, N., Husbands, P., and Harvey, I. (1995), "Noise and the Reality Gap: the Use of Simulation in Evolutionary Robotics", F. Morán, A. Moreno, J.J. Merelo and P. Cachou (Eds.), Advances in Artificial Life: Proceedings of the Third European Conference on Artificial Life (ECAL 95), Lecture Notes in Artificial Intelligence, Vol. 929, Springer-Verlag, pp. 704-720.
43. Kodjabachian, J. and Meyer, J-A (1995), "Evolution and Development of Control Architectures in Animats", Robotics and Autonomous Systems, Vol. 16, pp. 161-182.
44. Kodjabachian, J. and Meyer, J-A (1998), "Evolution and Development of Modular Control Architectures for 1-D Locomotion in Six-Legged Animats", Connection Science, Vol. 10, No. 3-4, pp. 211-37.
45. Kodjabachian, J. and Meyer, J-A (1998), "Evolution and Development of Neural Controllers for Locomotion, Gradient-Following, and Obstacle-Avoidance in Artificial Insects", IEEE Transactions on Neural Networks, Vol. 9, No.5, pp. 796-812.
46. Koza, J.R. (1989), "Hierarchical Genetic Algorithms Operating on Populations of Computer Programs", Proc. 11th Int. Joint Conf on Artificial Intelligence, San Mateo, California.
47. Koza, J.R. (1994), Genetic Programming II: Automatic Discovery of Reusable Programs, MIT Press, Cambridge, MA.
48. Lee, W-P., Hallam, J., and Lund, H.H. (1996), "A Hybrid GP/GA Approach for Co-evolving Controllers and Robot Bodies to Achieve Fitness-Specified Tasks". Proceedings of IEEE Third International Conference on Evolutionary Computation, IEEE Press.
49. Lund, H.H. and Hallam, J., Sufficient Neurocontrollers can Be Surprisingly Simple, Research Paper 824, Department of Artificial Intelligence, University of Edinburg, 1996.
50. Lund, H.H. and Miglino, O. (1998), "Evolving and Breeding Robots", Proceedings of First European Workshop on Evolutionary Robotics, Springer-Verlag.

51. Lund, H.H., Miglino, O., Pagliarini, L., Billard, A., and Ijspeert, A. (1998), "Evolutionary Robotics – A Children's Game", Proceedings of IEEE 5th International Conference on Evolutionary Computation.
52. Maes, P. (1990), "A Bottom-up Mechanism for Behavior Selection in an Artificial Creature", Proceedings of the First International Conference on Simulation of Adaptive Behavior (SAB90).
53. Mataric, M. and Cliff, D. (1996), "Challenges in Evolving Controllers for Physical Robots", Robotics and Autonomous Systems, Vol. 19, No. 1, pp. 67-83.
54. Mataric, M.J. (1992), "Integration of Representation into Goal Driven Behavior Based Robotics", IEEE Transactions on Robotics and Automation, Vol. 8, No. 3, pp. 304-312.
55. Matellán, V., Fernández, C., and Molina, J.M. (1998), "Genetic Learning of Fuzzy Reactive Controllers", Robotics and Autonomous Systems, Vol. 25, pp. 33-41.
56. Meeden, L. (1996), "An Incremental Approach to Developing Intelligent Neural Network Controllers for Robots", IEEE Transactions on Systems, Man and Cybernetics Part. B: Cybernetics, Vol. 26, No. 3, pp. 474-485.
57. Menczer, F. and Belew, R.K. (1994), "Evolving Sensors in Environments of Controlled Complexity", Artificial Life IV, R. Brooks and P. Maes (Eds.), Cambridge, MA, MIT Press, pp. 210-221.
58. Miglino, O., Lund, H.H., and Nolfi, S. (1995), "Evolving Mobile Robots in Simulated and Real Environments", Artificial Life, Vol. 2, No. 4, pp 417-434.
59. Miglino, O., Nafasi, K., and Taylor, C. (1995), "Selection for Wandering Behavior in a Small Robot", Artificial Life, Vol. 2, pp. 101-116.
60. Mondada, F. and Floreano, D. (1995), "Evolution of Neural Control Structures: Some Experiments on Mobile Robots", Robotics and Autonomous Systems, Vol. 16, pp. 183-195.
61. Nolfi, S. (1997), "Evolving Non-Trivial Behaviors on Real Robots: A Garbage Collecting Robot", Robotics and Autonomous Systems, Vol. 22, pp. 187-198.
62. Nolfi, S. (1997), "Using Emergent Modularity to Develop Control Systems for Mobile Robots", Adaptive Behavior, Vol. 5, No. 3/4, pp. 343-363.
63. Nolfi, S. and Parisi, D. (1997), "Learning to Adapt to Changing Environments in Evolving Neural Networks", Adaptive Behavior, Vol. 5, No. 1, pp. 75-98.
64. Nolfi, S., Elman, J., and Parisi, D. (1994), "Learning and Evolution in Neural Networks", Adaptive Behavior, Vol. 1, pp. 5-28.
65. Nolfi, S., Floreano, D., Miglino, O., and Mondada, F. (1994), "How to Evolve Autonomous Robots: Different Approaches in Evolutionary Robotics". R. Brooks and P. Maes (Eds.), Proceedings of Fourth International Conference on Artificial Life, Cambridge, MA, MIT Press.
66. Nordin, P., Banzhaf, W., and Brameier, M. (1998), "Evolution of a World Model for a Miniature Robot Using Genetic Programming", Robotics and Autonomous Systems, Vol. 25, pp. 105-116.
67. Patel, M.J., Colombetti, M., and Dorigo, M. (1995), "Evolutionary Learning for Intelligent Automation: a Case Study", Intelligent Automation and Soft Computing Journal, Vol. 1, No. 1, pp. 29-42.
68. Rechenberg, I. (1965), Cybernetic Solution Path of an Experimental Problem, Library Translation 1122, Royal Aircraft Establishment, Farnborough, UK.
69. Reynolds, C.W. (1994), "Evolution of Corridor Following Behavior in a Noisy World", From Animals to Animats 3, MIT Press, pp. 402-410.

70. Rumelhart, D.E., Mclelland, J.L. (1986), Parallel Distributed Processing 1, 2. MIT Press.
71. Salomon, R. (1997), "The Evolution of Different Neuronal Control Structures for Autonomous Agents", Robotics and Autonomous Systems, Vol. 22, pp. 199-213.
72. Santos, J. and Duro, R.J. (1998), "Evolving Neural Controllers for Temporally Dependent Behaviors in Autonomous Robots", Tasks and Methods in Applied Artificial Intelligence, A.P. del Pobil, J. Mira and M. Ale (Eds.), Lecture Notes in Artificial Intelligence, Vol. 1416, Springer-Verlag, Berlín, pp. 319-328.
73. Schwefel, H.P. (1975), Evolutionsstrategie und Numerische Optimierung, Ph. D. Thesis, Technische Universität Berlin.
74. Sharkey, N.E. (1997), "The new Wave in Robot Learning", Robotics and Autonomous Systems, Vol. 22, pp. 179-185.
75. Stone, J.V. (1995), "Evolutionary Robots: Our Hands In Their Brains?", Artificial Life IV, pp. 400-405.
76. Walter, G. (1953), The Living Brain, Norton, New York.

Chapter 4. A Simulation Environment for the Manipulation of Naturally Variable Objects

George Coghill

Department of Electrical and Electronic Engineering, University of Auckland, New Zealand

g.coghill@auckland.ac.nz

Abstract. One area where the use of robots has been impractical, up to the present time, is where the objects they handle are of an irregular shape. Robots are now very effective in manufacturing industries where their precision operations can be preprogrammed to produce machined parts of known dimensions to required tolerances. However, it is difficult to use robot arms to manipulate objects that are irregular and unpredictable. For example, in the food processing industry it is necessary to carry out operations such as shelling seafood, or filleting fish. The major problems are caused by inconsistencies in size, shape and texture. This work describes the possibility of using adaptive robot controllers to learn the correct operations by trial and error. The adaptive element is provided by a modified CMAC neural network, which implements a kind of reinforcement learning to gradually improve the robots actions. Rather than build a physical robot to carry out such a task, it was felt that a cheaper and more effective approach would be to create a realistic computer simulation environment in which to test out these ideas. This avoids spending a large amount of effort trying to maintain a real robot, which may eventually turn out to be inadequate to successfully execute the tasks required of it. By building an effective model, we may learn about the desired characteristics of such a robot and at the same time have a re-useable system with which we may tackle similar problems. We describe the system basics and our current progress towards these goals.

Keywords. Natural variability, 3D, simulation environment, neural networks, virtual robot

1 Introduction

The motivation for this project came from two different areas of previous research. The first was the development of a very promising intelligent motion control system, which was carried out at the University of Auckland[3]. The climax of this work was the creation of a 3-D simulation environment in which a biped robot was able to learn to walk. While this was impressive, it was difficult to think of really useful applications. Applications such as Mars Rovers, or undersea robot motion control (where communication band-width is low) seem fairly exotic as far as practical industrial applications are concerned. One of our previous research experiences with building robots was frustrating in that most of the effort was found to be involved in mechatronics and just keeping the robot maintained. The prospect of actually using these robots to carry out conceptual investigations always seemed to be on the horizon.

The success and the realistic nature of the simulations carried out by the walking robot led to a realization that many other simple human tasks might be carried out by the same approach. The following description outlines the component parts of such an effort and where, surprisingly, existing research results can fill in the missing parts of the puzzle.

The eventual aim of the project then, is to realistically simulate a simple operation such as de-boning a fish. A problem which is easily carried out by a human expert, but which is too variable in its nature, to be handled easily by an automatic system.

2 Learning Movement

The aim of this project is to be able to pick up and carry out a sequence of useful actions on a natural object. Of course, since the object is in a simulation environment, these objects will have to be modelled. In addition the manipulator "hands" will also be augmented by various tools to carry out specific tasks. This means that movements will involve interacting with the object to be manipulated, either with two hands, or a hand and a tool.

Visualisation of this task is obviously an important factor, but in this section we concentrate largely on the learning and control of movement of an arm and gripper. The nature of these components is discussed later.

The ultimate aim is to provide a realistic working model, so that the migration from the simulation environment to the real world is as painless as possible.

Most existing robot motion controllers are based upon precision movements in an engineering environment such as car assembly. Here the motion is just a sequence of precision steps, which may be complex in outward appearance, but essentially consist of a finite sequence of precision movements that were programmed into the controlling computer.

In our case, high precision is not the most important factor. The need is for the motion controller to learn to adapt to handle objects that have a variable shape, etc. Movements therefore have to cater for unanticipated changes with each new object. The area of most interest then, is that of adaptive control [1].

2.1 Reinforcement Learning

There are many variants to this concept; here we try to concentrate on the work carried out by Russell Smith [2][3]. Essentially, a reinforcement learning system is one where the success of an action is rewarded (or a failure is penalised). The problem is to try to select suitable strategies so that this reward is maximised. This is a well-known problem in the AI area [4], where it is known as the *credit assignment* problem. The difficulty is to determine which parameters in a complex system contribute most to the outcome. Again, there are many approaches, but here we are interested in the use of a modified CMAC [5] neural network as the adaptive element in the system. This means that the parameters of interest are the network weights. In the CMAC, as we shall see, the problem lies in identifying the most *eligible* weights as far as contribution to outcomes is concerned.

2.2 The CMAC

The Cerebellar Model Articulation Controller may be viewed, in its simplest terms, as a look-up table. Consider figure 1, where we have two analogue input channels. These inputs are quantised into discrete levels through digital to analogue converters. Each quantised state serves as an address to look up a weight value.

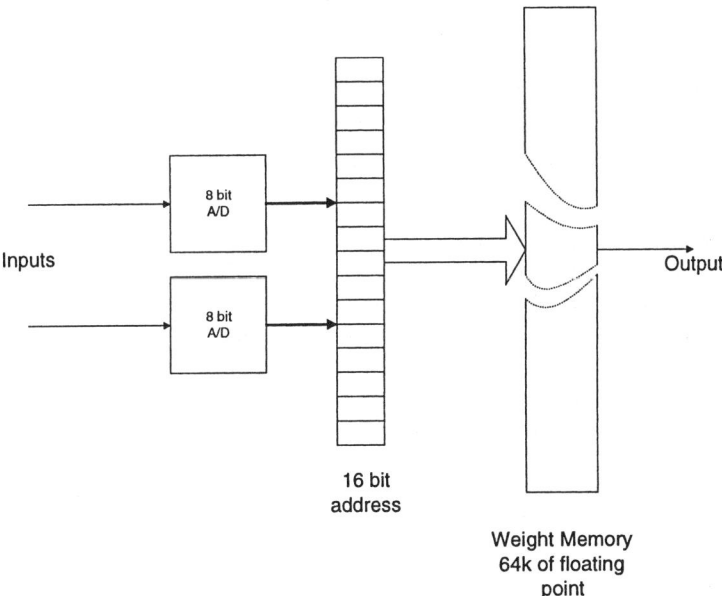

Figure 1: A very simple CMAC

The output will be a weight value and this is compared with some desired value resulting in a modification to that stored weight which is aimed at reducing the system error. There are two serious problems with this model. The first is noise immunity. If there is a noise spike in the input, the weight value selected will be inappropriate. This problem is solved at the cost of increased memory requirement by having more than one (always a power of 2) weight tables, slightly offset to each other with respect to the input value. If we imagine the weight memory in figure 1 as a two dimensional table (one dimension for each input), we could view four weight tables, slightly offset to each other, as shown in figure 2.

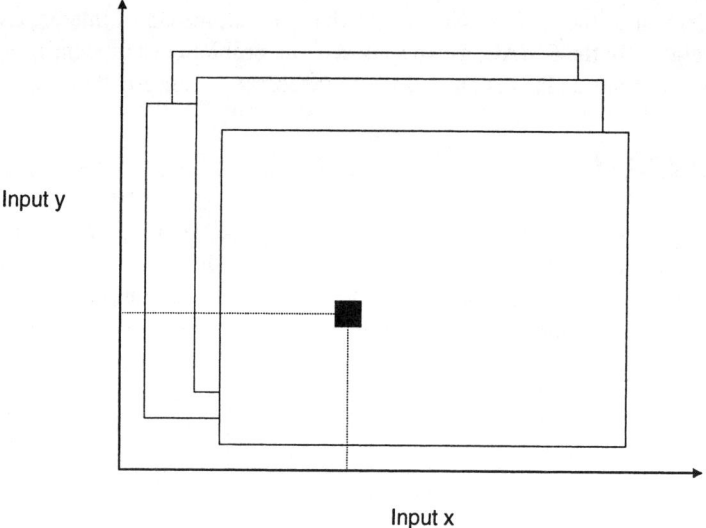

Figure 2: Overlapping weight tables

This improves the situation with regard to input noise immunity. We just need to create enough weight tables to obtain a more stable output. The system output is now the average of the selected weights (one from each table). Unfortunately, we have actually made our second problem even worse. With two input dimensions of only eight bit resolution and four weight tables, we require 256k weights. Any increase in input dimensionality, table overlapping, or input quantisation and the memory requirements explode to unacceptable levels. The CMAC solution to this problem is to use address hashing. The large virtual address space is hashed down to a much smaller physical memory space. The down side of this solution is that there is the possibility of hashing collisions due to two virtual addresses mapping to the one physical address. However, judicious selection of a suitable hashing function and a careful design should reduce collisions to an insignificant level.

2.1 The FOX

The FOX network [3], is a CMAC type network with the addition of a novel eligibility scheme. As described, due to the nature of the CMAC network, only a small subset of weights is updated with each learning step. These weights must have a high likelihood of being relevant to the outcome. This is especially true of movement learning systems where each successive point is physically close to the previous. The CMAC's behaviour, that of local generalisation, is well suited to this type of task. The FOX network has added functionality in that it assigns an eligibility value to recently adapted sets of weights, in addition to the current set. This eligibility function decays with time. Eligibility has been used before in the CMAC[6]. However, the FOX scheme has a number of advantageous properties that enable it to be tuned to specific applications. The FOX network has been incorporated into a modular control architecture, which may be used to build components for complex control systems that require some form of intelligent motion control.

3 The Graphical Environment

One of the decisions made in the design of the graphical environment was to try to use a very high-level 3D graphics system. At present, our needs are fulfilled by the povray system [7], which is available as free-ware. We are not currently aiming to produce hard commercial systems and so we are more interested in exploring new graphical environments at the least expense of development effort.

POV is effectively a language, which may be used to describe the specifics of a desired image, and the povray program interprets this specification. This may be used to create rendered images with a realistic simulation of lighting interaction (figure 3). Animations may also be created.

4 Creating Objects

It is intended to gradually build up a library of tools together with a library of natural objects, which emulate the physical properties of real objects. This is the area where most of the effort is required, both in skill sets and in computational activity.

Fortunately, this work is related to a field of research, which is already well established with some good results. This is in the disciple of medicine. One of the earliest recognised application areas for virtual reality was seen to be that of virtual surgery. It can provide a realistic training environment without risk to human patients. One of the more spectacular developments has been the work carried out by a group at Auckland University and the University of California[8][9].

Figure 3: An example povray image

This system, in fact, incorporates more features than required for our objects. Their work has produced a very realistic biomechanical model of the heart. They have developed a computational model, which uses specialised *finite element methods* to produce a virtual artefact that demonstrates the physical behaviour of soft tissue and its interactions with surgical instruments. Their model provides for much more than this, but these particular features are of great interest to this project's endeavours. Their model also has to be of a much higher resolution than ours. Slight inaccuracies when processing food items is likely to have an economic cost, rather than a human cost. Most of the food items envisaged would lend themselves well to a finite element approach; however, there are areas where different methods must be used. For example, serious work has been done in the area of modelling the behaviour of fabric[10]. The crinkling behaviour of folding cloth is very different from that of deformable, 3D soft tissue. The first problem we are studying is that of creating a realistic model of a fish which has to be deboned.

5 Image Understanding

One of the main problems in manipulating objects is that of reducing the dimensionality of images, while at the same time, preserving the essential properties. Many possible approaches might be considered. However, the following work [11], which was carried out in our own department, provides suitable outputs for mapping to a finite element model.

The *Mind's Eye Network* is a modular backpropagation (BP) style network. Each module is small having typically 3x3 inputs fed from input receptive fields. A module is in itself, a backpropagation model with a hidden layer. It also has outputs, which are fed (in addition to the feed-forward connection), as inputs to the neighbouring modules on the same layer. This single layer of BP modules feeds to an output slab, which forms a grid of output points. The network, as shown in figure 4 for a two dimensional object, is able to extract structure from a natural object and correct for noise corruption due to poor image reproduction.

A three dimensional version of the network has to be trained on images which show underlying bone structure.

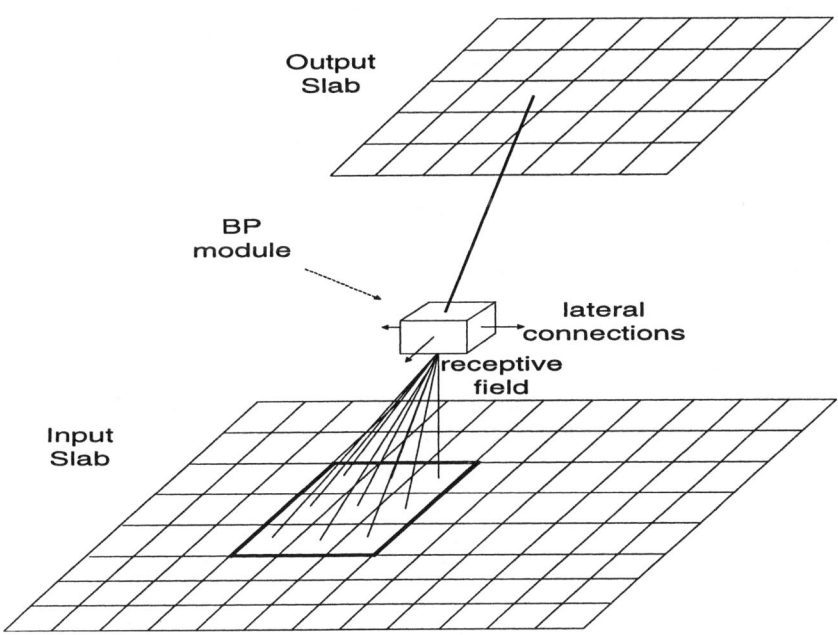

Figure 4: Mind's Eye Network

6 Manipulating Objects

As the human hand has been well evolved to the manipulation of natural objects, it seems sensible to design a gripper, which captures the main features. Research has already been published in this area [12][13][14]. It is agreed that a robot gripper with two fingers and a thumb can simulate the practical functionality of the hand to a satisfactory degree. A rough schematic of such a hand is shown in figure 5.

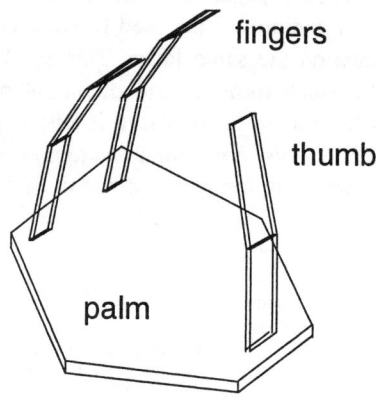

Figure 5: Gripper

Details of the best relative dimensions for such a gripper still have to be worked out. In fact research is currently being carried out independently on this aspect [15]. Again, it is intended that the FOX system will learn the adaptive control behaviour necessary for successful operation of this hand. The other important tool needed for successful processing of soft-tissue is a scalpel. However, the simulation of such a simple instrument should be relatively straightforward.

7 Intelligent Motion Control and Task Automation

The simulation environment, which is currently under development, will have two main aspects, the intelligent movements of the manipulators and their interactions with deformable "natural" objects. The FOX is intended to handle adaptive motion control issues requiring trajectory and joint movement. However, the general sequence of desired sub-tasks has to be entered into and understood by the system. These task descriptions will be generated as part of ongoing departmental research into *instructable systems* [16].

8 Future Research

As may be appreciated from the above description, this project requires a fairly multidisciplinary approach. It has many complex components, but we believe that all of the conceptual problems have already been solved. The amount of computing power required to complete this project is not beyond what is currently available. One of the strengths of this approach is the possibility of solving many food processing problems with enough confidence to enable real automated process lines to be developed.

References

1. Petros A. Ioannou, "Robust adaptive control"; Prentice-Hall, 1996.

2. Russell Smith, "An autonmous robot controller with learned behavior"; The Australian Journal of Intelligent Information Processing Systems, 3(2), Winter 1996.

3. Russell Smith, "Intelligent motion control with an artificial cerebellum"; PhD Thesis, Department of Electrical ans Electronic Engineering, University of Auckland, New Zealand, 1999.

4. L. P. Kaebling & A.W. Moore, "Reinforcement learning: a survey". Journal of Artificial Intelligence Research, 4(Jan-Jun):237-285, 1996.

5. W. Thomas Miller, Filson Glanz, & Gordon Kraft. "CMAC: An associative neural network alternative to backpropagation"; Proceedings of the IEEE, 78(10), 1990.

6. Chun-Shin Lin & Hyongsuk Kim, "CMAC-based adaptive critic self-learning control"; IEEE Transactions on Neural Networks, 6(3), 1995.

7. http://www.povray.org/ , the Persistence of vision raytracer, copyright © 1995-1998 Hallam Oaks Pty Ltd.

8. K D Costa, P J Hunter, J M Rogers, J M Guccione, L K Waldman, and A D McCulloch, "A three-dimensional finite element method for large elastic deformations of ventricular myocardium: Part I – Cylindrical and spherical polar coordinates"; ASME J. Biomech. Eng., 118(5):452-63, 1996.

9. K D Costa, P J Hunter, J M Rogers, J M Guccione, L K Waldman, and A D McCulloch, "A three-dimensional finite element method for large elastic deformations of ventricular myocardium: Part II – Prolate spherical coordinates"; ASME J. Biomech. Eng., 118(5):464-72, 1996.

10. D.E. Breen, D.M. House & M.J. Wozny, "Predicting the drape of woven cloth using interacting particles"; Computer Graphics (Proc. SIGGRAPH'94), 28(4), pp 365-372, 1994.

11. Stucke, T.J., Coghill, G.G. and Creak, G.A., " The mind's eye: extracting structure from naturally variable objects"; Neural, Parallel and Scientific Computations, Vol. 2, No 1, March, 1994, pp. 93-103.

12. L. D. Harmon, "Automated touch sensing: A Brief Perspective and Several New Approaches"; Proceedings-IEEE International Conference on Robotics and Automation, pages 326-331, March 1984.

13. E. Y. Chao, J. D. Opgrande, & F. E. Axmear, "Three-Dimensional force analysis of the finger joints in selected isometric hand functions"; Journal of Biomechanics, 9:387-396, 1976.

14. F. E. Hazelton, G. L. Smidt, A. E. Flatt, and R. J. Stephens," The Influence of Wrist Position on the Force Produced by the Finger Flexors"; Journal of Biomechanics, 8:301-306, 1974.

15. Angela Nugent, "Two finger with opposing thumb anthropomorhic robotic gripper with minimum gripping force"; Research Thesis, Texas Tech. Univ.

16. B. Macdonald's home page, http:/linux.ele.auckland.ac.nz/~macdon/

Chapter 5. Behavior-Decision Fuzzy Algorithm for Autonomous Mobile Robot

Yoichiro Maeda

Faculty of Information Science and Technology
Osaka Electro-Communication University
18-8 Hatsu-cho, Neyagawa, Osaka 572, Japan
maeda@mdlab.osakac.ac.jp

Abstract. The ambiguous state recognition when an autonomous mobile robot moves around a complex environment is one of the most important problems in the robot control. We propose a construction method of the behavior-decision system using fuzzy algorithms capable of expressing sequence flow which includes a mixture of both crisp and fuzzy processing. We also propose in this paper a method of tuning algorithms for giving robots the autonomous ability to judge purposes of actions like human. In this method, we try to express ambiguous situations which a robot will encounter and decision algorithm flows by using fuzzy algorithms with fuzzy branch controlled threshold, which we call it the behavior-decision fuzzy algorithm. Furthermore, we introduce a fuzzy inference shell and a mobile robot simulator developed for this research. Finally, we report some results of computer simulations and experiments concerning an evaluation of this method supposed simple in-door environments.

Keywords. Fuzzy algorithms, behavior decision, ambiguous state recognition, autonomous mobile robot, fuzzy shell.

1 Introduction

Recently the use of fuzzy logic in control applications in recent years has been remarkable. It is no exaggeration to say that fuzzy control has become recognized as the best model available for representing the knowledge of an experienced or skilled professional. Of the many features of knowledge representation methods which utilize fuzzy logic, the most advantageous include the ability to represent human knowledge, simply, using knowledge structures and the fact that rule maintainability is excellent due to the availability of compact representations of knowledge. However, in spite of the recent widespread use of fuzzy technology, certain remaining weak points such as insufficient learning ability, explanatory capacity, delays in logical analysis research and so on are beginning to require attention.

In any attempt to produce a high-order intelligent robot with autonomous behavior-decision capabilities, it is important to develop some new methodology [1][2], based on intelligent processing and model building methods, which is at the same time capable of describing complex behavioral sequences both flexibly and using macros. Unfortunately, even though sequence flow of this kind can be described easily using a representation method based on conventional AI production rules, such a method cannot be used to handle ambiguous states. On the other hand, use of a representation method based on conventional fuzzy inference can handle ambiguities, but cannot be used where sequence flow is required.

On this research [3][4][5], we therefore try to express the macro behavior-decision algorithm close to the one which humans use every day by utilizing fuzzy algorithms. Fuzzy algorithms can both describe ambiguous situations and represent sequence flows which mix fuzzy as well as non-fuzzy (crisp) processing. Furthermore, we also propose in this paper a method of tuning algorithms for giving robots the autonomous ability to judge purposes of actions like humans, where behavior is self-determined based on the situation in which they find themselves.

2 Fuzzy Branch Control in Fuzzy Algorithms

Inference methods based on MAX-MIN CG method used in conventional fuzzy control systems are insufficient from the standpoint of providing a methodology for writing more human-like macro judgement and decision-support algorithms. On the other hand, it has already been reported that fuzzy algorithms, capable of representing sequence flow including both crisp and fuzzy operations, are effective in expressing decision-support level control rules of this kind.

2.1 Original Fuzzy Algorithms

Fuzzy algorithms[6][7][8] are a concept first introduced by L. A. Zadeh as one possible tool for approximation analysis in decision making. According to Zadeh, fuzzy algorithms are "ordered sets of fuzzy instructions" and has the four following types of expressional capacity. These types are fuzzy definitional, generational, relational and decisional algorithms. The former in these types, the more primitive the expressiveness of the algorithm. The later in these types, the more macro representations can be used. An example of the rule representation with fuzzy algorithms is shown as follows (quoted from paper [8]).

IF x is small and y is large THEN z is very small ELSE z is not small.

IF x is large THEN (IF y is small THEN z is very large ELSE z is small)
ELSE z and y are very very small.

Fuzzy algorithms can be used not only for fuzzy inference processing which occurs in conventional fuzzy control, but can be used for coding algorithms which deal with both crisp and fuzzy quantities using a very natural representation. Further, since they allow the processing of fuzzy branch, which a sequence of flow branches according to fuzzy processing, they can be used in all types of algorithms including sequence flow.

On the other hand, with original fuzzy algorithms, processing after fuzzy branches occur is supposed to be done in parallel and the consequent part (THEN or ELSE part) executing after the branch are necessary to inherit the weight (fuzziness) of grades in the previous part (IF part). Since executing an algorithm using this method in practice results in large amounts of parallel processing with the possibility of the algorithm growing out of control, it comes to need the pruning processing to insure that algorithms remain a manageable size.

Branch processing found in algorithms is typically of one of the three following types.

a) IF part before branch is crisp processing,
   the execution of the THEN or ELSE part after branch is uni-fired.
b) IF part before branch is fuzzy processing,
   the execution of the THEN or ELSE part after branch is uni-fired.
c) IF part before branch is fuzzy processing,
   the execution of the THEN or ELSE part after branch is multi-fired.

Covering branching of types b) and c), fuzzy algorithms can describe human-like decision algorithms which include concepts such as worrying about a judgement and thinking about the possibility, both of which can occur in actual decision making. However, it is fairly difficult to realize a true fuzzy algorithm c) since such algorithms often grow out of control due to the countless branches and allow GOTO and CALL statements within rules. In response to these problems, even for branching type of b), it is possible to execute branches which include a level of possibility if the complement (1-M) of matching grade, given by M, for the THEN part is reflected in the level of matching of the ELSE part.

The effectiveness of fuzzy algorithm through a simple experiment for the fuzzy control of model cars was demonstrated by Murofushi et al.[9]. For the fuzzy algorithm discussed in this paper, a constant threshold value has been placed on the grade values of the previous part in the fuzzy branch. Branching is controlled by whether or not the matching value in the previous part has exceeded this threshold, so that processing of algorithms does not grow out of control. Although this method is very effective in control applications, it results in crisp branching that loses some of the characteristics of the original fuzzy algorithm.

2.2 Modified Fuzzy Algorithms

Fig.2.1 shows the general idea about the control method of fuzzy branches in this paper. We call it the modified fuzzy algorithm (MFA) hereafter. If a previous part

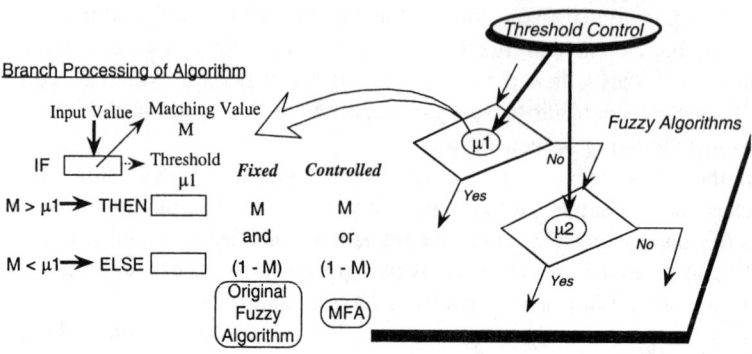

Figure 2.1: General idea of modified fuzzy algorithm (MFA)

has been represented by only crisp processing, the consequent part (THEN part meaning "yes" branch or ELSE part "no" branch) is executed with perfect matching when the previous part is satisfied. If a previous part has included fuzzy processing, the THEN part is executed only when a matching grade value in the previous part is greater than a given threshold ($\mu 1$ in the figure) and the ELSE part in the other case.

In the MFA, this threshold value itself is tuned in accordance with the control purpose determined autonomously (as described later). And the processing of a branch in the previous part can be described with fuzzy labels in this method, but rule-executions in the consequent part are not performed with parallel firing processing as original fuzzy algorithm. However, when a matching value in a previous part is M, the THEN part is executed with the matching value M and the ELSE part with the complement of matching value (1-M).

By using this method, an operator can describe "worried" situation on the decision making like human with simple rules. We therefore propose the hierarchical control method with the expressive function of macro decision making, which we call it the behavior-decision fuzzy algorithm. This method applied in the intelligent robot control is shown in Fig.2.2. According to the purpose which was decided in a planning layer (planner), the system recognizes the fuzzy purpose and controls the threshold of fuzzy branches for tuning the algorithm. The fuzzy algorithm tuned by the fuzzy purpose controls a robot and its sensing data are given to the fuzzy algorithm layer and the fuzzy purpose layer. This method can supply an available tool for complex decision making which includes both ambiguous recognition processes expressed by fuzzy rules and sequential decision processes expressed by AI production rules.

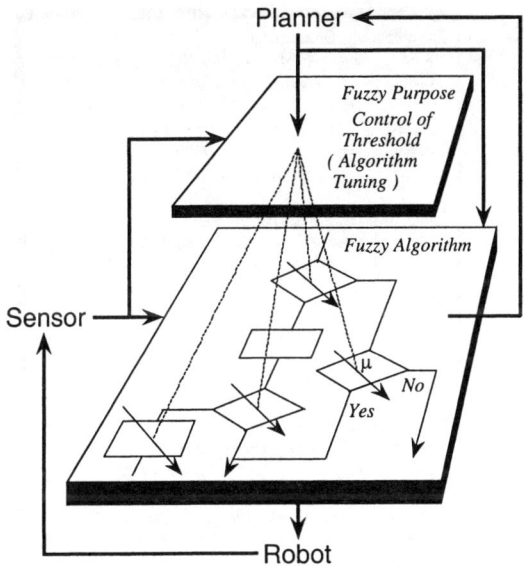

Figure 2.2: Behavior-decision fuzzy algorithm

3 Representation of Ambiguous States

Before discussing how to code the main flow with the behavior-decision fuzzy algorithm, we first tried to define ambiguous condition which a robot will encounter during motion. We call it fuzzy concepts and fuzzy states, using the fuzzy algorithms as previously described.

3.1 Fuzzy Concepts

First, as an example of fuzzy concepts which will be used in coding fuzzy states, attributes concerning obstacles (OB), walls (WA), forward free space (FS) and the robot itself (RO) are defined in the frame format as shown in Fig.3.1. We call it fuzzy-based frame, whose slot value can be defined with fuzzy labels as shown below left. The attributes with these slot values can be expressed by using both fuzzy variables or crisp variables. Here, each set of attributes corresponds to a linguistic variable for a human-like representation. For abbreviated labels with parentheses in fuzzy concepts, italic labels represent fuzzy values and all others represent crisp values.

```
Fuzzy concept  Obstacle (OB)
Kind     human (HM) / animal (AN) / not_creature (NC)
Char     very careful  (VC) / careful (C) / normal (NO)
Dist     close  (CL) / far (FR) / out of range (OR)
Dire     S_forward  (SF) / forward (FW) / side (SD)
Speed    very fast  (VF) / fast (F) / slow (SL)
S_dire   approach  (AP) / not_approach (NA)

Fuzzy concept  Wall (WA)
Dist     close  (CL) / far (FR) / out of range (OR)
Dire     S_forward  (SF) / forward (FW) / side (SD)

Fuzzy concept  Free-space (FS)
Width    very narrow  (VN) / narrow (N) / wide (W)
         / very wide  (VW) / out of scan (OS)
Dist     close  (CL) / far (FR) / out of range (OR)

Fuzzy concept  Robot (RO)
Wagon    with (WI) / without (WO)
Speed    very fast  (VF) / fast (F) / slow (SL)
Time     long  (LO) / little long (LL) / short (SH)
Energy   empty  (EM) / little empty (LE) / full (FU)
```

Figure 3.1: Definition of fuzzy concepts

3.2 Fuzzy States

It is one of the biggest problem in the locomotion of a robot that is the recognition by the robot of relatively ambiguous states such as whether it can move forward (passable) or avoid obstacles (avoidable). Since the explicit coding of ambiguous states of this kind is fairly difficult, we try to use an approximate representation of the fuzzy states as expressed with fuzzy algorithms. The expressions in Fig.3.2 show the fuzzy states of PASSABLE and AVOIDABLE represented by using above-mentioned fuzzy concepts.

Here, PASSABLE, IMPASSABLE, AVOIDABLE and UNAVOIDABLE are fuzzy labels in fuzzy states. In these expressions, a concept such as PASSABLE is not defined directly, but rather IMPASSABLE is first defined by several rules and then all remaining cases are taken PASSABLE. Previous parts of rules can include a mixture of both fuzzy values and crisp values, allowing for the definition of concepts in human-like manner. In this expression, the minimum grade value about matching in the previous part is found for each rule and only rules for which this value exceeds a certain threshold are executed.

To determine if a state is passable in the example being presented, the minimum grade value about matching rules would be found for each in order of top to bottom, and as soon as any exceeded threshold, the fuzzy state IMPASSABLE would result. If none of the rules exceeded the threshold, the fuzzy state PASSABLE would result. Since resulting states such as IMPASSABLE and others have also been represented by fuzzy labels, it is possible to get the degree of satisfaction about the states. It is possible to make fine adjustments to fuzzy state determinations by controlling the threshold value described above. Further, fuzzy state determinations can also be controlled indirectly by actually changing the slot value of fuzzy concepts which are used in fuzzy state definitional rules.

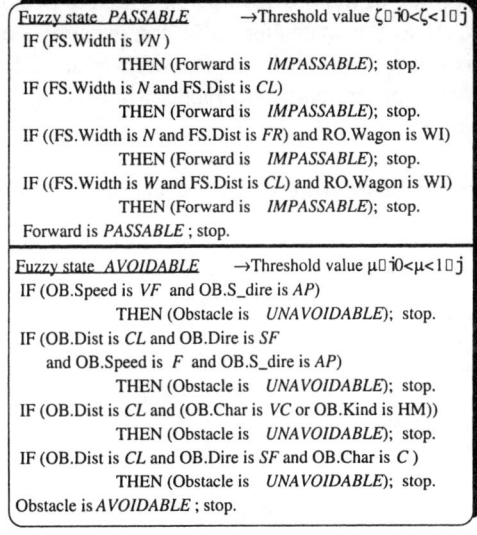

Figure 3.2: Representation of fuzzy states

4 Supervised Algorithm Control Based on Fuzzy Purpose

Since this part of the algorithm is for fine tuning the entire algorithm as required by a control purpose in that time, it must be of a higher order than the fuzzy algorithm. By configuring a hierarchical structure in this way, we consider constructing a supervised control method of the algorithm based on macro-judgments with an evaluation function of the fuzzy purposes. In the method being proposed, the configuration of the purpose judgement part includes two hierarchical determination parts of "strategy evaluation" and "tactic determination".

4.1 Strategy Evaluation

In this part, modes of motion is evaluated by information currently being obtained from surroundings. We supposed that a robot has three types of motion modes SAFETY, ECONOMY and MIN-TIME as shown in Fig.4.1. Here, RO.Energy and RO.Time are fuzzy labels representing current energy consumption versus total energy possessed by the robot and current time cost versus scheduled time to arrival. S, E and T are fuzzy labels representing the degree of satisfaction about the motion modes SAFETY, ECONOMY and MIN-TIME, respectively.

Expressive rules about these modes are described with the modified fuzzy algorithm. The grade value of the previous parts, which is decided by AND and OR operations, is equal to the matching value for execution of the consequent part. Degrees of satisfaction about each mode, Ms, Me and Mt, are set for each rule respectively. In this paper, only one mode whose this value is highest is selected as the purpose in that time.

```
SAFETY Mode
   IF ((OB.Dist is CL and OB.Dire is SF ) or (OB.Char is VC )
       or (WA.Dist is  CL and WA.Dire is SF ))
                           THEN (Mode is    S );   Ms = Mout.
ECONOMY Mode
   IF (RO.Energy is  EM ) THEN (Mode is E );   Me = Mout.
MIN-TIME Mode
   IF (RO.Time is  LO )    THEN (Mode is T );   Mt = Mout.
Mode Selection
   Select_Mode = max (Ms, Me, Mt).
```

Figure 4.1: Example of strategy evaluation rules

```
SAFETY Mode
   IF (Mode is S ) THEN
     1) ( ζ↓) and (η↑) and (μ↓)
        and (FS.Width ( N)→, OB.Char ( VC, C)←)
     2) and ( α∞↑) and (αw↑)
     3) and (Vlim ←) and (Slim←) .
```

Figure 4.2: Example of tactic determination rule

4.2 Tactic Determination

In this section, we explain about operations of the tactics suited for carrying out the strategy selected in the strategy evaluation part. This part is capable of modifying the algorithm mainly according to the three following methods.

1) Boundary control on fuzzy branches
   - modifying threshold values in the previous part of the fuzzy branch rules (direct control)
   - tuning membership parameters in fuzzy concepts (indirect control)
2) Modification of the control scheme

- modifying the degree of significance between path tracking and obstacle avoidance
- tuning the inference output value in danger evaluation
3) Changing control parameters
    - modifying the universe of discourse about velocity and steering membership functions
    - tuning the command output value of velocity and steering

Rules for these operations are described for each strategy. For example, a tactic determination rule concerned with SAFETY mode is shown in Fig.4.2. Here, the previous part is represented by a single fuzzy label which is same to the grade value output in the strategy evaluation part. The consequent part of this rule is executed according to the matching value of the previous part and the algorithm is modified. In this rule, the up and down arrows represent operations to raise or lower the threshold value within the fuzzy algorithm. The left and right arrows represent modification operations which slide values of membership functions on fuzzy concept to left or right. ζ, η and μ is threshold values for the fuzzy state evaluations and so on. ao and aw represent degrees of danger calculated by the danger evaluation using fuzzy reasoning. Vlim and Slim are the limit values about velocity and steering of a robot.

Tactics operations described above are determined based on the strategy selected every inference cycle. By applying such tactics for the behavior-decision fuzzy algorithm it is possible to perform online tuning of the fuzzy algorithm. In this method, it is possible to autonomously determine the control purpose on upper level of the algorithm and flexibly adjust the algorithm itself as compared with conventional fuzzy algorithms.

5 Overall Flow of Behavior-Decision Fuzzy Algorithm

According to the above-mentioned definition, we constructed the behavior-decision fuzzy algorithm for the autonomous mobile robot which moves in the complex environment. Fig.5.1 shows a flow chart representing the behavior-decision fuzzy algorithm based on the MFA. In the figure, the upper left is the start of the algorithm and the lower center the finish. This processing is repeated for every sampling time. In this algorithm, behavior phases have been divided into six major processing parts: STOP, MANUAL, PLAN, REPLAN, OBSTACLE and NORMAL, and supervised purpose judgement part: FUZZY_PURPOSE.

Meta rules for supervised control of overall algorithm flow in the expression using fuzzy algorithms are described as shown in Fig.5.2 (See Fuzzy-Algorithm META). Here, IMPASSABLE represents a fuzzy state as defined in section 3.2. Gothic letters occurring within the algorithm represent the fuzzy subalgorithms.

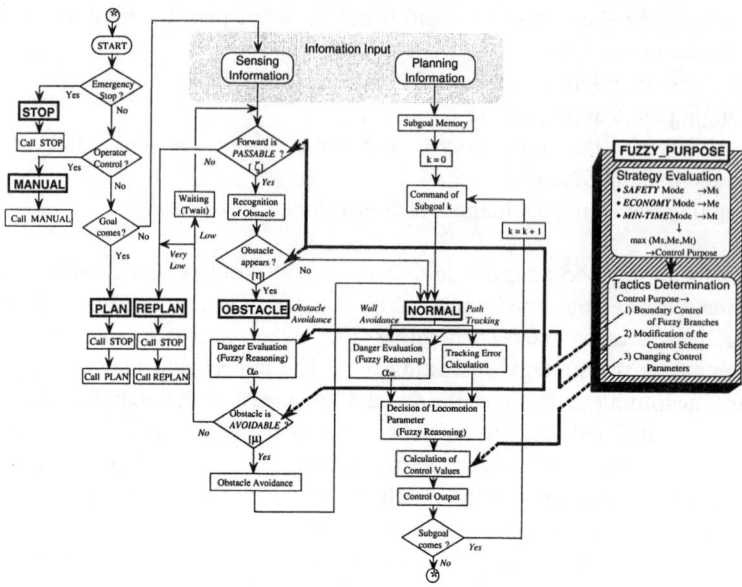

Figure 5.1: Processing flow of behavior-decision fuzzy algorithm

PLAN and REPLAN means processing routines in the path planning part [10]. In this simulation, we used stop motions in place of PLAN and back-track motions in place of REPLAN since the simulator in this paper has not path planning routine. The ability of using "call" and "goto" statements and handling both fuzzy and crisp values is remarkable feature of fuzzy algorithms. The gray part at the top of the figure shows the processing part for information input provided by the sensing and planning parts. Subgoal information given to the robot each time are assumed here to be input to memory as a target value while the robot is in offline.

The lower rule in Fig.5.2 shows a sample expression for the fuzzy subalgorithm OBSTACLE as it might be written for the sequence shown in Fig.5.2. It is assumed for this expression that the control purpose SAFETY has already been selected by the purpose judgement part. Here, italic letters represent fuzzy labels, Gothic letters fuzzy subalgorithms and all others crisp data. The phrase: "Obstacle is AVOIDABLE with Very Low", means that an obstacle is avoidable with very low grade in the possibility of avoiding it. In the step 5), after the recognition that an obstacle is unavoidable with low grade, the system is waiting for a while to look at the obstacle motion. After this waiting period ends, step 6) makes a return to the main algorithm META so as to attempt a retry. For the obstacle avoidance scheme called in step 7), we used algorithms [11][12] have been already proposed, but these will not be covered here.

```
Fuzzy-Algorithm META
1) Input Subgoals (k=1,...,m).
2) k = 0.
3) IF Emergent Stop happens THEN call STOP;
4) ELSE IF Operator interrupts THEN call MANUAL;
5) ELSE IF Goal comes THEN call STOP; call PLAN;
6) ELSE call FUZZY_PURPOSE;
        IF Forward is IMPASSABLE
            THEN call STOP; call REPLAN;
7) ELSE IF Obstacle appears THEN call OBSTACLE.
8) call NORMAL.
9) Output Control commands.
10) IF Subgoal comes THEN k = k + 1.
11) go to 3).

Fuzzy-Subalgorithm OBSTACLE
1) Decide α₀ [Fuzzy Reasoning].
2) IF Mode is S THEN α₀↑ and μ↓.
3) IF Obstacle is AVOIDABLE with Very Low THEN
        call STOP; call REPLAN;
4) ELSE IF Obstacle is AVOIDABLE with Low THEN
5)      Waiting; IF (Waiting time<Twait) THEN go to 5);
6)      go to META 6);
7) ELSE Call Obstacle_Avoidance_Rule. RETURN.
```

Figure 5.2: Example rules of behavior-decision fuzzy algorithm

6 Fuzzy Real-Time Advanced Shell

Problems involved in programming which uses fuzzy inferences include dealing with the complexity of managing shared data between standard membership functions and redundancy resulting from writing multiple subroutines using the same inference engine. We believe that it is possible to increase programming efficiency by utilizing a fuzzy shell possessing inference functions capable of dealing with such problems. However, although human interface capabilities such as found in rule editors using multiwindows can be of a comparatively high level, fuzzy expert systems which support conventional fuzzy inferences suffer from problems including difficulty with linking user's programs with the shell when running the shell and slow inference speeds because the shell systems are generally of a scale so large that realtime control is impossible.

For these reasons, the authors developed a fuzzy shell, capable of auto tuning user's programs. We call this shell FRASH (Fuzzy Realtime Advanced SHell) [13][14]. In this section, we describe the basic configuration of FRASH followed by an evaluation of the functionality of the shell in writing fuzzy inferences and fuzzy algorithms.

6.1 Features of FRASH

FRASH possesses the following main features.

- Features in FRASH Ver1.0 [FRASH1]
  1) The inference engine written in a library format
  2) The online rule editor written in a library format
  3) The offline rule editor using multiwindows (Rule editor)
  4) Inference display using multiwindows (Inference Monitor)
- Additional features in FRASH Ver2.0 [FRASH2]
  5) Frame-type data structures with slot values defined by using fuzzy numbers (Frame editor)
  6) C language source generator allowing the use of macro coding of fuzzy algorithms (Fuzzy algorithm Compiler)

1) and 2) are the most basic features of FRASH. In conventional systems which include a running inference shell, it has been necessary for the user to provide the shell with the processing of programs already written and then carry on further program development within the shell. In this system, the user's program is allowed to make the necessary library calls for the same results. Not only does this make program development simpler, but it allows for high-speed inferences as well since shell functions are separated into library units. Multiwindows are used in features 3) and 4) to reduce complexity during the input of rules and membership functions and to increase operability during rule debugging. For the additional feature 5), FRASH2 supports hierarchical programming using data attributes. The representability of data and maintainability of knowledge have been increased in FRASH2. by including frame-type data structures (hereafter referred to as fuzzy frame) capable of handling not only conventional crisp numbers, but fuzzy numbers as well for slot values. The programming of rules is also simplified by allowing modified fuzzy algorithms. Fuzzy algorithms allow the coding of fuzzy branch and can be used in decision support systems requiring sequence flow including both crisp and fuzzy operations.

In FRASH2, modified fuzzy algorithms mentioned in section 2.2 are coded using macro codes which are interpreted by a C language source code generator (fuzzy algorithm compiler). This feature can be used not only in the coding of conventional fuzzy control applications, but in sequential decision support control algorithms as well.

6.2 Basic Configuration of FRASH

Fig.6.1 shows the basic configuration of FRASH2. This system is made up of the four main parts, the scheduler, the editor process (offline rule and frame editor), the inference process (inference engine and online editor) and the inference display process. Scheduler mode is entered first when the shell is started. The editor, inference, or inference display processes can be selected from this mode. Any of these processes may be operated independently or simultaneously.

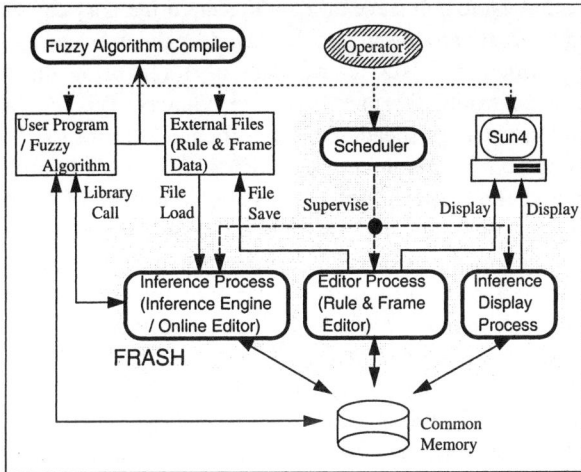

Figure 6.1: Basic configuration of FRASH

Simple descriptions of the features of each part of this shell are given below. Inference Engine supports typical inference methods such as the Mamdani MAX-MIN CG method. Inference Display process can perform the realtime graphic display showing the status of all membership functions and so on for every inference cycle. Offline Editor is a man-machine interface feature so that the user may input all data (rule sets, rules, variables, fuzzy labels, and membership values) in a multiwindow environment based on Open-Windows using a mouse. Online Editor provides a library so that all data may be modified in realtime during inference execution.

Typically, the user first enters the editor process and creates rules and frames, etc. using the multiwindow and graphic input features of the process. If this code is to be saved in an external file, rule-related data can be loaded from the user's program at run time. Next, the user can utilize inference functions (an inference engine) and online tuning functions (an online editor) by including the proper library calls in his program. Within the configuration of the shell, only inference process functions are all currently provided in a library format such that they may be called from a user's program. When the user described fuzzy algorithm in his program, a user program and an external data file are linked by a fuzzy algorithm compiler.

All three processes access shared memory for the output of data. Of this internal data, realtime values found within inferences (such as membership functions, fuzzy labels, and rules) can be displayed if the user has selected the inference display process. By using this feature, the user can revise rules offline while monitoring the results of inferences being made. Once rules have been finalized, rule data can be saved in a file and the rules can be created automatically at run time by having the user's program use the proper library function to specify that this file be loaded. Note that since all data in shared memory can be both read and written, it can be freely processed and displayed by the user's program.

FRASH is written in UNIX-C. All data is stored using the structures of C language and is controlled through the use of pointers. This data structure allows for the realization of a high-speed, memory-efficient realtime shell.

Inference Engine supports inference methods such as the Mamdani MAX-MIN weighted method (hereafter referred to as the Mamdani Method), a simplified fuzzy inference method in which post-inference parts of programs carry singleton values (the Singleton Method), a fuzzy inference method based on the Takagi-Sugeno model (the T-S Model Method). Therefore, the organization of data differs depending on each inference method. Membership data can be revised by changing the contents of data structures, while variables and fuzzy labels used by rules can be revised by changing pointers to data structures.

FRASH1 provides about 80 basic libraries. Standard fuzzy rules can be written easily by combining functions from these libraries. Furthermore, FRASH2 has additional libraries for fuzzy frame operations.

6.3 Fuzzy Frame Editor

In FRASH2, the organization of data into structures and representation of ambiguous knowledge is also possible since the system includes the ability to handle frame-type data structures with fuzzy slot values.

First, the user must define class or subclass structures and its attributes. An subclass inherits all attributes of its parent classes. For the attributes, the user can select either crisp or fuzzy. Next, the instances for an subclass are produced and defined its slot values for each attributes. Crisp attributes have only one variable, but fuzzy attributes have several fuzzy labels defined with a membership function. Membership functions for the slot value of fuzzy attributes are easily produced with graphic function editor. Changed data become available in another processes at realtime because all processes of FRASH use a common memory.

6.4 Fuzzy Algorithm Compiler

FRASH2 supports the modified fuzzy algorithm (MFA) mentioned in section 2.2 and therefore became powerful for expressive capacity of knowledges. FRASH1 was modified to incorporate this simple method of expressing fuzzy algorithms, in which branching is controlled by a threshold, into FRASH2. However, as already pointed out, since fuzzy algorithms allow CALL and GOTO statements within fuzzy rules, only offline editor of FRASH1 is not enough in the expressive function. It was therefore decided to incorporate a C language source code generator into FRASH2 so that fuzzy algorithms could be written using macros.

With this configuration, the user is allowed to write fuzzy algorithms using the offline editor with the feeling that he is writing C source code.

Fig.6.2 shows the fuzzy algorithm compiler of FRASH2. First, the rule and frame data files, whose name are "---.fzr", are made by offline editor (or online libraries). And source codes of a user program with the fuzzy algorithm, whose

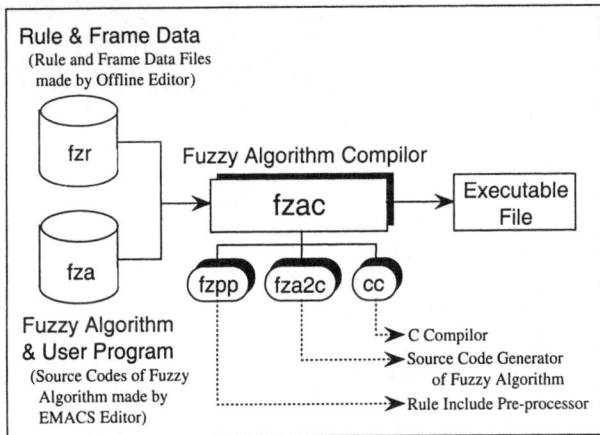

Figure 6.2: Fuzzy algorithm compiler

name are "---.fza", are made by EMACS editor. These files are linked with fuzzy algorithm compiler "fzac". The compiler "fzac" is constructed by three functions, that is a rule including pre-processor "fzpp", a fuzzy algorithm source code generator "fza2c" and a C compiler "cc". All files are compiled to a executable file through this compiler.

6.5 Operational Windows of FRASH

With FRASH2, coding can be made by using expressions nearly identical to those of actual fuzzy algorithms. Organization of data into structures and representation of fuzzy knowledge is also possible since the system includes the ability to handle fuzzy frames.

Editor windows of FRASH2 are shown in Fig.6.3. In this figure, a lower right window shows the top window, a upper right the frame editor window and a lower left the rule editor window. And a upper left window shows fuzzy algorithm editor window provided with the EMACS editor. In the top window, the user can select both rule and frame editor. In all editors except fuzzy algorithm, the user can build their frame data, rules, membership functions easily by crick buttons, slide levers, pull-down menus and so on. And the user can enter the editor mode of membership functions by double clicking the area of a fuzzy slot value in the frame editor window.

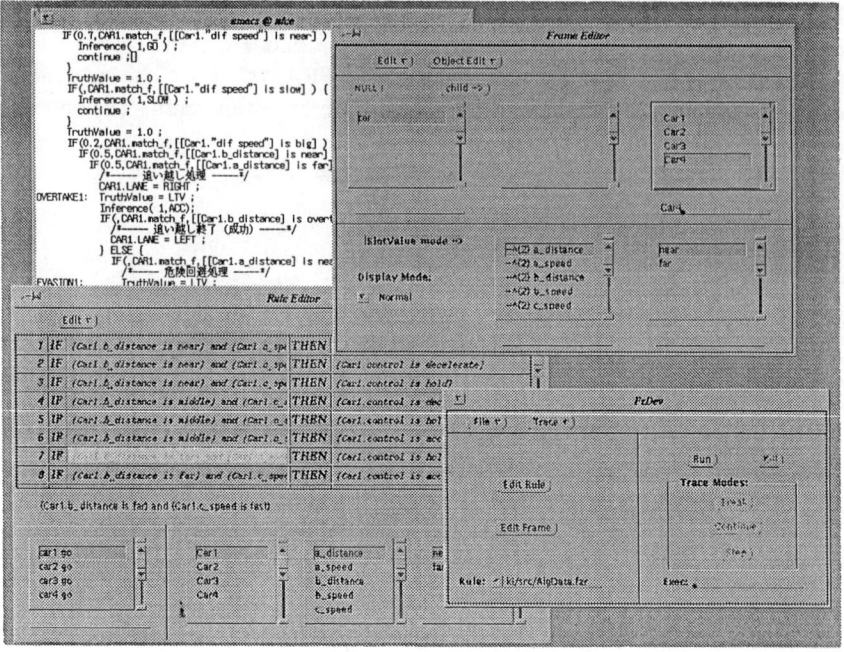

Fig.6.3. Editor windows of FRASH2

7 Simulations

This section describes the results of computer simulations of the proposed method supposed in a two-dimensional in-door area. These simulations in this section were mainly performed to check the representability of fuzzy states and the capabilities of the purpose judgement part expressed by FRASH2 which mentioned in the previous section.

7.1 Mobile Robot Simulator

We constructed the robot simulator to check the decision and motion of a robot. Fig.7.1 shows the locomotive model of a mobile robot. In this simulator, it suppose that a robot inputs following informations.
1) Planning informations (Offline data)
   (Psx, Psy, δ)k: Position of subgoals and permitted arrival limit (k=1,...,m)
2) Sensing information (Online data)
   (Prx, Pry, θr):   Present position and direction of a robot
                       [from CCD color image tracker (outside of the robot)]

di (i=1,...,8): Distance between a robot and an object
[from 8 sonar sensors covered 120 degrees area]

θo: Direction for an obstacle in the robot coordinates
[from CCD color image tracker (on the robot)]

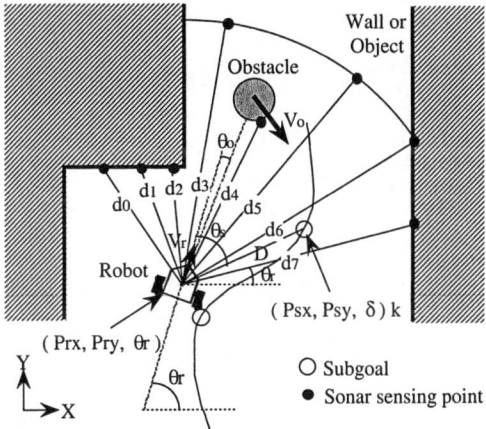

Figure 7.1: Locomotive model of the mobile robot

Information data 1) are given to the simulator in advance. In the final control system, these informations are sent from the path-planning part. Information data 2) are obtained by each sensor of a robot in on-line. From these informations, a robot can recognize each condition of the subgoals, obstacles and walls. Especially, position and velocity of an obstacle are decided by above sensing informations and its differential.

7.2 Threshold Control in the Simulation

More simplified rules than that one described before has been used in this simulation. Rules in the simulation are shown in Fig.7.2. The first rule about the fuzzy state PASSABLE is described based on fuzzy labels representing the width and distance of the forward space of the robot. Output of this rule is the degree of impassability ($0 < IM < 1$) of the forward space. If IM becomes higher than the threshold TH_imp, the forward space is decided as IMPASSABLE and a robot performs back-track in this simulation. In the second rule about the fuzzy state AVOIDABLE, it gives the degree of unavoidability ($0 < UN < 1$) for an obstacle. If UN exceeds the threshold TH_una, the obstacle is decided as UNAVOIDABLE and a robot performs back-track as well as IMPASSABLE. And if UN is more than half of TH_una and less than TH_una, a robot performs waiting motion (stopping for a while). The third rule about the control purpose SAFETY will get the degree of danger for control purpose evaluation. The robot is made to recognize the degree of danger ($0 < DA < 1$) according to the this evaluation rule.

The fuzzy labels CL and SF are set separately for both obstacles and surrounding objects.

```
Fuzzy state PASSABLE
IF P = [(FS.Width is VN ) or (FS.Width is N and FS.Dist is CL)] is high
   THEN (Forward is IMPASSABLE)              → TH_imp     <Back Track>
   ELSE (Forward is PASSABLE) .

Fuzzy state AVOIDABLE
IF A = [(OB.Speed is VF and OB.S_dire is AP ) or
       (OB.Dist is CL and OB.Dire is SF and OB.Speed is F and OB.S_dire is AP )] is high
   THEN (Obstacle is UNAVOIDABLE )  □@       → TH_una     <Back Track>
   ELSE IF A is little high
        THEN (Obstacle is LITTLE UNVOIDABLE )  → TH_una / 2  <Waiting>
        ELSE (Obstacle is AVOIDABLE ) .

Fuzzy purpose SAFETY
IF S = [(OB.Dist is CL and OB.Dire is SF ) or (WA.Dist is CL and WA.Dire is SF )] is high
   THEN (Mode is DANGER ) .
```

Figure 7.2: Fuzzy states and fuzzy purpose used in the simulation

In these rules, fuzzy branches for THEN or ELSE are controlled by thresholds of IMPASSABLE and UNAVOIDABLE. For example, the threshold of IMPASSABLE is decided every sampling time as follows.

$$TH\_imp = TH\_imp + G\_imp \{ TH\_imp\_0 (1 - DA^2) - TH\_imp \} \quad (7.1)$$

Here, TH_imp and TH_imp_0 shows the present and default threshold of IMPASSABLE, respectively. G_imp shows the control gain of the threshold. And the velocity of a robot is controlled by output value IM of the first rule as follows.

$$Vr = (1 - IM^2) Vr\_0 \quad (7.2)$$

In this equation, Vr_0 means initial value of the robot velocity. Actual velocity Vr of a robot always becomes less than Vr_0. These control operations correspond to the first and the third tuning method of the tactic determination rule in section 4.2.

7.3 Simulation Results

Fig.7.3 to 7.5 shows the trails of a robot and a moving obstacle in this simulation. The still objects within the environment (shown as black polygons) and the moving obstacle (shown as a large white circle going down at lower right area in the map) are detected using virtual ultrasonic sensors for a rough measurement of the distance. Table 7.1 shows the parameters of this simulation. We assume that the environment robot moves is a 4 by 4 meters room with 12 nodes of subgoals. During the simulation, the robot clears two subgoals (filled circles #3 and #11 in

Table 7.1: Parameters of the simulation

Calculating sample time : 0.1 sec
Drawing sample time : 0.3 sec
Environment : 4m × 4m room
Number of nodes : 12 nodes
        (small circle in the simulator)
Initial Threshold :
        $TH\_imp\_0 = 0.9$
        $TH\_una\_0 = 0.8$
        $G\_imp = 0.02$
        $G\_una = 0.05$
	Robot → autonomous mobile robot controlled speed and
	        steering (polygon with 2 wheels in the simulator)
	        speed : 10cm/s
	        size : about 30 cm in width
	Sonar → 8 lines from center of a robot
	        covering range : 120 degrees ahead
	        sensing limit : within 2 m
	Obstacle → moving object (a large circle in the simulator)
	        speed : 40 cm/s
	        size : 20 cm in diameter
	        moving time : 130 sec ~ 165 sec

the figure) moving from the starting point #0 in the lower left, to the final goal #11 in the upper right. The simulation time is shown in the left upper corner on the map.

The right side of a simulation window indicates each condition of fuzzy states, threshold values and a fuzzy purpose. Two bar graphs on left side shows degrees of IMPASSABLE(IM) and UNAVOIDABLE(UN). In these graphs, black arrows means threshold values of TH_imp and TH_una, and a white arrow in the graph of UN indicates the threshold TH_una/2 for determining the waiting mode. And a black triangle shows the point when the value of fuzzy states exceeded against the threshold value. The right bar indicates the evaluating value DANGER(DA) on fuzzy purpose SAFETY. The threshold values of IM and UN are controlled by this value DA.

Fig.7.3 shows the simulation result which indicated the robot motion with the threshold values TH_imp and TH_una controlled by the value DA. In the condition a), a robot recognized impassability of his front space and decided back-track motion to the node #3. In this period, we can find that the value of IM exceeded the threshold value TH_imp dropping down slightly. In the condition b), a robot entered the next little wide passage, but encountered an obstacle in right front. Also in this case, a robot goes back to the node #4 because the value UN exceeded the threshold value TH_una dropped by high degree of danger DA. In the condition c), the threshold values are fairly dropped for a robot encountered an moving obstacle from #6 to #7, but the value UN indicates between black and white arrows because the forward passage is very wide. Therefore, a robot goes on waiting for the time when an obstacle passes by unless the value UN exceeds the TH_una. And condition d) shows the state of a robot near the final goal #11.

On the other hand, the robot motion without threshold control is shown in Fig.7.4. In this case, each threshold does not change at any time and is kept the default value. Therefore, the value IM can not exceed the threshold value TH_imp and a robot passes the narrow passage from node #3 to #8 by force as shown in Fig.7.4 b). Fig.7.5 shows the motion of a robot controlled only the threshold value

94

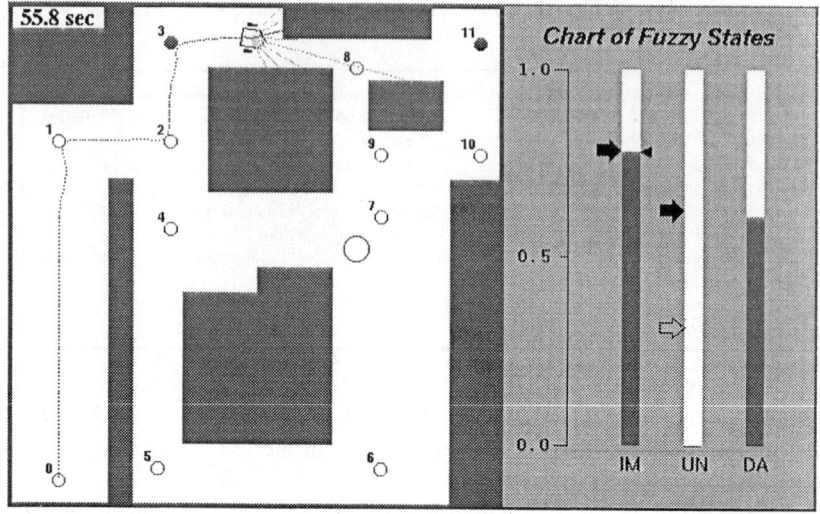

a)

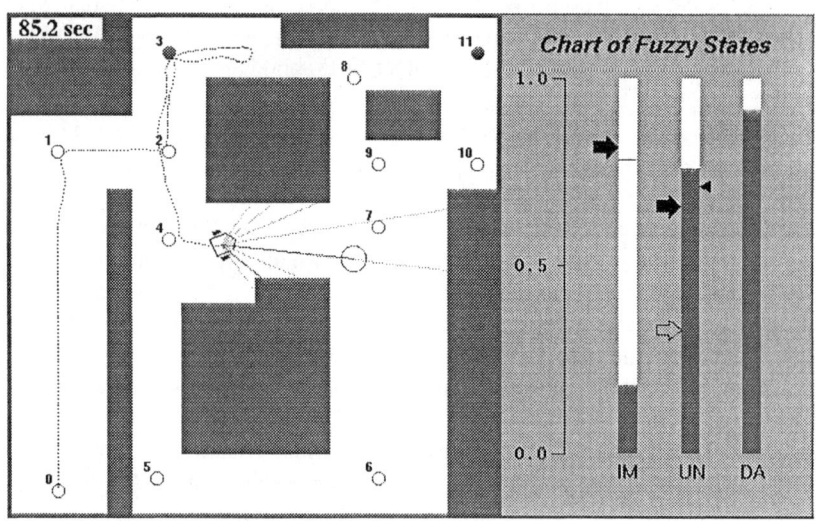

b)

Fig.7.3. Simulation results (1) [TH_imp : controlled, TH_una : controlled]

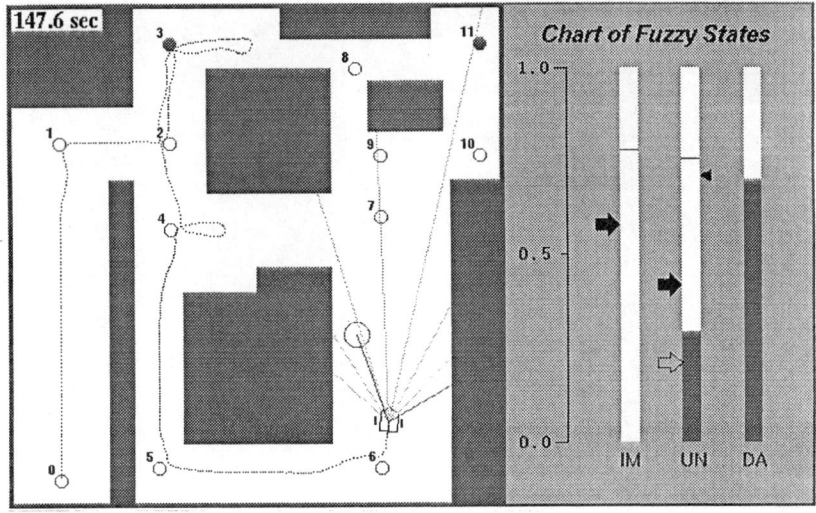

c)

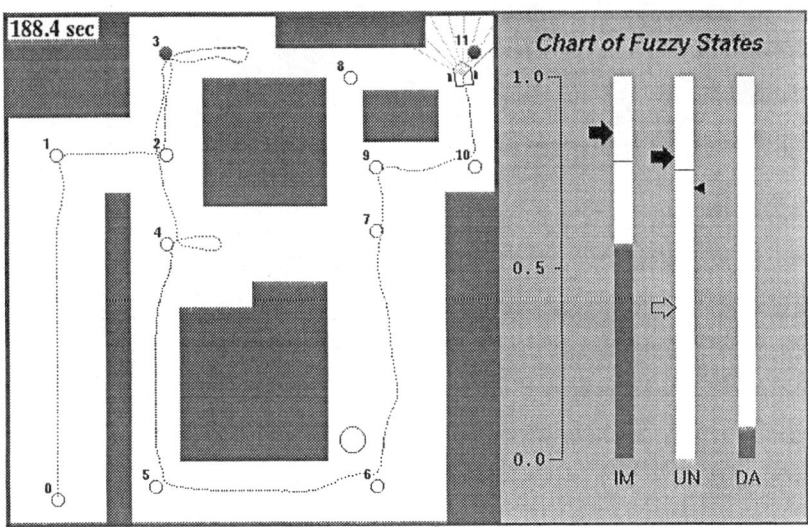

d)

Fig.7.3. Simulation results (1) [TH_imp : controlled, TH_una : controlled] (Continued)

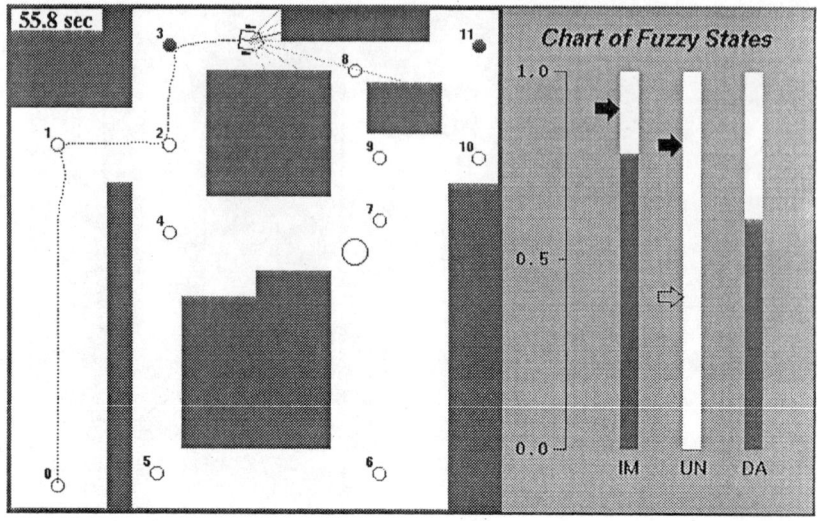

a)

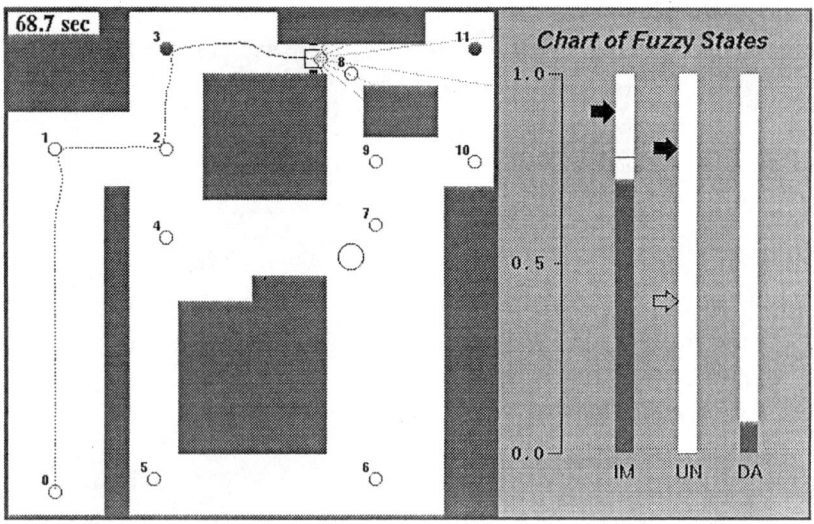

b)

Fig.7.4. Simulation results (2) [TH_imp : uncontrolled, TH_una : uncontrolled]

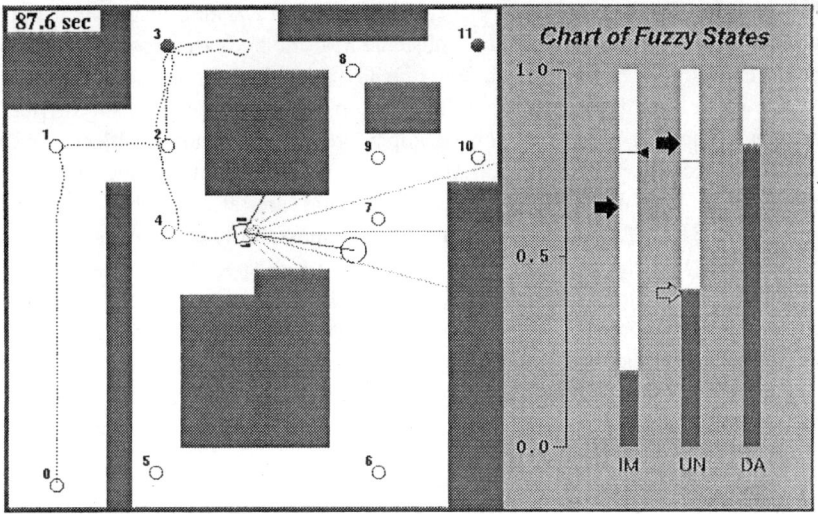

a)

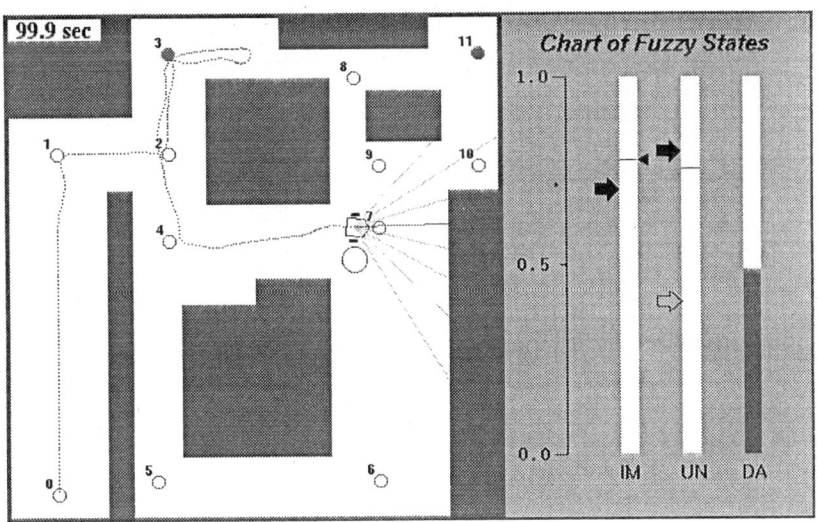

b)

Fig.7.5. Simulation results (3) [TH_imp : controlled, TH_una : uncontrolled]

TH_imp. In this case, a robot performed the back-track motion between node #3 and #8, but he passed by a standing obstacle near the node #7 because the value UN can't exceed the constant threshold value TH_una without control.

As shown in these figures, a robot clearly recognizes passage as being difficult whenever it enters a narrow region or approaches a surrounding objects or an obstacle. Especially, changing of thresholds in the graph shows that the IMPASSABLE and UNAVOIDABLE recognition are affected by the fuzzy purpose SAFETY. Although the degree of impassability here was actually determined only by the attribute value for forward free space and obstacles, it is clear that the evaluation of degree of safety which also took into account another attribute values or control purposes on upper level would realize fairly complex recognition. Furthermore, the motion of a robot will become still more similar to that of human by switching several control purposes of SAFETY, ECONOMY, MIN-TIME and so on.

8 Experiments

System constitution of our autonomous mobile robot is shown in Fig.8.1. The total experimental system is mainly constructed with three parts : robot, controller and supervisor. The robot (HERO2000) is controlled by the controller (PC98RX) which is commanded by the supervisor (SUN4 workstation). And every control signal and sensor data between the robot and the controller are sent through RS232C wireless communication for the robot is necessary to move around.

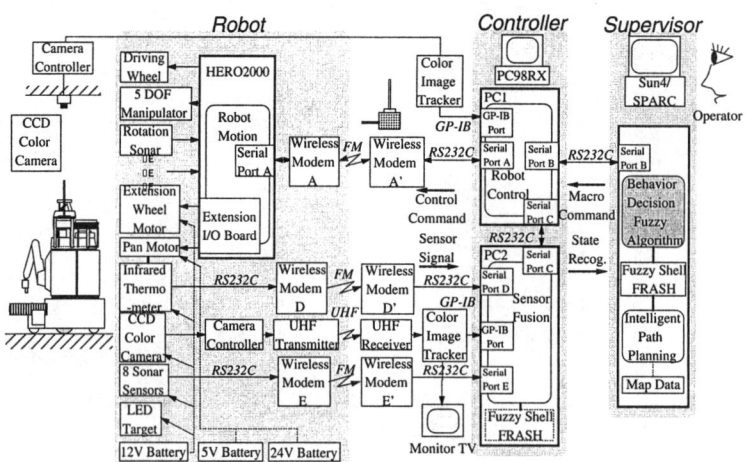

Figure 8.1: System constitution of autonomous mobile robot

In all systems, fuzzy reasoning and fuzzy algorithms are described with our original fuzzy shell FRASH which has the inference engine and the online rule/frame editor in the library format, and the offline rule/frame editor and the real-time inference display process using multiwindows. This shell has frame-type data structures with slot values defined by using fuzzy numbers and C language source code generator allowing the use of macro coding of fuzzy algorithms as mentioned in section 6.

The outward appearance of our autonomous mobile robot is shown in Fig.8.2. This robot has the sensor fusion system, which is composed by ultrasonic sensors, position detector system (color image tracker) and noncontact infrared thermometer, can recognize a human or an animal in front of a robot by evaluating the degree of living using fuzzy reasoning[15]. And the robot recognizes the objects with the eight sonar system and its position and direction are detected by the color image tracker outside of the robot.

Fig.8.3 shows the appearance of this experiment. We performed experiments about autonomous locomotive function of our robot. A white curved line shows the tracking route of the robot. In the experiment whose environment is equal to the simulation, our robot exhibited almost similar motion to the simulation results in a previous section. Therefore, the utility of this method was proved by this practical experiment with an autonomous mobile robot.

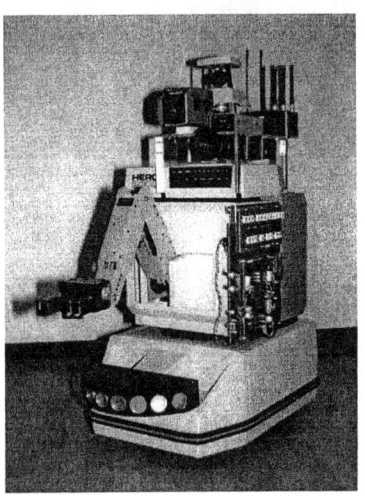

Figure 8.2: Configuration of mobile robot

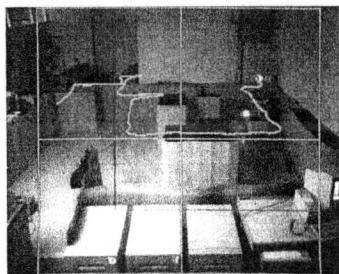

Figure 8.3: Robot trajectory in the experiment

9 Conclusion

It desirable to be able to express operational sequences which include ambiguous states when trying to represent "human-like" behavior-decision abilities in the intelligent robot. In this paper, we proposed the modified fuzzy algorithm (MFA) which has the algorithm tuning function based on fuzzy algorithms capable of expressing sequence flow and handling both crisp and fuzzy processing. The construction method of the behavior-decision system for autonomous mobile robots using the MFA was also described in this paper.

The main features of this method are as follows.

1) Allows the flexible and macro knowledge representation including behavior-decision sequences through the utilization of the MFA.
2) Allows the definition of fuzzy concepts expressed by using fuzzy-based frames and fuzzy states expressed by fuzzy algorithms.
3) Possesses algorithm tuning capability, which has flexible adaptation to the environment, based on the threshold value controlled by the fuzzy purpose.

We also developed a fuzzy inference shell FRASH for our research. This shell possesses the following functions: inference engines, online tuning libraries, an offline editor and an inference display for the fuzzy inference, and C language source generator to express fuzzy algorithms and frame-type data structures allowing fuzzy numbers. In this research, we developed the mobile robot simulator by using FRASH. Some results of computer simulations supposed a simple in-door environment and experiments using a real autonomous mobile robot were also shown to prove the effectiveness of our method proposed in this paper.

Acknowledgements

This research was carried out in the Laboratory for International Fuzzy Engineering Research (LIFE), Japan. The authors would like to express our deepest gratitude to Dr. Kaoru Hirota, Prof. in Tokyo Institute of Technology for very useful advices and Dr. Toshiro Terano, LIFE Director for giving us the opportunity to undertake this research.

References

1. T.Sawaragi, O.Katai and et.al. Opportunistic path-planning and fuzzy locomotion control for an autonomous mobile robot based on the vision information. Proc. of 12th Intelligent System Symposium, 63-68, 1990.
2. C.Isik and A.M.Meystel. Pilot level of a hierarchical controller for an unmanned mobile robot. IEEE Jour. of Robotics and Automation, Vol.4, No.3, 241-255, 1988.
3. Y.Maeda, M.Tanabe, M.Yuta and T.Takagi. Control purpose oriented behavior-decision fuzzy algorithm with tuning function of fuzzy branch. Proc. of International Fuzzy Engineering Symposium (IFES'91), 694-705, 1991.

4. Y.Maeda, M.Tanabe, M.Yuta and T.Takagi. Hierarchical control for autonomous mobile robots with behavior-decision fuzzy algorithm. Proc. of IEEE International Conference on Robotics and Automation, 117-122, 1992.
5. Y.Maeda, M.Tanabe and T.Takagi. Behavior-Decision Fuzzy Algorithm for Autonomous Mobile Robots. Information Sciences, Vol.71, No.1, 145-168, 1993.
6. E.S.Santos. Fuzzy algorithm. Information and Control, 17, 326-339, 1970.
7. L.A.Zadeh. A new approach to system analysis. Proc. of Conference on Man and Computer, 55-94, 1971.
8. L.A.Zadeh. Outline of a new approach to the analysis of complex systems and decision process. IEEE Trans. on Systems, Man, and Cybernetics, Vol.SMC-3, No.1, 28-44, 1973.
9. M.Sugeno, T.Murobushi and et.al. Fuzzy algorithmic control of a model car by oral instructions. Fuzzy Sets and Systems, 32, 207-219, 1989.
10. M.Tanabe, Y.Maeda, M.Yuta and T.Takagi. Path planning method for mobile robot using fuzzy inference under vague information of environment. Proc. of International Fuzzy Engineering Symposium (IFES'91), 758-769, 1991.
11. Y.Maeda. Fuzzy obstacle avoidance method for a mobile robot based on the degree of danger. Proc. of the North American Fuzzy Information Processing Society Workshop (NAFIPS'90), 169-172, 1990.
12. H.Koyama, Y.Maeda and S.Fukami. Obstacle avoidance using the fuzzy production system. Proc. of Joint Hungarian-Japanese Symposium on Fuzzy Systems and Applications, 102-105, 1991.
13. Y.Maeda, J.Murakami and T.Takagi. FRASH - A fuzzy real-time auto-tuning shell with expressive function of fuzzy algorithms. Proc. of Singapore International Conference on Intelligent Control and Instrumentation (SICICI'92), 1363-1368, 1992.
14. Y.Maeda. Fuzzy Real-time Advanced Shell for Intelligent Control with Fuzzy Algorithm Compiler. Journal of Robotics and Mechatronics, Vol.8, No.1, 49-57, 1997.
15. Y.Maeda, M.Tanabe and M.Yuta. Demonstration of Macro Behavior-Decision on Intelligent Autonomous Mobile Robot. Proc. of International Fuzzy Engineering Symposium (IFES'91), p.1110-1111, 1991.

Chapter 6. Modelling the Emergence of Speech and Language Through Evolving Connectionist Systems

John Taylor[1] and Nikola Kasabov[2]

[1]Department of Linguistics, John_Taylor@stonebow.otago.ac.nz
[2]Department of Information Science, NKasabov@otago.ac.nz
University of Otago, P.O Box 56, Dunedin, New Zealand

Abstract. *This chapter presents the hypothesis that language knowledge evolves in the human brain through incremental learning and that the process can be modelled with the use of evolving connectionist systems - a recently introduced neural network paradigm. Several assumptions have been hypothesised and proven through simulation: (a) the learning system evolves its own representation of spoken language categories (phonemes) in an unsupervised mode through adjusting its structure to continuously flowing examples of spoken words (a learner does not know in advance which phonemes there are going to be in a language, nor, for any given word, how many phonemes segments it has); (b) learning words and phrases is associated with supervised presentation of meaning; (c) it is possible to build a 'life-long' learning system that acquires spoken languages in an effective way, possibly faster than humans, provided there are fast machines to implement the evolving, learning models.*

Keywords. *Evolving spoken languages; evolving phonemic structures; on-line adaptive learning; neural networks.*

1 Introduction

This chapter presents a hypothesis, a computational model, and some preliminary experimental results. We investigate the possibility of creating a computer, connectionist-based model of the process of phoneme knowledge acquisition. The task is concerned with the process of learning in humans and how this process can be modelled in a program [1,7,12,15,44,54,55]. The following questions will be attempted: How can continuous learning in humans be modelled? What conclusions can be drawn from the modelling results, that may be used in the long term to improve learning and teaching processes, especially with respect to languages? How is learning a second language related to learning a first language?

The research aim of the project is computational modelling of processes of phoneme category acquisition, using natural, spoken language as training input to a connectionist system. A particular research question concerns the characteristics of an "optimal" input and optimal system parameters that are needed for the phoneme categories of the input language to emerge in the system in the least time. Also, by tracing how the target behaviour actually emerges on the machine, hypotheses about what learning parameters might be critical to the language-acquiring child could be made.

By attempting to simulate the emergence of phoneme categories in the language learner, the chapter addresses some fundamental issues in language acquisition, and will input into linguistic and psycholinguistic theories of acquisition. It will have a special bearing on the question, whether language acquisition is driven by general learning mechanisms, or by innate knowledge of the nature of language (Chomsky's Universal Grammar). Moreover, the research presented here generates hypotheses about the optimal input to a language learning system, and the actual algorithms that are used by the language-learning child.

From an information processing point of view, the research throws light on how the evolving cognitive system deals with and adapts to environmental constraints, which in this case are represented by the structures imposed upon perception by the language system.

The basic methodology consists in the training of an evolving connectionist structure (a modular system of neural networks) with mel-scale transformations of natural language utterances. The basic research question is whether, and to what extent, the network will organise the input in clusters corresponding to the phoneme categories of the input language. We will be able to trace the emergence of the categories over time, and compare the emergent patterns with those that are known to occur in child language acquisition.

The model is of evolving connectionist systems (ECOS) recently developed [21-25]. ECOS evolve as they learn from examples. The learning process is on-line, incremental, continuous, 'life-long'. Biologically plausible principles have been used in ECOS, such as: learning and forgetting, creating and pruning connections and neurons, sleep learning, 'echo'-type learning and others. The principles of ECOS have been successfully applied on small, bench-mark data and problems including adaptive speech recognition.

In preliminary experiments, it may be advisable to study circumscribed aspects of a language's phoneme system, such as consonant-vowel syllables. Once the system has proved viable, it will be a relatively simple matter to proceed to more complex inputs, involving the full range of sounds in a natural language, bearing in mind that some languages (such as English) have a relatively large phoneme system compared to other languages (such as Maori) whose phoneme inventory is more limited.

Moreover, it will be possible to simulate acquisition under a number of input conditions:
- input from one or many speakers;

- small input vocabulary vs. large input vocabulary;
- simplified input first (e.g. CV syllables) followed by phonologically more complex input;
- different sequences of input data.

The research project presented here is at its initial phase, but the results are expected to contribute to a general theory of human/machine cognition.

Technological applications of the research concern the development of self-adaptive systems. These are likely to substantially increase the power of automatic speech recognition systems.

2 The Dilemma "Innateness versus Learning" Reconsidered

2.1 A General Introduction to the Problem

A major issue in contemporary linguistic theory concerns the extent to which human beings are genetically programmed, not merely to acquire language, but to acquire languages with just the kinds of properties that they have [43]. For the last half century, the dominant view has been that the general architecture of language is innate, the learner only requires minimal exposure to actual language data in order to set the open parameters given by Universal Grammar as hypothesised by Noam Chomsky [7]. Arguments for the innateness position include the rapidity with which all children (barring cases of gross mental deficiency or environmental deprivation) acquire a language, the fact that explicit instruction has little effect on acquisition, and the similarity (at a deep structural level) of all human languages. A negative argument is also invoked: the complexity of natural languages is such, that they could not, in principle, be learned by normal learning mechanisms of induction and abstraction.

Recently, this view has been challenged. Even from within the linguistic mainstream, it has been pointed out that natural languages display so much irregularity and idiosyncrasy, that a general learning mechanism has got to be invoked; the parameters of Universal Grammar would be of little use in these cases [9]. Moreover, linguists outside the mainstream have proposed theoretical models which do emphasise the role in input data in language learning [34,56,57,5]. On this view, language knowledge resides in abstractions (possibly, rather low-level abstractions) made over rich arrays of input data.

In computational terms, the contrast is between systems with a rich in-built structure, and self-organising systems which learn from data [15]. Not surprisingly, systems which have been pre-programmed with a good deal of language structure vastly outperform systems which learn the structure from input data. Research on the latter is still in its infancy, and has been largely restricted to modelling circumscribed aspects of a language, most notably, the acquisition of irregular verb morphology [44]. A major challenge for future research will be to model the acquisition of more complex configurations, especially the interaction of phonological, morphological, syntactic, and semantic knowledge.

Our focus in this paper is the acquisition of phonology, more specifically, the acquisition of phoneme categories. All languages exhibit structuring at the phoneme level. We may, to be sure, attribute this fact to some aspect of the genetically determined language faculty. Alternatively, and perhaps more plausibly, we can regard the existence of phoneme inventories as a converging solution to two different engineering problems. The first problem pertains to a speaker's storage of linguistic units. A speaker of any language has to store a vast (and potentially open-ended) inventory of meaningful units, be they morphemes, words, or fixed phrases. Storage becomes more manageable, to the extent that the meaningful units can be represented as sequences of units selected from a small, finite inventory of segments (the phonemes). The second problem, refers to the fact that the acoustic signal contains a vast amount of information. If language learning is based on input, and if language knowledge is a function of heard utterances, a very great deal of the acoustic input has got to be discarded by the language learner. Incoming utterances have got to be stripped down to their linguistically relevant essentials. Reducing incoming utterances to a sequence of discrete phonemes solves this problem, too.

2.2 Our Approach

Infant acquisition

Research by Peter Jusczyk [19] and others, has shown that new-born infants are able to discriminate a large number of speech sounds - well in excess of the number of phonetic contrasts that are exploited in the language an infant will subsequently acquire. This is all the more remarkable, since the infant vocal tract is physically incapable of producing adult-like speech sounds at all [36]. By about 6 months, perceptual abilities are beginning to adapt to the environmental language, and the ability to discriminate phonetic contrasts that are not utilised in the environmental language declines. At the same time, and especially in the case of vowels, acoustically different sounds begin to cluster around perceptual prototypes, which correspond to the phonemic categories of the target language, a topic researched by Kuhl [32]. Thus, the "perceptual space" of e.g. the Japanese- or Spanish-learning child becomes increasingly different from the perceptual space of the English- or Swedish-learning child. Japanese, Spanish, English, and Swedish "cut up" the acoustic vowel space differently, with Japanese and Spanish having far fewer vowel categories than English and Swedish. However, the emergence of phoneme categories is not driven only by acoustic resemblance. Kuhl's research showed that infants are able to filter out speaker-dependent differences, and attend only to the linguistically significant phoneme categories.

It is likely that adults in various cultures, when interacting with infants, modulate their language in ways which optimise the input for learning purposes. This is not just a question of vocabulary selection (though this, no doubt, is important). Features of "child-directed speech" include exaggerated pitch range and slower articulation rates [32]. These maximise the acoustic distinctiveness of the different vowels, in that features of co-articulation and target undershoot [37], characteristic of rapid conversational speech, are reduced.

Specific questions are:

1) Is it necessary to train the system with input that approximates its characteristics to those of "Motherese", with respect to both vocabulary selection and articulation?

2) Can the system organise the acoustically defined input without prior knowledge of the characteristics of the input language? If so, this would be a significant finding for language acquisition research! Recall that children learning their mother tongue do not know in advance how many phoneme categories there are going to be in the language, nor even, indeed, that language will have a phoneme level of organisation.

A further set of research questions concerns the simulation of bilingual acquisition [29]. We can distinguish two conditions:

1) Simultaneous bilingualism, i.e. the system is trained simultaneously with input from two languages (spoken by two different speakers, or two sets of speakers). Can the system keep the two languages apart? Comparisons with simultaneous acquisition by the bilingual child can be made.

2) Late bilingualism, i.e. training of an already trained system on input from a second language. Will the acquisition of the second language show interference patterns characteristic of human learners?

Phonemes

While it is plausible to assume that aspects of child-directed speech facilitate the emergence of perceptual prototypes for the different phonemes of a language's sound system, it must be borne in mind that phoneme categories are not established only on the basis of acoustic-phonetic similarity. Phonemes are theoretical entities, at some distance from acoustic events. (This, after all, is why automatic speech recognition is such a challenge!) The value of the phoneme concept for language description (and transcription) is, of course, obvious, as is the cognitive value of phoneme-sized units to the language user. As long as the child's vocabulary remains very small (up to a maximum of about 40 - 50 words), it is plausible that each word is represented as a unique pathway through acoustic space, each word being globally distinct from each other word. But with the "vocabulary spurt", which typically begins around age 16 - 20 [3], this strategy becomes less and less easy to implement. Up to the vocabulary spurt, the child has acquired words slowly and gradually; once the vocabulary spurt begins, the child's vocabulary increases massively, with the child sometimes adding as many as 10 words per day to his/her store of words. Under these circumstances, it is highly implausible that the child is associating a unique acoustic pattern with each new word. Limited storage and processing capacity requires that words be broken down into constituent elements, i.e. the phonemes. Rather than learning an open-ended set of distinct acoustic patterns, one for each word (tens of thousands of them!), the words come to be represented as a linear sequence of segments selected form an inventory of a couple of dozen distinct elements.

Phoneme analysis

Linguists have traditionally appealed to two procedures for identifying the phonemes of a language:

(a) the principle of contrast. The vowels [i:] and [I] are rather close (acoustically, perceptually, and in terms of articulation). In English, however, the distinction is vital, since the sounds differentiate the words *sheep* and *ship* (and countless others). The two sounds are therefore assigned to two different phonemes. In Spanish, the difference is linguistically irrelevant (there are no "minimal pairs" exploiting the contrast). Consequently, Spanish speakers tend to lump the two sounds together in the same category, and experience considerable difficulty in controlling the difference in their speech production (as when learning English as a foreign language).

(b) economy of distribution. The "h" sounds of English do not constitute an acoustically homogeneous category. Phonetically, an "h" sound is simply a voiceless version of a following vowel. The initial sounds in *he, head, hard, hut, who,* etc., are noticeably different. The differences, however, are irrelevant to the English sound system, and are, in fact, fully predictable, given the identity of the following vowel. The "h" sounds are in "complementary distribution", and can therefore be classified as variants, or "allophones", of a single /h/ phoneme. A similar state of affairs holds for the stop consonants (though in this case the situation is not open to introspection). Rapid changes, especially of the second formant, are a major cue for differentiating /ba, da, ga/. However, the second formant trajectory also depends on the identity of a following vowel: the perceptual identity of the initial sound in /di, du, da/ is not matched by acoustic identity [36]

A more dramatic case is given by certain consonants of Japanese, where the choice of consonant is highly dependent on the identity of a following vowel. Below are some syllables of Japanese, arranged according to the identity of the vowel phonemes (shown in the columns), and the identity of the consonant phonemes (shown in the rows):

	/a/	/i/	/u/	/e/	/o/
/s/	sa	shi	su	se	so
/t/	ta	chi	tsu	te	to
/z/	za	ji	zu	ze	zo
/h/	ha	çi	fu	he	ho

In terms of acoustic similarity, the [ç] of [çi] might cluster with either the [sh] of [shi] or the [s] of [se]. But in terms of the phonological system of Japanese (and the intuitions of Japanese speakers), [ç] has got to cluster with [h]: [h] "becomes" [ç] when it precedes [i].

A final point on phoneme systems: On the ideal scenario, a language has a small, fixed set of phonemes. Any word in the language can be unambiguously represented as a linear sequence of these phonemes (or variants of them). However, careful analysis of a language often shows cracks in the ideal system. A characteristic feature of New Zealand English is the effect on vowel quality of a following "l". The general effect is to cause the vowel to retract and to lower, depressing both the first and second formants. At the same time, some vowel contrasts are neutralised: most New Zealanders do not distinguish *dole* and *doll*, *celery* and *salary*. It is actually far from clear, what the "correct" phoneme analysis is of words with post-vocalic "l". This means, that in assessing the results of our experiments, we will need to bear in mind the possibility of alternative analyses. The results may even cause us to query some aspects of tranditional phonemic analyses.

Modelling of phoneme acquisition

We can conceptualise the sound of a word as a path through multi-dimensional acoustic space. Repeated utterances of the same word will be represented by a bundle of paths which follow rather similar trajectories. As the number of word types is increased, we may assume that the trajectories of different words will overlap in places; these overlaps will correspond to phoneme categories.

It is evident that an infant acquiring a human language does not know, a priori, how many phoneme categories there are going to be in the language that s/he is going to learn, nor, indeed, how many phonemes there are in any given word that the child hears. (We should like to add: The child learner does not know in advance that there are going to be such things as phonemes at all! Each word simply has a different global sound from every other word.) A minimum expectation of a learning model, is that the language input will be analysed in terms of an appropriate number of phoneme-like categories. Of special interest, is the success of the system in handling the more problematic cases mentioned above, such as cases of complementary distribution.

The earlier observations on language acquisition and phoneme categories suggest a number of variables which can be testing in the modelling of phoneme acquisition:

(a) Does learning require input which approximates the characteristics of "motherese", e.g. with regard to careful, exaggerated articulation, also with respect to the frequency of word types in the input language?

(b) Does phoneme learning require lexical-semantic information? The English learner will have evidence that *sheep* and *ship* are different words, not just variants of one and the same word, because *sheep* and *ship* mean different things. Applying this to our learning model, the question becomes: Do input utterances need to be classified as tokens of word types?

(c) Critical mass. It would be unrealistic to expect stable phoneme categories to emerge after training on only a couple of acoustically non-overlapping words. We might hypothesise that phoneme-like organisation will emerge only when a critical mass of words types have been extensively trained, such that each phoneme has been presented in a variety of contexts and word positions. First language acquisition research suggests that the critical mass is around 40 - 50 words.

(d) The speech signal is highly redundant, in that it contains vast amounts of acoustic information that is simply not relevant to the linguistically encoded message. We hypothesise that a learning model will need to be trained on input from a variety of speakers, all articulating the "same words". The system must be introduced to noise, in order for noise to be ignored.

Underlying our experiments is the basic question: Can the system organise the acoustic input with minimal specification of the anticipated output? (In psycholinguistic terms, this is equivalent to reducing to a minimum the contribution of innate linguistic knowledge). Indeed, our null hypothesis will be, that phoneme systems emerge as organisational solutions to massive data input. If it should turn out, that learning can be modelled with minimal supervision, this would have very significant consequences for linguistic and psycholinguistic theories of human language learning.

Bilingual acquisition

Once satisfactory progress has been made with modelling phoneme acquisition within a given language, a further set of research questions arises concerning the simulation of bilingual acquisition. We can distinguish two conditions:

1) Simultaneous bilingualism. From the beginning, the system is trained simultaneously with input from two languages (spoken by two sets of speakers) [29].

2) Late bilingualism. This involves training an already trained system with input from a second language [29].

It is well-known that children manage bilingual acquisition with little apparent effort, and succeed in speaking each language with little interference from the other. In terms of an evolving system, this means that the phoneme representations of the two language are strictly separated. Even though there might be some acoustic similarity between sounds of one language and sounds of the other, the distributions of the sounds (the shape of the trajectories associated with each of the languages) will be quite different.

Late acquisition of a second language, however, is typically characterised by interference from the first language. The foreign language sounds are classified in terms of the categories of the first language. (The late bilingual will typically retain a "foreign accent", and will "mis-hear" second-language utterances.) In terms of our evolving system, there will be considerable overlap between the two languages; the acoustic trajectories of the first language categories are so entrenched, that second language utterances will be forced into the first language categories.

3 Methods and Tools for Modelling Speech and Language Acquisition

3.1 Traditional AI Methods

The theory and practice of intelligent systems (IS) for modelling the acquisition of speech and language is still in its infancy despite of the rapid development of the information technologies in the recent years. However several techniques have been attempted, such as statistical techniques, e.g. Hidden Markov Models, neural networks (NN) [40,47,38], fuzzy systems (FS) [20], rule-based systems [11], genetic algorithms (GA) [14], artificial life [42,53], hybrid systems [20]. Several NN theories, models and methods for adaptive learning and dynamical modification of NN structures have been introduced so far [38,2,6,7,21-28] but only few of them are tested on the problem of modelling the speech emergence [1,8].

Even though the learning algorithms of NNs strongly relate to the NN structures, the dualism of learning and structure still exists and many connectionist methods deal with the "learning only" issue, others - with the "structure only" issue. Some of the references below are given in the respective aspect, i.e. learning, or structure. In the human brain the two processes are strongly inter-related.

Here we introduce briefly some concepts that have already been modelled in computers and that are related to the problem of speech acquisition.

Adaptive learning aims at solving the well-known stability/plasticity dilemma [6]. Methods for adaptive learning fall into three categories, namely incremental learning, lifelong learning, on-line learning. In the case of the NN structure, the bias/variance dilemma has been acknowledged by several authors [6]. The dilemma is that if the structure of a NN is too small, the NN is biased to certain patterns, and if the NN structure is too large there is too much variance that results in over-training, poor generalisation, etc. In order to avoid this problem, a NN (or an intelligent system - IS) structure should change dynamically during the learning process, thus better representing the patterns in the data and the changes in the environment. In terms of dynamically changing IS structures, there are three approaches taken so far: constructivism, selectivism, and a hybrid approach [21,23].

Constructivism is about developing NNs that have a simple initial structure and grow during its operation. This theory is supported by biological facts [39,50,13].

The growth can be controlled by either a similarity measure (similarity between new data and already learned ones), or by the output error measure, or by both. A measure of difference between an input pattern and already stored ones is used to insert new nodes [21-25]. *Selectivism* is concerned with pruning unnecessary connections in a NN that starts its learning with many, in most cases redundant, connections. The hybrid evolving approach to adaptive learning and structure modification involves both growing and shrinking during the learning process [21-25]. This method is used in the present approach.

On-line learning is concerned with learning data as the system operates (usually in a real time) and the data might exist only for a short time. NN models for on-line learning are introduced and studied in [2,21-25]. Several investigations proved that the most popular neural network models and algorithms are not suitable for adaptive, on-line learning, including multi-layer perceptrons trained with the backpropagation algorithm, radial basis function networks, self-organising maps SOMs and fuzzy neural networks. These models either operate on a fixed size connectionist structure, that limits its ability to accommodating new data, or require both new data and the previously used ones to adjust to the new data, or require many iterations of passing data through the connectionist structure in order to learn it, which could be unacceptably time-consuming. Some of them require a global optimisation algorithm where during the learning of each data item all the elements of the connectionist structure have to change. Problems of choosing the optimum initial structure, of arriving at a local minimum, of catastrophic forgetting, of lack of meaningful explanation of the information stored in the connections, and others, are often experienced (see [40,20,28,]).

Despite the successful development and use of NN, rule-based systems, FS, GA, hybrid systems, and other AI methods for adaptive training, radically new methods and systems are required both in terms of learning and structure development in order to model the emergence of spoken languages and for continuous learning in a computer system.

In this paper a paradigm called ECOS (Evolving COnnectionist Systems) along with two of its implementations - evolving fuzzy neural networks for supervised learning (EFuNNsu) and evolving fuzzy neural network for unsupervised learning (EFuNNun) - are explored. The major principles of ECOS and EFuNNs are presented in [21-25] and are briefly discussed below.

3.2 ECOS - Evolving Connectionist Systems

ECOS are systems that evolve in time through interaction with the environment, i.e. an ECOS adjusts its structure with a reference to the environment [21-25].

ECOS are multi-level, multi-modular structures where many modules have inter-, and intra- connections. The evolving connectionist system does not have a clear multi-layer structure. It has a modular open structure. The functioning of the ECOS is based on the following general principles [21-25]:

(1) learn fast from a large amount of data, e.g. through one-pass training;

(2) adapt in an on-line mode where new data is incrementally accommodated;
(3) have an 'open' structure where new features (relevant to the task) can be introduced at any stage of the system's operation, e.g., the system creates "on the fly" new inputs, new outputs, new modules and connections;
(4) memorise data exemplars for a further refinement, or for information retrieval;
(5) learn and improve through active interaction with other IS and with the environment in a multi-modular, hierarchical fashion;
(6) adequately represent space and time in their different scales; have parameters that represent short-term and long-term memory, age, forgetting, etc.;
(7) deal with knowledge in its different forms (e.g., rules; probabilities); analyse themselves in terms of behaviour, global error and success; "explain" what the system has learned and what it "knows" about the problem it is trained to solve; make decisions for further improvement.

3.3 Evolving Fuzzy Neural Networks (EFuNN) for Supervised and Unsupervised Learning

EFuNNs are first introduced in [24] where preliminary results were given. EFuNNs have a five-layer structure where nodes and connections are created/connected as data examples are presented. An optional short-term memory layer can be used through a feedback connection from the rule (also called, case) node layer. The layer of feedback connections could be used if temporal relationships between input data are to be memorised structurally. The third layer of neurons (rule nodes) in EFuNN evolves through either supervised (EFuNNsu) or unsupervised (EFuNNun) learning. The fourth layer of neurons in EFuNNsu represents fuzzy quantisation for the output variables, similar to the input fuzzy neurons representation in layer two of neurons. The fifth layer represents the real values for the output variables (fig.1).

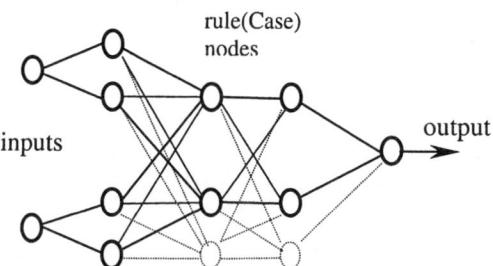

Figure 1: A simplified, exemplified (2 inputs, 1 output, 2 membership functions) diagram of an EFuNNsu. In the EFuNNun the fourth and the fifth layer of neurons are not present.

3.4 Learning Modes in EFuNNsu and EFuNNun

Different learning, adaptation and optimisation strategies and algorithms can be applied on an EFuNN structure. Some of them are: (a) Active learning - learning is performed when a stimulus (input pattern) is presented and kept active; this is the main learning mode. (b) Passive (inner, ECO) learning mode - learning is performed when there is no input pattern presented to the EFuNN. In this case the process of further elaboration of the connections in EFuNN is done in a passive learning phase, when existing connections, that store previously fed input patterns, are used as "echo" (here denoted as ECO) to reiterate the learning process (see for example [25]).

Different structure optimisation techniques can be applied during the learning process:
(a) Pruning and forgetting - the nodes and connections that are not actively participating in the learning process get pruned according to set criteria;
(b) Aggregation and abstraction - rule nodes that are close in the problem space (accommodate similar examples or exemplars) get merged together with the centre of the cluster they form being kept as a new rule node.

4 Modelling the Emergence of Speech and Language

The areas of the human brain that are responsible for the speech and the language abilities of humans evolve through the whole development of an individual [1,4,46,45,48,49,51,52]. Computer modelling of this process, before its biological, physiological and psychological aspects have been fully discovered, is an extremely difficult task. It requires flexible techniques for adaptive learning through an active interaction with a teaching environment.

It can be assumed that in a modular spoken language evolving system, the language (languages) modules evolve through using both domain text data and spoken information data fed from the speech recognition part. The language module produces final results as well as a feedback for adaptation in the previous modules. This idea is currently being elaborated with the use of ECOS.

4.1 An EFuNNun Model of the Emergence of Phonemes – a Simple Example

An EFuNNun is used here with inputs that represent features taken from a continuous signal - spoken words. In the experimental results shown in fig.2, three time lags of 26 mel scale coefficients are used taken from a window of 12 ms, with an overlap of 50%. In the other experiments 39 features are used (12 mel scale coefficients taken each 20ms and the power of the signal in this period, the 13 delta values and the 13 delta-delta values.

With each new input vector from one word the vector data is either associated with an existing rule node that is modified to accommodate this data, or a new rule node is created (see fig.2a). Regularly the rule nodes are aggregated (fig.2b), while reduces the number of the nodes and places them in the centres of the data clusters. After the whole word is presented the aggregated rule nodes represent the centres of the phoneme clusters without the concept of phonemes being presented to the system (fig.2c,d). The latter figures show clearly that three rule nodes were evolved (after aggregation) that represent the stable sounds as follows: frames 0-53 and 96-170 are allocated to rule node 1 that represents /silence/; frames 56-78 are allocated to rule node 1 that represents the phoneme /ei/; frames 85-91 are allocated to rule node 3 that represents the phoneme /t/; the rest of the frames represent transitional states, e.g. frames 54-55 the transition between /silence/ and /ei/, frames 79-84 - the transition between /ei/ and /t/, and frames 92-96 - the transition between /t/ and /silence/, are allocated to some of the closest rule nodes.

If the aggregation is done with more sensitivity, there will be rule nodes evolved to represent these short transitional sounds.

When more pronunciations of the word 'eight' are presented to the EFuNNun model the model refines the phoneme regions and the phoneme rule nodes. Words that represent all phonemes in English were presented to the EFuNNun model and the results of the emergence of all phoneme and some transitional sounds in English are presented in fig.3.

The EFuNNun model allows for experimenting with different strategies of elementary sound emergence:
(a)   increased sensitivity over time;
(b)   decreased sensitivity over time;
(c)   single language sound emergence;
(d)   multiple languages presented one after another;
(e)   multiple languages presented simultaneously (alternative presentation of words from different languages);
(f)   aggregation within word presentation;
(g)   aggregation after a whole word is presented;
(h)   aggregation after several words are presented
(i)   the effect of alternative presentation of different words versus the effect of one word presented several times and then the next one, etc.
(j)   the role of the transitional sounds and the space they occupy;
(k)   using forgetting in the process of learning;
(l)   other emerging strategies.

115

Figure 2: Experimental results with an EFuNNun model for phoneme acquisition - a single pronunciation of the word 'eight'. From top to bottom: the two dimensional input space of the first two mel scale coefficients of all frames taken from the speech signal 'eight' and numbered with the consecutive time interval, and also the evolved rule nodes (denoted with larger font) that capture each of the three phases of the signal: /silence/, /ei/, /t/, /silence/; the time of the emergence of the rule nodes; all the 78 element mel-vectors of the word 'eight' over time.

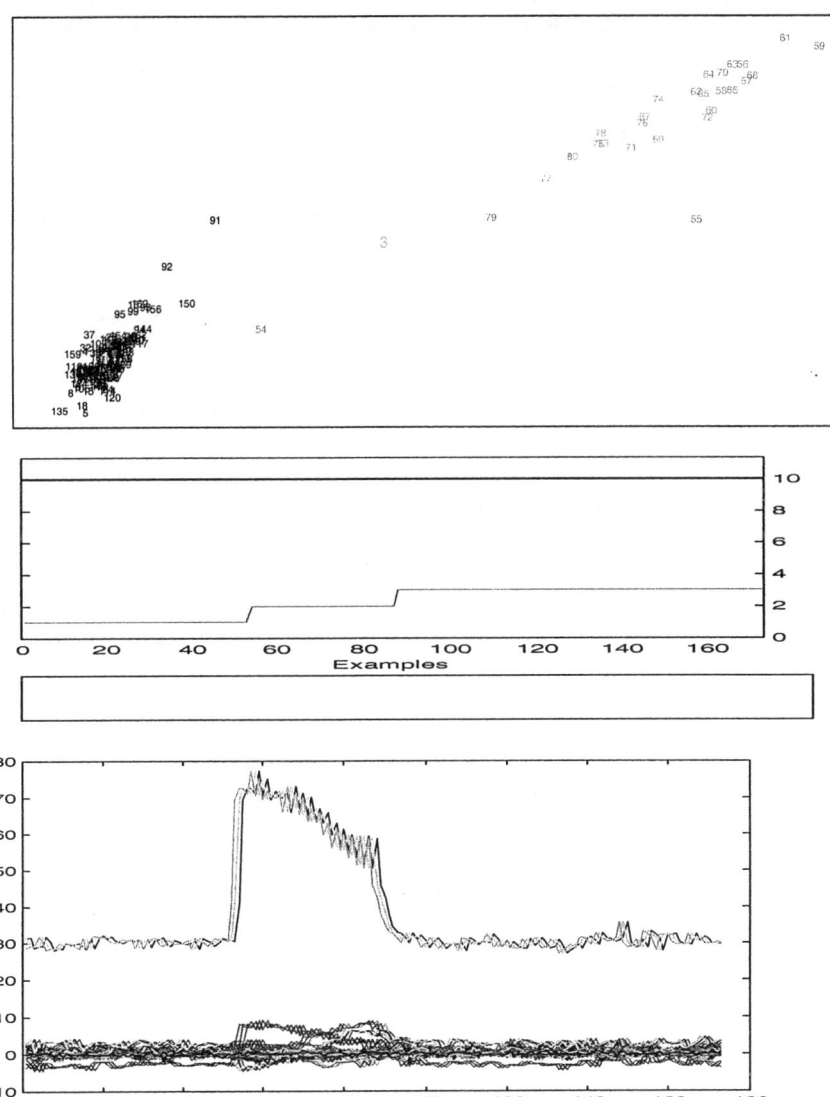

4.2 Eolving the Whole Phoneme Space of NZ English

To create the clustered model for New Zealand English, several speakers from the Otago Speech Corpus (http://kel.otago.ac.nz/hyspeech/corpus/) were selected to train the system. Here, 18 speakers (9 Male, 9 Female) each spoke 128 words three times. Thus, approximately 6912 utterances were available for training. During the training, a word example was chosen at random from the available words. The waveform underwent a Mel-scale cepstrum (MSC) transformation to extract 12 frequency coefficients, plus the log energy, from segments of approximately 23.2ms of data. These segments were overlapped by 50%. Additionally, the delta and delta-delta values of the MSC coefficients and log energy were extracted, for an input vector of dimensionality 39.

The EFuNN system was trained until the number of rules was constant for over 100 epochs. A total of 12000 epochs were performed – each on one of the 12,000 data examples (see [58] for detail). The sensitivity threshold parameter of the EFuNNun was set to $Sthr$ of 0.85. The aggregation threshold was allowed to change, with a target number of rule nodes of 100. The other parameters were as their default values.

Figure 3 shows: (a) The connections of the evolved EFuNNun (70 rule nodes are evolved that capture 70 elementary sounds) from spoken words presented one after another, each frame of the speech signal being represented by 13 features (12 MSC and the power), their delta features and the delta-delta features (all together 39 input features); (b) The evolved rule nodes from the EFuNNun and the trajectory of the word "zero" projected in the MSC1-MSC2 input space; (c) The trajectory of the word "zero" and the labelled rule nodes projected in the MSC1-MSC2 input space.

Figure 4 shows three representations of a spoken word from the corpus. Firstly, the word is viewed as a waveform. This is the raw signal as amplitude over time. The second view is the MSC space view. Here, the 12 frequency components are shown (Figure 4 bottom). This approximates a spectrogram. The third view (top) shows the activation of each of the 70 rule nodes created over time. Darker areas represent a high activation. Additionally, the winning rules are shown as circles. Numerically, these are: 1 1 1 1 1 1 2 2 2 2 22 2 2 11 11 11 11 11 24 11 19 19 19 19 15 15 16 5 5 16 5 15 16 2 2 2 11 2 2 1 1 1. Some further testing showed that recognition of words depended on not only the winning rule node, but also the path of the recognition. Additionally, an n-best selection of rule nodes may increase discrimination.

Trajectory plots

The trajectory plots, shown in figures 5, 6, 7, and 8 are in three of the 39 possible dimensions. Here, the first and seventh MSC are used for the x and y co-ordinates. The log energy is represented by the z-axis. A single word, 'sue', is shown in fig 5.

The starting point is shown as a square. Several frames represent the hissing sound, which has low log energy. The vowel sound has increased energy, which fades out toward the end of the utterance. Two additional instances of the same word, spoken by the same speaker, are shown in fig 6. Here, a similar trajectory can be seen. However, the differences in the trajectories represent the intra-speaker variation. Inter-word variability can be seen in fig.7, which shows the 'sue' from fig. 5 (dotted line) compared with the same speaker uttering the word 'nine'. Even in the three-dimensional space shown here, the words are markedly different. The final trajectory plot (fig.8) is of two similar words, 'sue' (dotted line) and 'zoo' (solid line) spoken by the same speaker. Here, there is a large overlap between the words, especially in the latter section, the vowel sound.

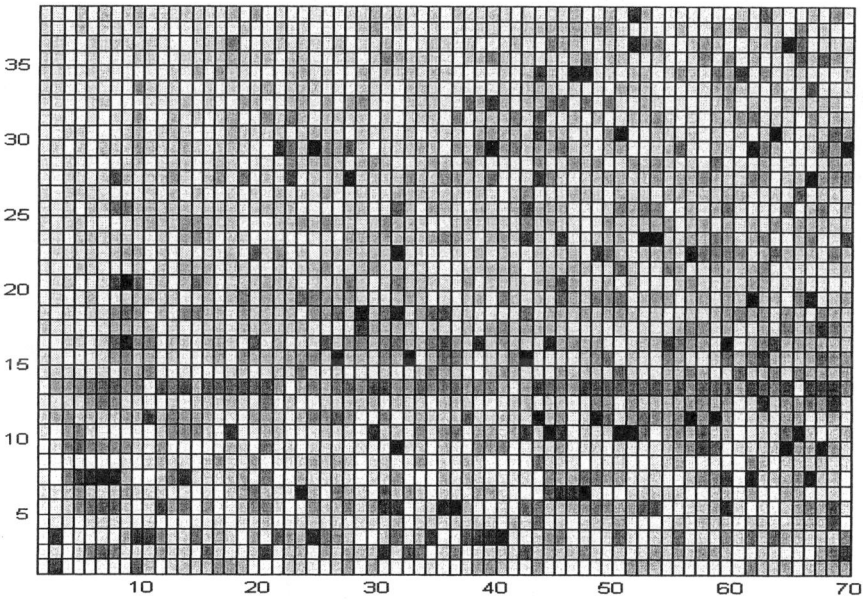

Figure 3a: The connections of the evolved EFuNNun from spoken words presented one after another, each frame of the speech signal being represented by 13 features (12 MSC and the power), their delta feature and the delta-delta features (all together 39 input features)

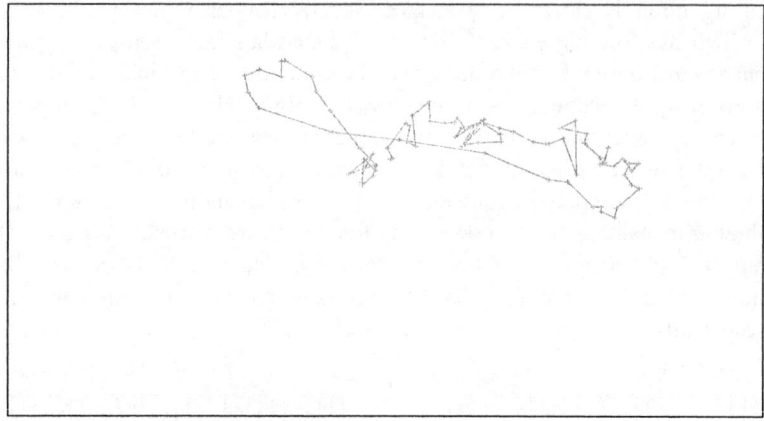

Figure 3b: The evolved 70 rule nodes from the EFuNNun and the trajectory of the word "zero" projected in the MSC1-MSC2 input space

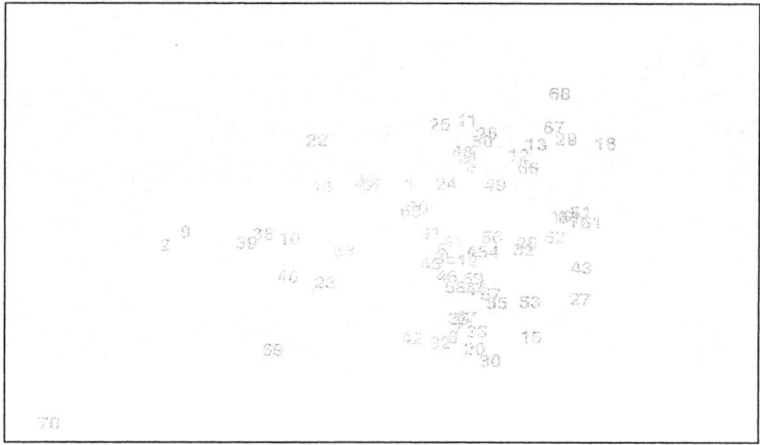

Figure 3c: The trajectory of the word "zero" (the consecutive time frames are numbers with the smaller font numbers) and the labelled rule nodes (larger font) projected in the MSC1-MSC2 input space.

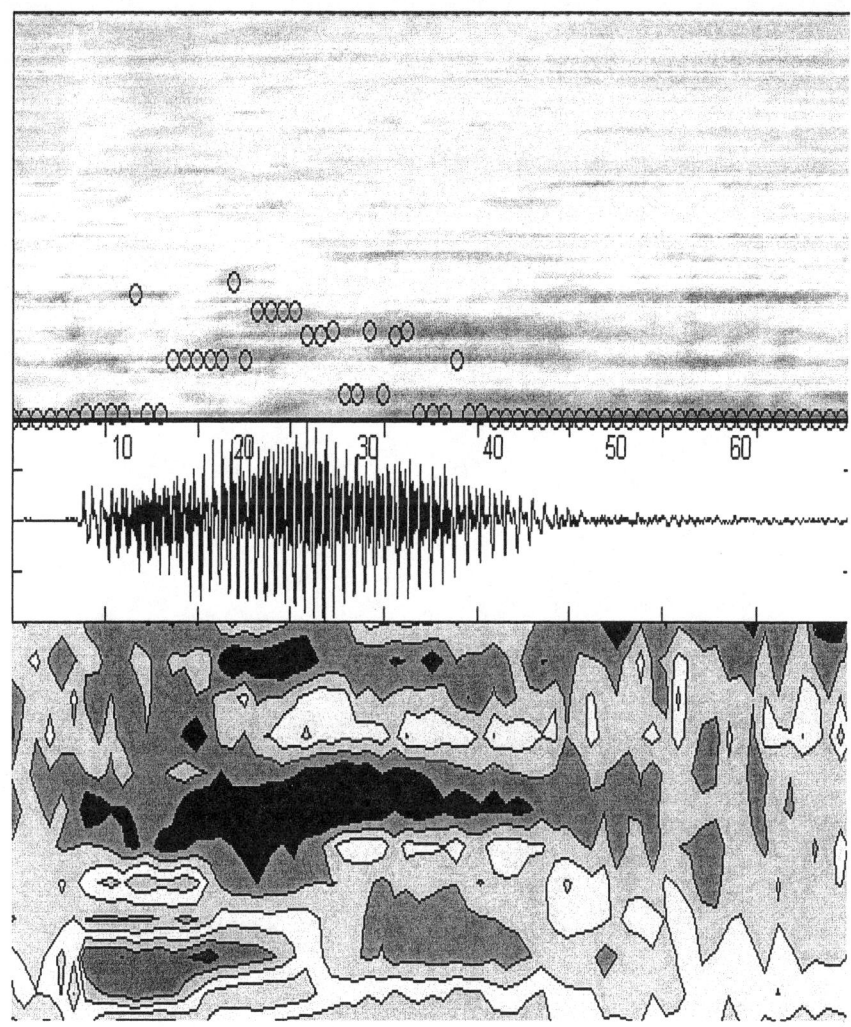

Figure 4: Representation of a spoken word: 'zero'

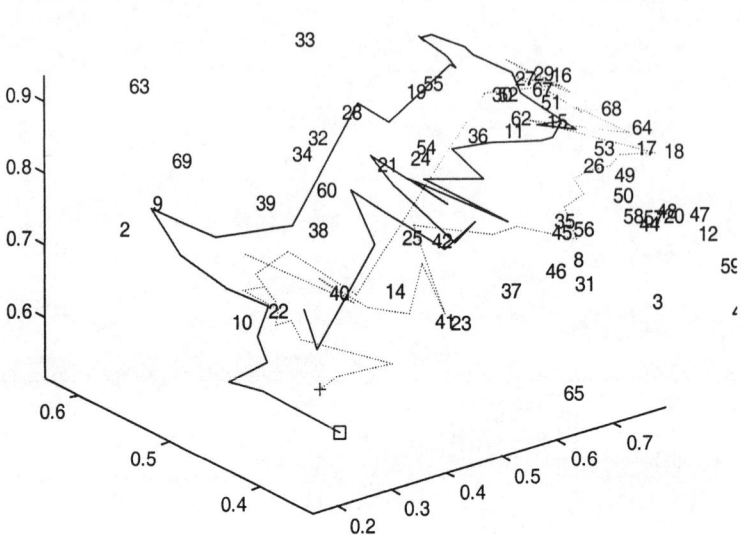

Figure 5: Two utterances of the word 'sue' along with the 70 rules nodes of the evolved EfuNNun shown in the MS1-MS7-logE space

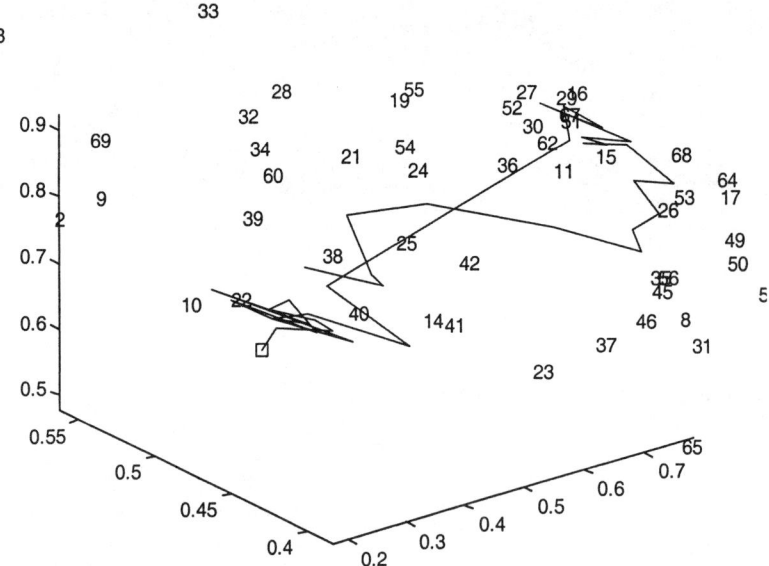

Figure 6: Trajectory of a spoken word 'sue' long with the 70 rules nodes of the evolved EFuNNun shown in the MS1-MS7-logE space

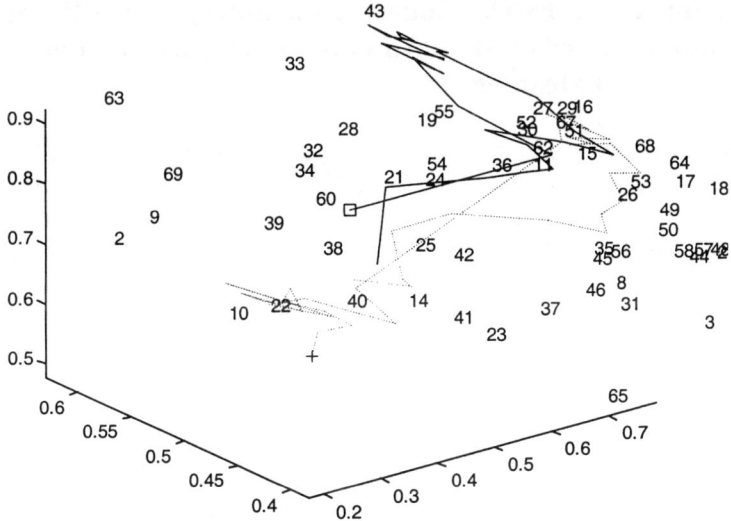

Figure 7: Trajectories of 'sue' and 'nine' in the input space MS1-MS7-logE, along with the 70 rules nodes of the evolved EFuNNun also presented

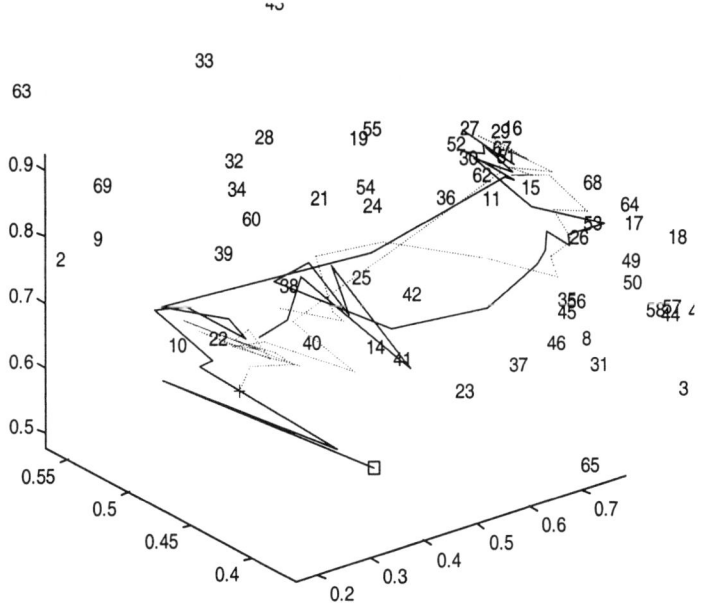

Figure 8: Trajectories of the words 'sue' and 'zoo' along with the 70 rules nodes of the evolved EFuNNun in MS1-MS7-logE space

4.3 A Supervised ECOS Model for the Emergence of Word Clusters Based on Both Auditory Traces and Supplied (Acquired) Meaning

The next step of this project is to develop a supervised model based on EFuNNun for phoneme cluster emergence, and EFuNNsu for word recognition. After the EFuNNun is evolved (it can still be further evolved) a higher-level word recognition module is developed where the inputs of the EFuNNsu are the activated rule nodes from the phoneme EFuNNun over a period of time. The outputs of the EFuNNsu are the words that are recognised. The number of words can be extended over time thus creating a new output node that is allowable in an EFuNNsu system. A new, sentence recognition layer, can be built on top of this model. The new layer will use the input from the previous layer (a sequence of recognised words over time) and will activate an output node that represents a sentence (a command, a meaningful expression, etc). At any time in the functioning of the system new sentences can be introduced to the system which makes the system evolvable over time.

5 Conclusions

Modelling the emergence of spoken languages is an extremely intriguing task that has not been solved so far despite existing papers and books. The simple evolving model that is presented in this chapter illustrates the main hypothesis raised in this material, that linguistic concepts such as phonemes are learned rather than inherited. The evolving of sounds, words and sentences can be modelled in a continuous learning system that is based on the evolving connectionist systems techniques ECOS.

Acknowledgements

The research was done as part of a research programme UOO808 funded by the Public Good Science (PGSF) Fund of the New Zealand Foundation for Research Science and Technology (FRST), and partially by a divisional research grant in the Division of Humanities, University of Otago, New Zealand. We would like to thank Richard Kilgour and Qun Song for the implementation of some of the tools and for conducting some of the experiments.

References

1. Altman, G., Cognitive Models of Speech Processing, MIT Press, 1990.
2. Amari, S. and Kasabov, N. eds, "Brain-like Computing and Intelligent Information Systems", Springer Verlag,1997.
3. Bates, E. and J. Goodman. On the emergence of grammar from the lexicon. In: B. MacWhinney (ed), The Emergence of Language, 29-79. Mahweh: Lawrence Erlbaum. 1999.
4. Brent, M. Advances in the computational stu dy of language acquisition, in: M. Brent (ed) Computational Approaches to Language Acquisition (1996)
5. Bybee, J. (1995). Regular morphology and the lexicon. Language and Cognitive Processes 10, 425-455.
6. Carpenter, G. and Grossberg S., Pattern recognition by self-organizing neural networks, The MIT Press, Cambridge, Massachusetts (1991)
7. Chomsky, N., *The Minimalist Program*, Cambridge, MA: MIT Press, 1995.
8. Cole, R. et al. "The Challenge of Spoken Language Systems: Research Directions for the Nineties", IEEE Transactions on Speech and Audio Processing, vol.3, No.1, 1-21, 1995
9. Culicover,P.Syntactic Nuts: Hard cases, Syntactic theory, and Language Acquisition. Oxford: Oxford University Press. 1999.
10. Damasio, H., Grabowski, T., Tranel, D., Hichwa, D., Damsio, A. A neural basis for lexical retrieval, Nature, 380, April, 499-505 (1996)
11. Deacon, T. The symbolic species (The co-evolution of language and the human brain) Penguin Press,1998
12. Drossaers, M. Little Linguistic Creatures (Closed Systems of Solvable Neural Networks for Integrated Linguistic Analysis), PhD Thesis, University of Twente, Enschede (1995)
13. Durand, G., Kovalchuk, Y., Konnerth, A. Long-term potentiation and functional synapse induction in developing hippocampus, Nature, 381, May, 71-75 (1996)
14. Edelman, G., Neuronal Darwinism: The theory of neuronal group selection, Basic Books (1992)
15. Elman, J., E.Bates, M.Johnson, A.Karmiloff-Smith, D.Parisi and K.Plunkett, Rethinking Innateness (A Connectionist Perspective of Development), The MIT Press, 1997
16. Fletcher and Macwhinney (1996). The Handbook of Child Language. Oxford: Blackwell
17. Fodor, J. and Pylyshyn, Z. (1988) "Connectionism and Cognitive Architecture: a Critical Analysis," Cognition 28, 3-71
18. Harris, C. Connectionism and Cognitive Linguistics, Connection Science, vol.2, 1&2 (1990), 7-33
19. Jusczyk, P. The Discovery of Spoken Language, The MIT Press, 1997
20. Kasabov, N. (1996) Foundations of Neural Networks, Fuzzy Systems and Knowledge Engineering, MIT Press, Cambridge, Massachusetts, 550 pages
21. Kasabov, N. Brain-like functions in evolving connectionist systems for on-line, knowledge-based learning, in: T.Kitamura (ed) Modelling brain functions for intelligent systems, World Scientific – FLSI Series (2000)
22. Kasabov, N. ECOS: A framework for evolving connectionist systems and the eco learning paradigm, Proc. of ICONIP'98, Kitakyushu, Oct. 1998, IOS Press, 1222-1235

23. Kasabov, N. Evolving fuzzy neural networks for on-line, knowledge based learning, IEEE Transactions on Systems, Man, and Cybernetics (2000)
24. Kasabov, N. Evolving Fuzzy Neural Networks - Algorithms, Applications and Biological Motivation, in: Yamakawa and Matsumoto (eds), Methodologies for the Conception, design and Application of Soft Computing, World Scientific, 1998, 271-274
25. Kasabov, N. The ECOS Framework and the ECO Learning Method for Evolving Connectionist Systems, Journal of Advanced Computational Intelligence, 2 (6) 1998, 195-202
26. Kasabov, N., "A framework for intelligent conscious machines utilising fuzzy neural networks and spatial temporal maps and a case study of multilingual speech recognition", in: ref.2, 106-126 (1998)
27. Kasabov, N., R. Kozma, R. Kilgour, M. Laws, J. Taylor, M. Watts, and A. Gray, "A Methodology for Speech Data Analysis and a Framework for Adaptive Speech Recognition Using Fuzzy Neural Networks and Self Organising Maps", in: Kasabov and Kozma (eds) Neuro-fuzzy techniques for intelligent information systems, Physica Verlag (Springer Verlag) 1999
28. Kasabov, N., R.Kilgour, and S.Sinclair, From hybrid adjustable neuro-fuzzy systems to adaptive connectionist-based systems for phoneme and word recognition, Fuzzy Sets and Systems, 103 (1999) 349-367
29. Kim, K., Relkin, N., Min-lee, K., Hirsch, J. Distinct cortical areas associated with native and second languages, Nature, 388, July (1997)
30. Kohonen, T., "The Self-Organizing Map", Proceedings of the IEEE, vol.78, N-9, pp.1464-1497, (1990).
31. Köpke, K.M.. (1988). Schemas in German plural formation. Lingua 74: 303-335.
32. Kuhl, P. (1994) Speech Perception, in: Introduction to Communication Sciences and Discorse, F.Minifie (ed), san Diego, Singular P., 77-142
33. Lakoff, G. and M. Johnson (1999). Philosophy in the Flesh. Basic Books, New York
34. Langacker, R. (1986, 1991). Foundations of Cognitive Grammar. 2 vols. Stanford University Press.
35. Li, Ping and Brian MacWhinney (1996). Cryptotype, over-generalisation and competition: A connectionist model of the learning of English reverse prefixes. Connection Science 8: 3-30.
36. Liberman, A., F. Cooper, D. Shankweiler and M. Studdert-Kennedy, Perception of the speech code. Psychological Review 74, 431-461. 1967.
37. Lindblom, B. Spectrographic study of vowel reduction. Journal of the Acoustic Society of America, 35, 1773-1781. 1963
38. Matsumotot, G., M.Ichikawa, Y.Shigematsu (1996) Brain Computing, In: Methodologies for Conception, Design, and Application of Intelligent Systems, World Scientific, 15-24
39. McClelland, J., B.L. McNaughton, and R.C. Reilly "Why there are Complementary Learning Systems in the Hippocampus and Neocortx: Insights from the Successes and Failures of Connectionist Models of Learning and Memeory", CMU Technical Report PDP.CNS.94.1, March, (1994)
40. McClelland, J., Rumelhart, D., et. al. Parallel Distributed Processing, vol. II, MIT Press (1986)

41. Nijholt, A. and J. Hulstijn (eds). Formal Semantics and Pragmatics of Dialogue. Proceedings Twendial'98 (TWLT13), Universty of Twente, The Netherlands, May 1998
42. Parisi, D. (1997). An artificial life approach to language. Brain and Language 59, 121-146
43. Pinker, S. (1994) The Language Instinct: How the Mind Creates Language. Penguin
44. Plunkett, K. (1996). Connectionist approaches to language acquisition. In Fletcher & MacWhinney (eds), The Handbook of Child Language. Oxford: Blackwell, 36-72
45. Port, R., and T.van Gelder (eds) Mind as motion (Explorations in the Dynamics of Cognition) , The MIT Press, 1995
46. Regier, T. (1996). The Human Semantic Potential: Spatial Language and Constrained Connectionism. MIT Press.
47. Rummelhart, D. and McClelland, J. (1985) "On Learning the Past Tenses of English Verbs," in McClelland and Rummelhart et. al. (1986) Ch. 18, pp. 216—271
48. Segalowitz, S.J. Language functions and brain organisation, Academic Press, 1983
49. Seidenberg, M. language acquisition and use: learning and applying probabilistic constraints, Science, 275, 1599-603 (1997)
50. Shastri, L. A biological grounding of recruitment learning and vicinal algorithms, TR-99-009, International Computer Science Institute, Berkeley, (1999)
51. Shastri, L. and C.Wendelken, Knowledge fusion in the large - taking a cue from the brain, Int. Conf. FUSION'99, Sunnyvale, June 1999
52. Snow, C. and C Ferguson. Talking to children: Language input and language acquisition. Cambridge: Cambridge University Press. 1977.
53. Stephens, C., Olmedo, I., Vargas, J., Waelbroack, H., Self adaptation in evolving systems, Artificial Life, 4(2) 183-201
54. Stork, D. (1991) Sources of neural structure in speech and language processing, Int. Journ. on Neural Systems, Singapore, 2 (3), 159-167
55. Taylor, J. R. & MacLaury, R. E. (eds.), Language and the Cognitive Construal of the World. Berlin, Mouton de Gruyter (1995) 406p.
56. Taylor, J. R. An Introduction to Cognitive Linguistics. Oxford, Clarendon Press, 1999, to appear
57. Taylor, J. R. Linguistic Categorization: Prototypes in Linguistic Theory. 2nd Edition. Oxford, Clarendon Press (1995)
58. Taylor, J., N.Kasabov and R.Kilgour, Modelling the emergence of speech sound categories in evolving connectionist systems, in: Proc. Joint. Conf. Information Sciences – JCIS, Atlantic City, 2000

Part II

Intelligent Human Computer Interaction and Scientific Visualisation

Chapter 7. Discovering the Visual Signature of Painters

H.Jaap van den Herik and Eric O. Postma

IKAT / Computer Science Department, Universiteit Maastricht, The Netherlands
herik@cs.unimaas.nl

Abstract. Recent developments in image classification have focused on efficient preprocessing of visual data to improve the performances of neural networks and other learning algorithms when dealing with content-based classification tasks. Given the high dimensionality and redundancy of visual data, the primary goal of preprocessing is to transfer the original data to a low-dimensional representation that preserves the information relevant for the classification. This contribution reviews modern preprocessing (dimension-reduction) techniques and discusses their advantages and disadvantages. The performance of the techniques is assessed on a difficult painting-classification task that requires painter-specific features to be retained in the low-dimensional representation. Evaluation of the results shows that domain-specific knowledge provides a rough albeit indispensable guideline for determining the appropriate type of preprocessing. Furthermore, the evaluation shows that neural-network techniques are most suitable for executing and fine-tuning the preprocessing and subsequent classification. It is argued that further improvements can be gained by the use of a content-based attentional selection procedure. Our conclusion is that preprocessing should be tailored to the task at hand by combining domain knowledge with neural-network techniques, and that within fifty years the visual signature of painters is as recognizable as is any handwritten signature.

Keywords. Image recognition, neural networks, visual art recognition

1 Introduction

Automatic image recognition poses a challenge for present-day and future research in artificial intelligence. The aim of automatic image recognition is to generate an appropriate label describing the contents of an arbitrary image. Two main types of (automatic) image recognition can be defined. In the first type, the label describes the object (or group of objects) present in the image. In this case the label refers to spatially-localized object features. For instance, an image containing a tree in a landscape is labeled as *tree* on the basis of spatially-localized tree features. In the second type, the label characterizes the overall contents of an image by referring to a feature that is not necessarily spatially localized or linked to an object. For instance, the image of a tree in a landscape is labeled as *landscape* on the basis of spatially-distributed landscape features. Of course, both types of image recognition represent extremes of a continuum. After all, in human vision, the spatially-distributed context (i.e., the landscape) enhances the recognition of appropriately-positioned objects within that context [3]. Nevertheless, it does make sense to define the two types of recognition because both psychological and biological studies reveal the initial step of human vision to be a position-independent parallel process that recognizes elementary features such as color, texture, orientation, and size [2, 13, 32]. This fast process precedes a slower position-dependent object-recognition process which requires the elementary features making up a perceptual object to be combined. The fast parallel process is called *preattentive*, because it does not require spatial attentional selection (i.e., it is spatially independent), the slow process is referred to as *attentive* because it does require spatial attentional selection.

This contribution reviews preprocessing techniques that perform a preattentive analysis by detecting the elementary features of images. The effectiveness of the preprocessing techniques is assessed on a suitable test bench consisting of impressionistic paintings. The "hidden signature" contained in the painter's idiosyncratic application of paint, if captured by the preprocessing techniques, can be recognized with standard neural-network or other classification techniques.

The outline of the contribution is as follows. Section 2 discusses how preprocessing transforms images from image space to feature space by probing a specific kind of feature. Then Section 3 describes the quest for good features (i.e., features that capture the information relevant for the task at hand) by discussing the painting-classification task. Section 4 describes some basic features and the associated preprocessing techniques chosen on the basis of expert knowledge. Section 5 presents the classification results obtained using the preprocessing techniques discussed in Section 4, and also discusses their implications. Finally, Section 6 draws conclusions and presents an outlook on the future of automatic image recognition.

2 Image Space and Feature Space

A straightforward way of representing images is to define them as vectors in which the elements correspond to the pixel values. Provided that all images are normalized to a standard size they can all be represented as points in the same space. A distance between two points in the image space reflects the average of the distances between matched pairs of pixel values. However, such an image space is not a good image representation for at least two reasons. The first reason is the high dimensionality of the representation, which makes it unsuitable for neural-network algorithms (or machine-learning algorithms in general). Assuming the number of possible discrete values along each dimension to be n, we see that with increasing dimensionality d, the total number of values grows as n^d. Even when assuming that the images give rise to points that are uniformly distributed over the image space, the learning algorithms still must cover the entire space in order to achieve good recognition performances. It would require a storage capacity which grows exponentially with the number of dimensions and is (given the dimensionality of the images) not feasible. This is known as the *curse of dimensionality* (see, e.g., [23]). The second reason is that the values of the individual pixels barely convey any information on the characteristics of the images. Their individual values do not correlate with the contents of the image, whereas local configurations of values do.

The (local) image features are much more informative than the individual pixels and, when appropriately combined, give rise to a representation with a much lower dimensionality. Both human and computer vision suggest to focus on specific types of spatial features. In preattentive vision, spatial and chromatic information is relevant. Feature-based approaches are therefore suitable for our purposes. They may rely on chromatic features [30], on (local) spatial features [29], or on both [6, 17, 21, 23]. Combined with histogramming [28] these features yield a suitable representation space for image recognition. Henceforth, we refer to this representation space as *feature space* to emphasize the central role of the features chosen.

Our conception of image classification is illustrated in figure 1. The input pattern is the (digital reproduction of an) image. A preprocessing stage reduces the dimensionality of the input to a low-dimensional feature-space representation which serves as input to a (neural-network) classifier.

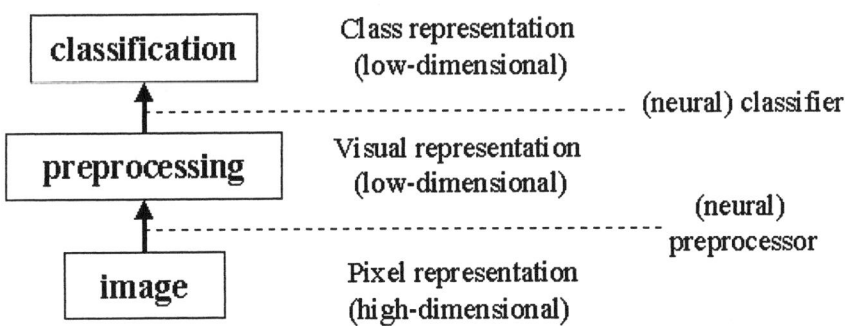

Figure 1. Illustration of image classification.

Domain knowledge, ranging from general knowledge to highly-specific knowledge has to be invoked for choosing the appropriate preprocessing technique(s) and the associated features. In the next section, domain knowledge is guiding our quest for good features to recognize impressionistic paintings.

3 The Quest for Good Features

The choice of a good feature is dependent on the task (e.g., classification of images) and the (preprocessing) technique involved. It requires a careful consideration of the domain investigated (e.g., impressionistic paintings) as well as a deep insight into the essence of the dimension reduction produced by the various preprocessing techniques. This section discusses the classification task, provides some relevant general domain knowledge, and describes the data set (see also the Appendix).

3.1 The Task: Impressionistic-Painting Classification

The effectiveness of a feature space is evaluated by applying it to a painting-classification task: paintings must be classified according to their maker. Determining the maker of an artwork is difficult; so far it requires human expertise. However, human subjectivity in the recognition process may give rise to mistakes or controversy.

One of the most famous examples of a mistake is the Van Meegeren forgery [19]. In 1937, the Dutch art expert Bredius identified the painting *The Disciples at Emmaus* (in Dutch: *De Emmaüsgangers*) as a genuine Johannes Vermeer. In 1938, the Boijmans Van Beuningen Museum in Rotterdam bought the painting and organized a special exhibit centered around the *Emmaüsgangers*. Just after the war, in 1945, the Dutch painter Han van Meegeren was arrested for selling another Vermeer to Göring. After two weeks of imprisonment, Van Meegeren confessed to have forged both the Vermeer sold to Göring and the *Emmaüsgangers*. His confession was corroborated by convincing evidence. Although experts, with the benefit of hindsight, have identified visual characteristics revealing the painting to be a forgery, these characteristics escaped the scrutiny of Bredius and his contemporaries.

Two recent examples illustrate the controversy associated with the human recognition of paintings once more. First, in the Rembrandt Research Project Van de Wetering [34] describes how his research group came to the decision that a large number of paintings originally credited to Rembrandt were not painted by the master, whereas two other paintings so far not recognized as Rembrandt's were identified as such and now belong to the Rembrandt collection. Second, there is the ongoing debate on the authenticity of Van Gogh's *Sunflowers*, bought by the Japanese insurance company Yasuda. Nowadays, art experts challenge the authenticity of the *Sunflowers* as a genuine *Van Gogh*, although it was generally accepted as an original Van Gogh for several decades.

The lack of objective criteria for determining the maker of a painting makes it a challenging image-recognition task. As a first step towards the automatic recognition of forgeries, we attempt to classify impressionistic paintings from six distinguished painters using preattentive recognition. Although this task may be considered to be much easier than the recognition of forgeries, it is a very hard task.

3.2 The Data Set

The data set used in all our experiments consists of 60 digital reproductions of paintings taken from the *WebMuseum* [20]. The 24-bit true-color images vary in size with an approximate average of 1000×1000 pixels and are all in JPEG format. Each pixel is defined as an *rgb* triplet, $r, g, b \in \{0,1,...255\}$. The data set contains 60 works by 6 (neo-)impressionistic painters (ten works per painter). The painters are: *Claude Monet, Vincent van Gogh, Paul Cézanne, Alfred Sisley, Camille Pissarro,* and *George Seurat*. The titles of the paintings included in the data set along with their creation dates are given in the Appendix. The paintings have been carefully selected so as to be as uniform as possible with respect to year of creation and contents. As a result, the years of creation of the paintings range from 1870 to 1894, and their contents are restricted to natural scenes (i.e., landscapes, townscapes, and sea views).

In our experiments with the data set, we have taken great care in preventing visual artefacts from affecting our results. In particular, we are well aware that the distortions introduced by the photographing and digitalization processes generate "signatures" in the spatial and chromatic characteristics of paintings. These signatures may be probed by our techniques and guide the recognition. Unfortunately, such artefacts cannot be prevented; they are unavoidable side-effects of the reproduction processes. Therefore, our aim is to ensure that the (unknown) artefacts are consistent throughout all digital reproductions in our data set. Only by ensuring this, the artefacts cannot be used for the recognition. In our data set consistency of artefacts is achieved by ensuring that all paintings are photographed by the same photographer using a single digitalization procedure.

4 The Features and the Preprocessing Techniques

Our choice of the feature set is guided by domain knowledge on visual art and impressionistic paintings in particular. Our knowledge stems from discussions with the art expert Frank Boom and some literature on art (e.g., [7, 19]). The main general characteristics known to be indicative of a painting's maker are: color, brushwork, and texture. These characteristics have determined our choice for preprocessing techniques. Since the experts seem to know only implicitly the more peculiar characteristics relevant for the recognition of the painting, a neural network classifier is used to collect analogous information from the feature-space representation. As a painter's brushwork yields a specific texture, we treat brushwork and texture as one characteristic. The following feature-space representations are studied.

- *Color:* color histograms.
- *Brushwork and texture:* oriented spatial features, (parts of) Fourier spectra, statistical descriptors, independent components, fractal dimension.

Each of these feature-space representations evokes a preprocessing technique. They are described in more detail below.

4.1 Color Histograms: RGB and HSI

Color histograms are straightforward low-dimensional representations of images. A color histogram is a vector containing the relative frequencies of pixel values as its elements. An RGB histogram consists of three such vectors, one for each color (red, green, blue).

Two examples of RGB color histograms are shown in the left part of figure 2. The upper histogram corresponds to Van Gogh's painting *Wheat Field under Threatening Skies* and the lower to his *The Church at Auvers-sur-Oise*. Although on casual visual inspection both these paintings have similar colors, the RGB histograms do not appear very similar.

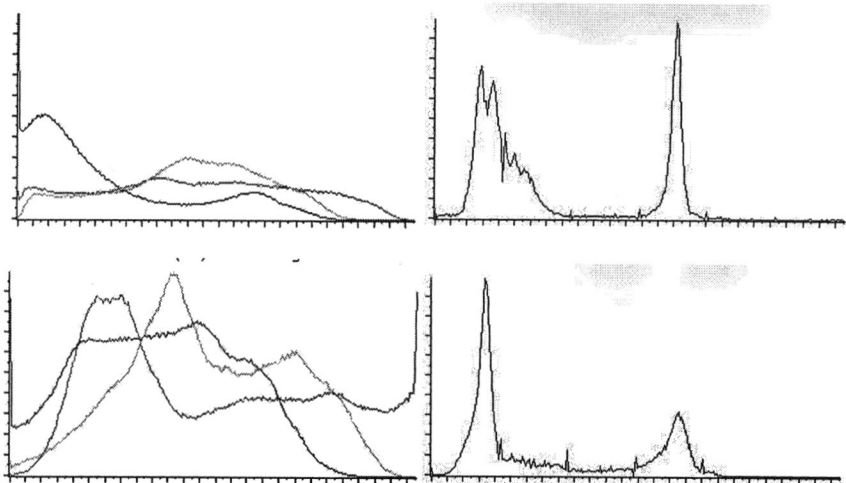

Figure 2. Left: two RGB histograms of similarly-colored Van Gogh paintings. Right: two *hue* histograms of the same paintings.

Apparently, the raw RGB values are not suitable for representing the perceptual appearance of color. The reason is that humans perceive color in terms of its hue, saturation, and intensity, rather than in terms of its three RGB components. The right part of figure 2 shows the *hue* histograms of both Van Goghs. (For clarity of presentation, the saturation and intensity components are not shown.) In this case, a clear correspondence in both histograms is observed, i.e., the two coinciding peaks.

The perceptually more appropriate hue (*H*), saturation (*S*), and intensity (*I*) values are obtained by measuring the individual RGB-color components, using the HSI model (*cf.* [8, 12]). The HSI representation is obtained from normalized RGB values as follows:

$$H = \cos^{-1}\left[\frac{\frac{1}{2}[(R-G)+(R-B)]}{\sqrt{(R-G)^2 + (R-B) \cdot (G-B)}}\right],$$

$$S = 1 - \frac{3}{(R+G+B)} \min(R,G,B), \text{ and}$$

$$I = \frac{1}{3}(R+G+B).$$

If $B > G$ then $H = 2\pi - H$ and H is normalized by dividing its value by 2π.

4.2 Oriented Spatial Features

To measure the presence of local spatially-oriented edges (e.g., as generated by the brushswork), we employ oriented Gaussian derivatives. Our choice is based on very general considerations regarding the nature of natural images [14]. A suitably normalized and scaled Gaussian function is defined as (*cf.* [24]):

$$G(x, y) = e^{-(x^2+y^2)}$$

with *x,y* the spatial Euclidian co-ordinates. The property of steerability [5] allows oriented derivatives of *G(x,y)* to be generated through linear combinations of the vertical and horizontal (first order) derivatives $dG^v(x,y)$ and $dG^h(x,y)$

$$dG^v(x,y) = \frac{\partial}{\partial x} G(x,y) = -2xe^{-(x^2+y^2)}$$

$$dG^h(x,y) = \frac{\partial}{\partial y} G(x,y) = -2ye^{-(x^2+y^2)}$$

using the interpolation equation [5]

$$dG^\theta(x,y) = \cos(\theta)dG^v(x,y) + \sin(\theta)dG^h(x,y)$$

with $\theta \in [0, \pi/2]$ the direction of the oriented derivative.

Figure 3 illustrates the feature set by showing eight directions, uniformly distributed over the full circular interval. The main two parameters are the feature size, i.e., the extent in number of pixels of the oriented Gaussian derivatives, and the number of orientations spanning the range from the vertical to the horizontal direction.

Figure 3. The oriented spatial features.

4.3 Fourier Spectra

Fourier's [4] analysis is probably the most frequently applied transformation in signal and image processing. Applied to a waveform, the Fourier transformation yields a spectral representation that expresses the amplitudes and phases of the sine-waveforms; when superimposed they make up the waveform. In two dimensions, the Fourier transformation yields a spectral representation that can be visualized as in figure 4. The center of the Fourier spectrum corresponds to the average "power" in the image, i.e., a spatial frequency of zero or the mean intensity of the image. Moving away from the center in any direction represents the presence and magnitude of waveforms of increasing spatial frequency r at the orientation perpendicular to the direction chosen. As illustrated in figure 4, keeping r constant and varying the angle ϕ, yields the orientation-dependent power at the spatial frequency r.

Figure 5 shows an example of a Fourier spectrum of one of the paintings in the data set. The bright center represents the zero-frequency component, the bright horizontal and vertical lines represent the strong presence of vertically- and horizontally-oriented edges in the painting, respectively. The strong vertical and horizontal components are (at least in part) caused by the square sampling grid of digital images.

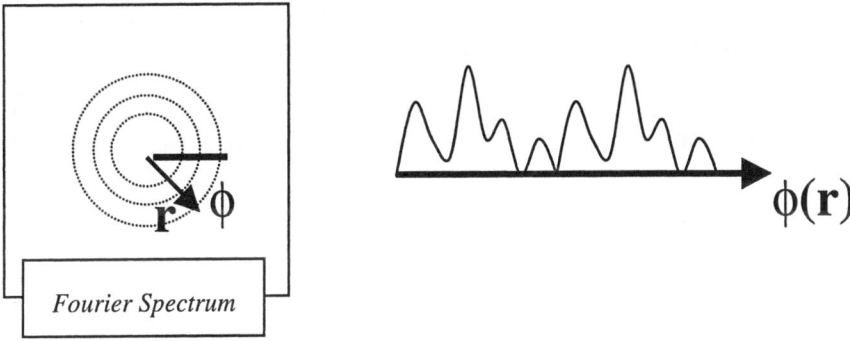

Figure 4. Illustration of the extraction of the orientation-sensitive features from the Fourier spectrum. *Left:* The distance r from the center of the spectrum is proportional to the spatial frequency. The angle φ corresponds to the orientation of the spatial components. *Right:* Keeping r constant and varying φ results in the power at spatial frequency r as a function the orientation.

Figure 5. Example of the Fourier spectrum of a painting.

4.4 Statistical Descriptors

Considering high-dimensional vectors (i.e., (parts of) images) as random samples we may use standard statistical descriptors for characterizing their distribution. Below, we apply four well-known statistical descriptors to nine parts of the painting. For this purpose all paintings are partitioned into nine non-overlapping regions with the same aspect ratio as the entire painting. Figure 6 illustrates the partitioning for three generic types of paintings, viz. landscape, square, and portrait paintings.

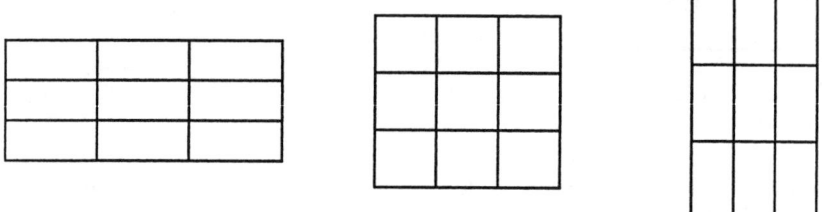

Figure 6. Illustration of the partitioning of the painting into nine regions for (from left to right) landscape, square, and portrait paintings.

Four well-known descriptors are the *mean*, the *standard deviation*, the *skewness*, and the *kurtosis*. These descriptors are defined as follows. The mean is simply the average of the pixel-intensity (or gray) values $I(x,y)$:

$$mean = \bar{I} = \frac{\sum_{x=1}^{N}\sum_{y=1}^{M} I(x,y)}{NM},$$

with N and M representing the width and the height of the image in pixels, respectively. The mean probes the average intensity of the nine parts of the painting which may reveal painter-specific distributions of lighter shades. For instance, an idiosyncratic characteristic of a painter may be his use of lighter shades for skies, i.e., the top three regions of the paintings.

The standard deviation (*sd*) is defined as:

$$sd = \sigma = \frac{\sqrt{\sum_{x=1}^{N}\sum_{y=1}^{M}(I(x,y)-\bar{I})^2}}{NM-1}.$$

The standard deviation probes the amount of variation in shading within each region which may reveal painter-specific shade variations.

The skewness and kurtosis describe the deviation of the distribution of the image vectors from a normal or Gaussian distribution. In case both descriptors are zero, the samples are normally distributed. The skewness describes the skew of the distribution and is defined as

$$skew = \frac{\sum_{x=1}^{N}\sum_{y=1}^{M}\left(\frac{I(x,y)-\bar{I}}{\sigma}\right)^3}{NM}.$$

The kurtosis describes the "peakedness" of the distribution:

$$kurt = \frac{\sum_{x=1}^{N}\sum_{y=1}^{M}\left(\frac{I(x,y)-\bar{I}}{\sigma}\right)^4}{NM} - 3.$$

The skewness and kurtosis descriptors may capture painter-specific non-Gaussian deviations of (parts of) the images.

4.5 Independent Components

Small patches of the paintings may be considered of being composed of mixtures of sources or components. For instance, a patch may be composed of artefacts caused by the JPEG compression, the texture of the painting's canvas, and various patterns related to the painting and the painter. From a statistical point of view, these components may be isolated by measuring the associated distributions in the image space. The technique for doing this is known as *independent component analysis (ICA)*.

ICA is a statistical method that transforms multidimensional vectors (such as images) into components that are as independent from each other as possible [10]. The basic idea of ICA is to use the non-Gaussian nature of the distributions to retrieve the underlying sources. The interested reader is referred to Hyvärinen and Oja [11] for a concise introduction to ICA.

We applied the *FASTICA* method [10] to parts of the paintings in our data set. The first 100 independent components so obtained are shown in Figure 7.

Although the components so obtained are hard to interpret, some textural structure (i.e., horizontal and vertical lines) is clearly visible in some of them.

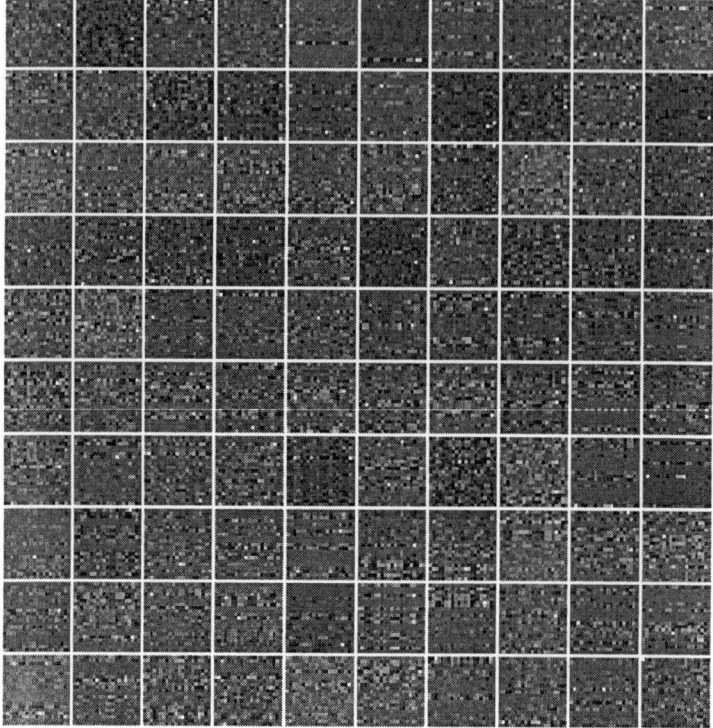

Figure 7. The first 100 independent components.

4.6 Fractal Dimension

The texture of a painting (as revealed in its digital reproduction) may reflect the painter's brushwork. Texture is therefore an interesting feature to probe. Our efforts to obtain effective textural feature-space representations were hampered by the fact that there are many different ways of representing texture. This is certainly due to the lack of a rigorous definition of "texture". In fact, the feature-space representations discussed in sections 4.3 and 4.5 can also be considered to be sensitive to the texture of paintings.

One relatively novel way of assessing the texture of an image is by measuring the fractal dimension of an image. The use of concepts from chaos theory is now widespread in image processing [15]. In preattentive image recognition, the concept of fractal dimension can be applied in, for instance, segmentation tasks (see, e.g., [18]). Fractal dimension is a measure of the self-similarity of (a part of) an image across a range of scales [16]. Many natural shapes, such as trees, clouds, and coastlines, have a high degree of self-similarity across spatial scales [1]. Zooming in on a tree reveals small-scale branching patterns that are similar to the large-scale branching patterns. In an analogous way, image textures can be characterized by their self-similarity across scales.

Taylor, Micolich, and Jonas [31] measured the fractal dimension of Jackson Pollock's drip paintings. They found the fractal dimension to increase with the date of creation of the painting. Although Pollock's paintings are more homogeneous in texture than the impressionistic paintings in the data set, probing their texture in terms of fractal dimension may be informative of the painting's maker.

To measure the fractal dimension of the paintings, we measured the Hurst coefficient ([27] cited in [18]) which is an approximation of the fractal (correlation) dimension. For each painting the average Hurst coefficient was determined for each of the nine regions displayed in figure 6. The results are shown in Figure 8.

5 Experiments

This section puts the features of section 4 to the test by training a standard neural-network classifier on the associated feature-space representations.

5.1 Generic Preprocessing

The raw images containing digital reproductions of the impressionistic paintings were submitted to a generic preprocessing stage involving two steps: resizing and low-pass filtering. In the resizing step the images were resampled to sizes commensurate with the actual dimensions of the paintings. As a result of the resizing, a pixel represents a fixed square area of the physical painting surface for all images included in the data set. In the low-pass filtering step, a Gaussian filter was applied to all images to remove the high spatial-frequency noise and artefacts generated by, e.g., the JPEG coding.

5.2 From Image Space to Feature Space

The (preprocessed) images contained in the data set were transformed to the following 11 types of feature-space representations: Color histograms (RGB, HSI, and hue), Oriented spatial-feature histograms, Fourier spectra (circular coefficients), Statistical descriptors (mean, standard deviation, skewness, kurtosis), Independent components, and Fractal histograms (Hurst coefficients).

The dimensionality of the feature spaces were chosen as small as possible without losing too much information. For color histograms, the dimensionality (i.e., the number of bins) varied from 1 to 256 per color component; So for RGB and HSI the maximum is 768. Histograms of the oriented spatial features contained 40 bins. The Fourier spectrum was kept at a fixed dimensionality of 50 (evenly sampled on the half-circle interval $0 \leq \phi \leq \pi$). Nine average fractal coefficients were included (one for each region). Also for each statistical descriptor the dimensionality equaled the number of regions, i.e., 9. The number of independent components was kept at 100.

5.3 Automatic Classification

A multilayer neural network is capable of mapping points from an input (i.e., feature) space towards an output space [26]. The length of the input vector matches the dimensionality of the feature space. The dimensionality of the output space depends on the number of categories needed for the classification task. In our painting-recognition task, six output nodes suffice, each active output represents a painter. The transformation from input to output space is achieved by an intermediate representation formed in a so-called hidden layer. The dimensionality of the representation space (i.e., the number of nodes in the hidden layer) is a parameter of the network structure.

The feature-space representations of the training set were used to train the multilayer neural network using the standard backpropagation algorithm [26]. To measure the generalization performance G, a 10-fold cross-validation procedure was performed (see, e.g., [33]). To avoid overfitting we used the early-stopping criterion (e.g., [25]).

6 Results

The image space containing the vectors representing the paintings in our data set is transformed by the preprocessing techniques into different feature spaces. Each of these feature spaces emphasizes certain characteristics of the paintings while ignoring others. Whether the feature spaces obtained capture the characteristics related to the hidden signatures of the painters is determined by a standard neural-network classifier.

The generalization performance is an estimate of the proportion correctly-classified unknown paintings (i.e., correctly assigned to one of the six painters in the data set). The graph in figure 8 shows the generalization performances obtained for the 11 feature-spaces representations.

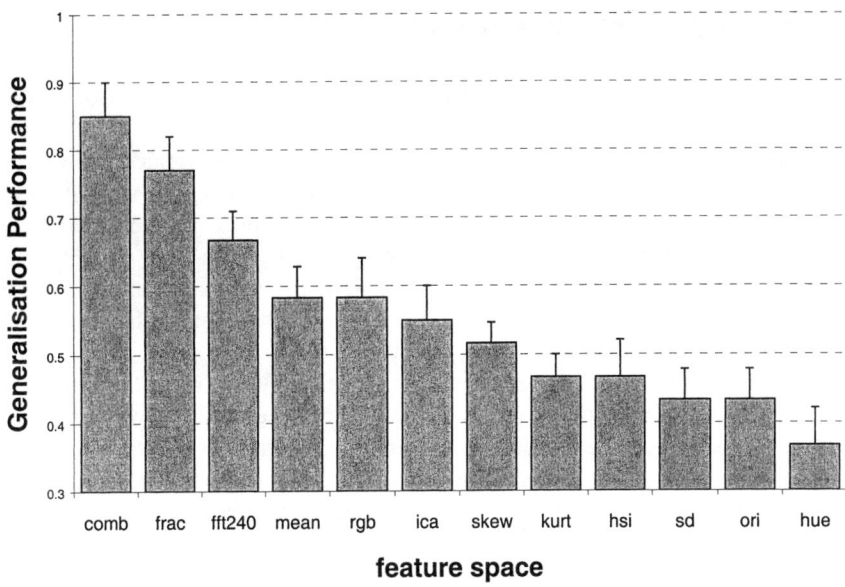

Figure 8. Generalization Performance for the feature-space representations tested. Error bars indicate the standard error of the mean (SE). Abbreviations: *comb*: combination of the best three feature-space representations; *frac*: fractal dimension; *fft240*: (Fast) Fourier Transform at 240 pixels from the center; *mean*: the average intensity in the nine regions of the painting; *rgb*: RGB color histogram; *ica*: independent components; *skew*: skewness (nine regions); *kurt*: kurtosis (nine regions); *hsi*: HSI color histogram; *sd*: standard deviation (nine regions); *ori*: oriented receptive-field histograms; *hue*: hue histogram.

All feature-space representations studied yield performances above chance level, i.e., the performance obtained by randomly assigning a painter to a painting yields a generalization performance of *0.167*. The results confirm the importance of texture and brushwork as measured by the fractal dimension and Fourier transformation. The mean intensity as measured over the nine regions turns out to be a simple but effective indicator which outperforms both color features. Surprisingly, the RGB histograms outperform the HSI histograms. By combining the three best feature-space representations, i.e., *frac*, *fft240*, and *mean*, a performance of *0.85* is obtained. Since we have not optimized the combined feature space, it is likely that the best performance will increase even further.

7 Conclusions and Outlook

Our knowledge-guided quest for good features yielded feature-space representations permitting faithful classification of a limited set of impressionistic paintings. Although not all features suggested by the expert yielded good results, domain knowledge proved to be indispensable for achieving our ultimate result. We have considered 11 types of features, but there are many more that can be tested. Gabor wavelets, principal components, and gray-level co-occurrence matrices, to mention a few, are certainly worth investigating. However, it was not our intention to be exhaustive. Rather, our main goal was to show the feasibility of knowledge-guided feature selection for neural networks applied to preattentive image recognition. Our quest focused on a task that is only a small representative of the general problem of automatic image labeling. Nevertheless, we believe that the success of our restricted study can be generalized to the preattentive recognition of larger sets of paintings. Attentive recognition techniques may be required to bridge the gap towards perfect generalization performance. We further believe that in the coming decades the combination of active vision [22] with ideas on subject modeling and efficient search algorithms [9] is a particular effective approach towards this goal.

Our conclusion is that preprocessing should be tailored to the task at hand by combining domain knowledge with neural-network techniques, and that within fifty years the visual signature of painters is as recognizable as is any handwritten signature.

Acknowledgements

The authors gratefully acknowledge that part of the project has been performed in the framework of TOKEN 2000, a project made possible by NAP funding, funding by NWO (The Netherlands Organization for Scientific Research), and supported by the Rijksmuseum. Moreover, they wish to thank Kees-Jan Verkerk, Chris Dobbelstein, and Armand Oprey for preparing the impressionistic-paintings data set and for performing some preliminary studies on preprocessing techniques. We also thank Frank Boom for providing us with expert knowledge on the paintings and their makers.

References

1. Barnsley, M.F. (1992). Fractals Everywhere. San Diego: Academic Press.
2. Beck, J. (1982). Textural segmentation. In J. Beck (Ed.), *Organization and Representation in Perception* (pp. 285-317). Hillsdale, NJ: Erlbaum.

3. Biederman, I., Rabinowitz, J.C., Glass, A.L., and Stacey, E.W., Jr. (1974). On the information extracted from a glance at a scene. *Journal of Experimental Psychology*, **103**, 597-600.
4. Fourier, J. (1888). *Théorie Analytique de la Chaleur*. Gauthiers-Villars.
5. Freeman, W.T. and Adelson, E.H. (1991). The design and use of steerable filters. *IEEE Transactions in Pattern Analysis and Machine Intelligence*, **13**, 891-906.
6. Funt, B.V. and Finlayson, G.D. (1995). Color constant color indexing. *IEEE Transactions on Pattern Analysis and Machine Intelligence*, **17**, 522-528.
7. Gage, J. (1999). Colour and Meaning. Art, Science and Symbolism. London: Thames and Hudson.
8. Gonzalez, R.C. and Woods, R.E. (1993). *Digital Image Processing*. Reading, MA: Addison-Wesley Publishing Company.
9. Herik, H.J. van den (1995). How to model thoughts and actions. *Nieuw Archief voor Wiskunde, Part IV*, **13** (3), 363-380.
10. Hyvärinen, A. (1999). Fast and robust fixed-point algorithms for independent component analysis. *IEEE Transactions on Neural Networks*, **10**, 626-634.
11. Hyvärinen, A. and Oja, E. (1999). Independent Component Analysis: A Tutorial. http://www.cis.hut.fi/projects/ica.
12. Jain, A.K. (1989). Fundamentals of Digital Image Processing. Prentice Hall.
13. Julesz, B. and Bergen, J.R. (1983). Textons, the fundamental elements in preattentive vision and perception of textures. *The Bell Systems Technical Journal*, **62**, 1619-1645.
14. Koenderink, J.J. and Doorn, A.J. Van (1988).The basic geometry of a vision system. In R. Trappl (Ed.), *Cybernetics and Systems'88* (pp. 481-485). Dordrecht: Kluwer Academic Publishers.
15. Lu, N. (1997). *Fractal Imaging*. San Diego: Academic Press.
16. Mandelbrot, B.B. (1977). *The Fractal Geometry of Nature*. W.H. Freeman and Company.
17. Mel, B. 1997). SEEMORE: Combining color, shape, and texture histogramming in a neurally-inspired approach to visual object recognition. *Neural Computation*, **9**, 777-804.
18. Parker, J.R. (1997). *Algorithms for Image Processing and Computer Vision*. New York: John Wiley & Sons, Inc.
19. Phillips, D. (1995). How do forgers deceive art critics? In R. Gregory, J. Harris, P. Heard, and D. Rose (Eds.), *The Artful Eye* (pp. 372-388). Oxford: Oxford University Press.
20. Pioch, N. (1996). *The Webmuseum*, Paris. http://sunsite.doc.ic.ac.uk/wm/
21. Postma, E.O., Herik, H.J. van den, and Hudson, P.T.W. (1997a). Image Recognition by Brains and Machines. In S. Amari and N. Kasabov (Eds), *Brain-like Computing and Intelligent Information Systems* (pp. 25-47). Singapore: Springer-Verlag.
22. Postma, E.O., Herik, H.J. van den, and Hudson, P.T.W. (1997b). SCAN: A scalable model of covert attention. *Neural Networks*, **10**, 993-1015.
23. Postma, E.O, Herik, H.J. van den, and Hudson, P.T.W. (1998). Spatio-chromatic Features for Image Recognition. In H. Prade (Ed.), *Proceedings of the European Conference on Artificial Intelligence, ECAI'98* (pp. 637-641). John Wiley & Sons, Chichester.
24. Rao, R.P.N. and Ballard, D.H. (1995). An active vision architecture based on iconic representations. *Artificial Intelligence*, **78**, 461-505.

25. Reed, R.D. and Marks II, R.J. (1999). Neural Smithing. Supervised Learning in Feedforward Artificial Neural Networks. Cambridge, MA: MIT Press.
26. Rumelhart, D.E., Hinton, G.E., and Williams, R.J. (1986). Learning internal representations by error propagation. In D.E. Rumelhart and J.L. McClelland (Eds.), *Parallel Distributed Processing: Explorations in the microstructure of cognition, vol. I: Foundations.* (pp.318-362). Cambridge, MA: MIT Press.
27. Russ, J.C. (1990). Surface characterisation: Fractal Dimensions, Hurst Coefficients, and Frequency Transforms. *Journal of Computer Assisted Microscopy*, **2**, 249-257.
28. Schiele, B. and Crowley, J.L. (1996). Object recognition using multidimensional receptive field histograms. In B. Buxton and R. Cipolla (Eds.) *Proceedings of the ECCV'96*, 610-619. Berlin: Springer-Verlag.
29. Schmid, C. and Mohr, R. (1997). Local grayvalue invariants for image retrieval. *IEEE Transactions on Pattern Analysis and Machine Intelligence*, **19**, 530-534.
30. Swain, M. and Ballard, D.H. (1991). Color indexing. *International Journal of Computer Vision*, **7**, 11-32.
31. Taylor, R.P., Micolich, A.P., and Jonas, D. (1999). Fractal analysis of Pollock's drip paintings. *Nature*, **399**, 422.
32. Treisman, A.M. (1982). Perceptual grouping and attention in visual search for features and objects. *Journal of Experimental Psychology: Human Perception and Performance*, **8** (2), 194-214.
33. Weiss, S. M. and Kulikowski, C. A. (1991). Computer systems that learn: classification and prediction methods from statistics, neural nets, machine learning and expert systems. San Mateo, CA: Morgan Kaufmann.
34. Wetering, E. van de (1997). *Rembrandt: The painter at work.* Amsterdam: Amsterdam University Press.

35. Appendix

In Table 1 we have listed by painter the titles of the paintings included in the data along with their creation dates.

painter	title of the painting	year
Monet	Impression, Sunrise	1873
	The Boat Studio	1876
	Saint-Lazare Station	1877
	The Artist's Garden at Vetheuil	1881
	Rock Arch West of Etretat (The Manneport)	1883
	Garden in Bordighera, Impression of Morning	1884
	Meule, Soleil Couchant	1891
	Poplars along the River Epte	1891
	Wheatstacks	1891
	Rouen Cathedral, the West Portal, Dull Weather	1892
Van Gogh	Olive Trees with the Alpilles in the Background	1889
	Road with Cypres and Star	1889
	Landscape with House and Ploughman	1890
	Thatched Cottages in the Sunshine: Reminiscences of the North	1890
	The Church at Auvers-sur-Oise	1890
	The White House at Night	1890
	Village Street and Stairs with Figures	1890
	Village Street in Auvers	1890
	Wheat Field under Threatining Skies	1890
	First Steps (after Millet)	1891
Cézanne	Study: Landscape at Auvers	1873
	Jas de Buffan, the Pool	1876
	Houses along the Road	1881
	Gardanne	1886
	Mountains in Provence	1888
	House and Farm at Jas de Bouffan	1890
	House and Trees	1892
	Well: Millstone and Cistern under Trees	1892
	The House with the Cracked Walls	1893
	The Great Pine	1894
Sisley	Bridge at Villeneuve-la-Garenne	1872
	Autumn: Banks of the Seine near Bougival	1873
	Garden Path in Louveciennes	1873
	Snow at Louveciennes	1874
	Flood at Port-Marly	1876
	Station at Sevres	1879
	The Chemin de By through Woods at Roches-Couraut, St. Martin's Summer	1880
	Provenchers's Mill at Moret	1883
	Moret-sur-Loing	1891
	The Church at Moret	1894
Pissarro	The Stage Coach at Louveciennes	1870
	Entrance to the Village of Voisins	1872
	The Orchard	1872
	Gelee Blanche (Hoarfrost)	1873
	The Chestnut Trees at Osny	1873
	Village Path	1875
	The Red Roofs	1877
	Landscape at Chaponval	1880
	The Shepherdess (Young Peasant Girl with a Stick)	1881
	Haymakers Resting	1891
Seurat	Forrest at Pontaubert	1881
	Alfalfa Fields, Saint-Denis	1885
	Boats, Low Tide, Grandcamp	1885
	English Channel at Grandcamp	1885
	Seine at Courbevoie	1885
	Lighthouse at Honfleur	1886
	Seine at Le Grande Jatte	1888
	Sunday at Port-en-Bessin	1888
	Channel at Gravelines, Evening	1890
	Channel at Gravelines, in the direction of the Sea	1890

Table 1. The paintings included in the data set.

Chapter 8. Multimodal Interactions with Agents in Virtual Worlds

Anton Nijholt and Joris Hulstijn

Centre for Telematics and Information Technology,
University of Twente, PO Box 217, 7500 AE Enschede, the Netherlands
anijholt@cs.utwente.nl

Abstract. *In this chapter we discuss our research on multimodal interaction in a virtual environment. The environment we have developed can be considered as a 'laboratory' for research on multimodal interactions and multimedia presentation, where we have multiple users and various agents that help the users to obtain and communicate information. The environment represents a theatre. The theatre has been built using VRML (Virtual Reality Modeling Language) and it can be accessed through World Wide Web (WWW). This virtual theatre allows navigation input through keyboard function keys and mouse, but there is also a navigation agent which tries to understand keyboard natural language input and spoken commands. Feedback of the system is given using speech synthesis. We also have Karen, an information agent which allows a natural language dialogue with the user. In development are several talking faces for the different agents in the virtual world. We investigate how we can increase the user's commitment to the environment and its agents by providing context and increasing the user's feeling of 'presence' in the environment.*

Keywords. *Human-computer interaction, virtual reality, embodied agents, talking faces, text-to-speech synthesis, speech recognition, presence.*

1 Introduction

World Wide Web allows interactions and transactions through Web pages using speech and language, either by inanimate or live agents, image interpretation and generation, and, of course the more traditional ways of presenting explicitly pre-defined information of text, tables, figures, pictures, audio, animation and video. In a task- or domain-oriented way of interaction current technology allows the recognition and interpretation of rather natural speech and language in dialogues. However, rather than the current two-dimensional web-pages, many interesting parts of the Web will become three-dimensional, allowing the building of virtual worlds inhabited by user and task agents, with which the user can interact using different types of modalities, including speech and language interpretation and generation. Agents can work on behalf of users, hence, human computer interaction will make use of 'indirect management', rather than interacting through direct manipulation of data made visible on the screen.

In this chapter we present our research on the development of an environment in which users can display different behavior and have goals that emerge during the interaction with this environment. Users who, for example, decide they want to spend an evening outside their home and, while having certain preferences, cannot say in advance where exactly they want to go, whether they first want to have a dinner, whether they want to go to a movie, theatre, or to opera, when they want to go, etc. During the interaction, goals, possibilities and the way they influence each other become clear. One way to support such users is to give them different interaction modalities and access to multimedia information. We discuss a virtual world for presenting information and allowing natural interactions about performances, associated artists and groups, availability of tickets, etc., for some existing theatres in the city of Enschede, the Netherlands, the home town of our university.

The interactions take place in a virtual theatre, a realistic model of one of our local theatres. This so-called 'Muziekcentrum' offers its potential visitors information about performances (music, cabaret, theatre, opera) by means of a brochure that is published once a year. In addition to this yearly brochure it is possible to get information at an information desk in the theatre (during office hours), to get (more recent and updated) information by phone (either by talking to a theatre employee or by using Interactive Voice Response Technology) and to get information from local daily and weekly papers and monthly announce-ments issued by the theatre. The central database of the theatre holds the information that is available at the beginning of the 'theatre season'. Our aim is to make this information about theatre and performances much more accessible to the general audience.

The interactions between user (the visitor) and system take place using different task-oriented agents. These agents allow mouse and keyboard input, but inter-actions can also take place using speech and language input. In the current system both sequential and simultaneous multi-modal input is possible. There is also multi-modal (both sequential and simultaneous) output available. The system presents its information through agents that use tables, chat windows, natural language, speech and a talking face. At this moment this talking face uses speech synthesis with associated lip movements. Other facial animations are possible (movements of head, eyes, eyebrows, eyelids and some changes in face color). These possibilities have been designed and associated with utterances of user or system, but not yet fully implemented.

In this chapter it is also discussed how our virtual environment can be considered as an interest community and it is shown what further research and development is required to obtain an environment where visitors can retrieve information about artists, authors and performances, can discuss performances with others and can be provided with information and contacts in accordance with their preferences. In addition, but this has not been realized yet, we would like to offer our virtual environment for others to organize performances, meetings and to present (video) art. We would like to offer this environment for experiments on mediated communication between visitors and for performances, done by avatars with or by avatars without user participation.

The virtual environment we consider is web-based and the interaction modalities that we consider confine to standards that are available or that are being developed for world wide web. From a more global point of view research topics that have been aimed at are:

- Modelling effective multi-modal interactions between humans and computers, with an emphasis on the use of speech and language
- Commercial transactions, (local, regional and global) information services, education and entertainment in virtual environments
- Web-based information and transaction services, in particular interactions in virtual environments

Indexing and retrieval of multi-media information (in our case, multi-media information about artists and performances) available in theatre databases and on the world wide web is an issue that will receive more attention in the near future. We assume that it will become clear how to generalize our approaches and how to tune them to domains different from our (virtual) theatre domain.

2 History and Motivation

Some years ago, the Parlevink Research Group of the University of Twente started research and development in the area of the processing of (natural language) dialogues between humans and computers. In order to do so, one can choose to take a general, domain-independent approach. This allows general research in syntactic analysis, semantic and pragmatic interpretation and the modelling of dialogues in general. Hence, research has to embedded in current state-of-the-art research on parsing, unification, grammar formalisms, semantics and representation of dialogue utterances, discourse representation and the representation of 'common sense' and world knowledge. That is, knowledge that has to be represented and made accessible in order to get our system to understand user utterances and to generate intelligent system utterances.

Although this domain-independent approach has been followed, we also started research where the domain of application and the user interaction associated with the domain is a basic assumption of interaction modelling and dialogue management. In our case, the domain is that of a theatre information service and related transactions (e.g., to make a reservation). In order to be able to model 'natural' interaction, Wizard of Oz experiments have been designed and a corpus of (keyboard) natural language dialogues in this domain has been obtained.

Our dialogues involve transactions. Such dialogues display a more complex structure than mere inquiry or advisory dialogues. Two tasks are executed in parallel: obtaining information and ordering (Jönsson [8]). In our corpus similar complex behaviour related to these tasks can be found. Users browse, inquire and retract previous choices, for instance when tickets are too expensive. Hence, we allow interactions where the goal is not set before; it develops during the dialogue and the user will update goals depending on the information obtained. It is also the environment in which the dialogues are embedded and the possibility to explore environment and interaction modalities that invites users to browse through the available information just like leafing through a brochure.

Our research led to the development of a (keyboard-driven) natural language accessible information system (SCHISMA), able to inform users about theatre performances and to allow users to make reservations for performances. The system made use of the database of performances in the local theatres of the city of Enschede. In the next sections we will give more information about the design of this theatre information system. The system is far from perfect. However, if a user really wants to get information and has a little patience with the system, he or she is able to get this information. A more general remark should be given: When we offer an interface to the general audience to access an information system, do we want to offer an intelligent system that knows about the domain, that knows about users, their preferences and other characteristics, etc., or do we assume that any user will adapt to the system that is being offered? This is not an exclusive or, nevertheless the point is extremely important. It has to do with group characteristics (men, women, old, young, naive, professional, experienced, etc.), but also with facilities and alternatives provided by the owner of the system. As an example, consider a transport and railway information system. Human operators are available to inform about times and schedules of busses and trains. However, the number of operators, from a user's point of view, is insufficient. Callers can wait (and pay for the minutes they have to wait) or choose for a computer-operated system to which they can talk in natural speech, but possibly have to accept that they need more interactions in order to get themselves understood. Hence, it really depends on the application, the situation and the users involved (do they want to pay for the services, do they want to adapt to the interface, does the provider offer an alternative, etc.), whether we can speak of a successful natural language accessible dialogue information system.

We do not really disagree with a view where users are expected to adapt to a system. On the other hand, wouldn't it be much more attractive (and interesting from a research point of view) to be able to offer environments, preferably on worldwide web, where different users have different assumptions about the available information and transaction possibilities, have different goals when accessing the environment and have different abilities and experiences when accessing and exploring such an environment? We like to offer a system such that we can stimulate and expect users to adapt to it and find effective, efficient, but most of all enjoyable ways to get or to get done what they want.

In the next section we discuss how we can add 'context' to our dialogue system. With 'context' we mean that we would like to add visual and auditory cues in the presentation of information and to allow users to choose the (combination of) interaction modalities that best suits his or her preferences for performing the 'task' that has to be done. With 'context' we also refer to the possibility that users come to consider the environment as an interest community, where they can exchange information with other users.

3 VR Context-Embedded Interaction

3.1 Environment Visualization

In order to add context to our natural language dialogue system we decided to visualize the environment in which people can get information about theatre performances, can make reservations and can talk to theatre employees and other visitors. VRML, agent technology, text-to-speech synthesis, talking faces, speech recognition, etc., became issues after taking this decision. They will be discussed in the next sections. Visualization allows users to refer to a visible context and it allows the system to disambiguate user's utterances by making use of this context. Moreover, the visualization can make it possible for the system to influence the interaction behavior of the user in such a way that more efficient and natural dialogues with the system become – in principle - possible.

Our virtual theatre (http://parlevink.cs.utwente.nl/) has been built according to the design drawings made by the architects of our local theatre. Part of the building has been realized by converting AutoCAD drawings to VRML97. Video recordings and photographs have been used to add 'textures' to walls, floors, etc. Sensor nodes in the virtual environment activate animations (opening doors) or start events (entering a dialogue mode, playing music, moving spotlights, etc.). Visitors can explore the environment of the building, hear the carillon of a nearby church, look at a neighboring pub and movie theatre, etc. and they can enter the theatre (cf. Figure 1) and walk around, visit the concert hall, admire the paintings on the walls, go to the balconies and, take a seat in order to get a view of the stage from that particular location. When the performance hall is entered, the lights dim, spot lights are moving over the stage and some music starts playing. Information about today's performances is available on an information board that is automatically updated using information from the database with performances. In addition, as may be expected, visitors may go to the information desk in the

Figure 1: Entrance of the theatre

theatre, see previews of performances and start a dialogue with an information and transaction agent called 'Karen'. This agent has a 3D talking face (see section 4).

Apart from navigating, clicking on interesting objects (resulting in access to web pages with information about performances, access to web magazines, etc.) and interacting with person-like agents we allow a few other interactions between visitors and virtual objects. For example, using the mouse, the visitor can play with the spotlights and play notes on a keyboard that is standing in some far away part of the building. There is a floor map near the information desk where people can click on positions in order to be 'transported' to their seat in the performance hall so they can see the view they have. On the desk is also a monitor on which they can see pictures or video previews of performances. Unfortunately, most performances do no have a video preview available yet, so we can not display them for every performance that is in the database.

3.2 Visualizing Agents

We assign natural tasks in our environment to agents. It can be useful to visualize them using talking faces and animated 3D avatars. From several studies (cf. Friedman [6]) it has become clear that people engage in social behavior toward machines. It is also well known that users respond differently to different 'computer personalities'. It is possible to influence the user's willingness to continue working even if the system's performance is not perfect. Users can be made to enjoy the interaction and they can be made to perform better, all depending on the way the interface and the interaction strategy have been designed. People behave differently in the presence of other people than they do when they are alone. Similarly, in experiments it has been shown that people display different behavior when interacting with a talking face than they do with a text-display interface. This behavior is also influenced by the facial appearance and the facial expressions that are shown. People tend to present themselves in a more positive light to a talking face display and they are more attentive when a task is presented by a talking face (cf. Sproull et al. [12]). From these observations we conclude that introducing talking faces can help make interaction more natural and shortcomings of technology more acceptable to users.

There is another reason to introduce visualized task-oriented agents. The use of speech technology in information systems will continue to increase. Most currently installed information systems that work with speech, are telephone-based systems where callers can get information by speaking aloud some short commands. There are exceptions. For example, there are applications where customers can call a specific telephone number which connects them to a website. Data at the site can be accessed and retrieved (using speech recognition and speech synthesis). This is especially useful when a caller needs to order, to book, to fill in forms, etc., or to access information that is available in tables. Another example is the demonstrator for the SNCF (French railways) train travel information kiosk, where train passengers have access to information services using multimodal input and multimedia output. Speech and touch are among the input modalities. Sound, video, text and graphics are the output modalities. Speech is becoming more and more important in these multimodal interfaces. One of the main problems in speech-only dialogue systems is the limitation of the context. As long as the context is narrow they perform well, but wide contexts are causing problems. One reason to introduce task-oriented agents is to restrict user expectations and utterances to the different tasks for which agents are responsible. Obviously, this can be enhanced if the visualization of the agents helps to recognize the agents tasks.[1]

3.3 Multi-modal and Multi-agent Approach

When a user has the possibility to change easily from one modality to an other, or can use combinations of modalities when interacting with an information system, then it is also more easy to deal with shortcomings of some particular modality. Multi-modality has two directions. That is, the system should be able to present multi-media information and it should allow the user to use different input modalities in order to communicate with the system. Not all communication devices that are currently available for information access, exploration of information and for transaction allow more than one modality for input or output. This is especially true if we look at world wide web interfaces.

---

[1] It may be the case that different specialist agents are taken more seriously than one generalist agent. See Nass et al. [11] who report about different appreciation of television programs depending on whether they were presented in a 'specialist' or in a 'generalist' setting.

More serious, however, hardly any research has been done to distinguish discourse and dialogue phenomena, let alone to model them, for multi-modal tasks. The same holds for approaches to funnel information conveyed via multiple modalities into and out of a single underlying representation of meaning to be communicated (the cross-media information fusion problem). Similarly, on the output side, there is the information-to-media allocation problem.

In addition to the issue of multi-modality, there is a need for a multi-agent approach. Our system deals with actors that present information, reason about information, communicate with each other and that realize transactions (e.g. through negotiation). In this case agents can take roles ranging from presenting windows on a screen, reasoning about information that might be interesting for a particular user, and being recognizable (and probably visible) as being able to perform certain tasks.

3.4 Virtual Communities

Today there are examples of virtual spaces that are visited and inhabited by people sharing common interests. These spaces can for example, represent offices, shops, class rooms, companies, etc. However, it is also possible to design virtual spaces that are devoted to certain themes and are tuned to users (visitors) interested in that theme or to users (visitors) that not necessarily share common (professional, recreational or educational) interests, but share some common conditions (driving a car, being in hospital for some period, having the same therapy, belonging to the same political party, etc.).

In the previous subsections we have looked at possibilities for theatre visitors to access information, to communicate with agents designed by the provider of the information system and to explore an environment with the goal to find information or to find possibilities to enter into some transaction. Hence, we have a community of people interested in theatre, in music, in performers and their environment has been modeled along the lines of an existing theatre. We need to investigate how we can allow communication between users or visitors of this web-based information and transaction system. Users can help each other to find certain information, they can inform each other (especially when they know about the other's interests), they can have conversations about common interests and they can have domain-related collaboration (e.g., in our case, they can decide to perform a certain play where the actors are distributed among different web sites but sharing the same virtual stage).

As a (not too complicated) example we mention a virtual world developed by the virtual worlds group of Microsoft in co-operation with The Fred Hutchinson Cancer Research Center in Seattle. This so-called "Hutch World" enables people struggling with cancer to obtain information and interact with others facing similar challenges. Patients, families and friends can enter the password protected three-dimensional world (a rendering of the actual outpatient lobby), get information at a reception desk, visit a virtual gift shop, etc. Each participant obtains an avatar representation. Participants can engage in public chat discussions or invitation-only meetings. A library can be visited, its resources can be used and participants can enter an auditorium to view presentations.

4 Agents in the Virtual Theatre

4.1 An Agent Platform in the Virtual Environment

In the current prototype version of the virtual theatre we distinguish between different agents: We have an information and transaction agent, we have a navigation agent and there are some agents under development. An agent platform has been developed in JAVA to allow the definition and creation of intelligent agents. Users can communicate with agents using speech and natural language keyboard input. Any agent can start up other agents and receive and carry out orders of other agents. Questions of users can be communicated to other agents and agents can be informed about each other's internal state. Both the information & transaction agent and the navigation agent are in the platform. But also the information board, presenting today's performances, has been introduced as an agent. And so can other objects in the environment.

4.2 The Information & Transaction Agent

Karen, the information & transaction agent, allows a natural language dialogue with the system about performances, artists, dates, prices, etc. Karen (Figure 2) wants to give information and to sell tickets. Karen is fed from a database that contains all the information about performances in the (existing) theatre.

Our current version of the dialogue system of which Karen is the face uses a 'rewrite and understand' approach. User utterances are simplified using a great number of rewrite rules. The resulting simple sentences are parsed. The output can be interpreted as a request of a certain type. System response actions are coded as procedures that need certain arguments. Missing arguments are subsequently asked for. The system is modular, where each 'module' corresponds to a topic in the task domain. For example, a module has to take care of a date a user is referring to (next Wednesday, over two weeks, tomorrow).

There are also modules for each step in the understanding process: the rewriter, the recognizer and the dialogue manager. The rewrite step can be broken down into a number of consecutive steps that each deal with particular types of information, such as names, dates and titles. The dialogue manager initiates the first system utterance and goes on to call the rewriter and recognizer process on

Figure 2: Karen at the information desk

the user's response. Also, it provides an interface with the database management system. Results of queries are represented as bindings to variables, which are stored in the global data-structure, called context. Based on the user utterance, the context and the database, the system has to decide on a response action, consisting of database manipulation and dialogue acts. The arguments for the action are dug out by the dedicated parser, associated with the topic. All arguments that are not to be found in the utterance are asked for explicitly. More information about this approach can be found in Lie et al. [9].

Presently the input to Karen is keyboard-driven natural language and the output in our for the general audience WWW accessible virtual world allows for a mix of speech synthesis and information presentation on the screen. As mentioned earlier, Karen's spoken dialogue contribution is presented by visual speech, that is, a 'talking face' on the screen, embedded in the virtual world, mouths the questions and part of the responses. If necessary, information is given in a window on the screen, e.g., a list of performances or a review of a particualar performance. The user can click on items to get more information or can type in further questions concerning the items that are shown.

4.3 The Navigation Agent

Navigation in virtual worlds is a well known problem. Usually, navigation input is done with keyboard and mouse. This input allows the user to move and to rotate, to jump from one location to an other, to interact with objects and to trigger them. There is more to say about navigation:

- *Input tools*: Which tools do we use to navigate? Keyboard, mouse, 3D stick, bookmarks, menus, speech, gestures, gaze control?
- *Physical*: How do we move in the virtual world? How can we avoid collissions with physical objects and with other agents? How can we employ knowledge of the environment and knowledge of tasks in order to control the movements of our agents?
- *Orientation*: How does a user know where s/he is and how s/he should move to get to another location in the virtual environment. How do we represent spatial knowledge and how do we assist users to think, reason and argue about objects, locations and routes in the virtual world?
- *Social*: How can we improve social navigation? There are so many users who know how to get at certain information that suits their interests. So why not exploit their knowledge to identify routes, to filter and to select information?

Related issues to navigation in multi-user environments are personal space, group space and privacy. Including these issues involves also the study of social norms, both in virtual and physical environments.

We developed a navigation agent that helps the user to explore the environment by means of speech commands. It is left to the user to choose between interaction modes for navigation. A smooth integration of the pointing devices and speech in a virtual environment requires that the system has to resolve deictic references that occur in the interaction. Moreover, the navigation agent should be able to reason (in a modest way) about the geometry of the world in which it moves. It knows about the user's coordinates in the virtual world and it has knowledge of the coordinates of a number of objects and locations. This knowledge is necessary when a visitor refers to an object close to the navigation agent in order to have a starting point for a walk in the theatre and when the visitor specifies an object or location as the goal of a route in the environment. The navigation agent is able to determine its position with respect to nearby objects and locations and can compute a walk from this position to a position with coordinates close to the goal of the walk.

Verbal navigation requires that names have to be associated with different parts of the building, objects and agents. Users may use different words to designate them, including references that have to be resolved in a reasoning process. The current agent is able to understand command-like speech or keyboard input. However, it hardly knows how to communicate with a visitor. The phrases to be recognized must contain an action (go to, tell me) and a target (information desk, synthesizer). It tries to recognize the name of a location in the visitor's utterance. When the recognition is successful, the agent guides the visitor to this location. When the visitor's utterance is about performances the navigation agent makes an attempt to contact Karen, the information and transaction agent. In progress is an implementation of the navigation agent in which the navigation agent knows more about (or is able to compute) current position and focus of gaze of the user, geometric relations between objects and locations, knowledge of previously visited locations or routes and knowledge of the previous communication with the visitor.

4.4 Language Skills of the Agents

At this moment our agents have different language skills. On the one hand we have Karen and a grammar specification of the input for Karen based on a corpus of WoZ obtained keyboard-based dialogue utterances. On the other hand we have a navigation agent with language skills that are based on the current limitations of speech input. We would like to automate the process of assigning language skills to agents in our environment as much as possible. Therefore we hope to see speech recognition technology move forward from keyword spotting, to finite state utterance specification to (word-graph) based context-free language specification. At this moment we follow the developments of Philips speech recognition software when looking at the recognition of spoken dialogue utterances for different agents. More fundamental, however, is our approach to induce grammars (context-free, probabilistic, unification constraints) from corpora of utterances (see Ter Doest [5]) collected in Wizard of Oz experiments. In this approach we tag a corpus with syntactic categories and superficial structure using Standard Generalised Markup Language (SGML). From this tagged data grammar rules, unification constraints and probabilities are derived. We have tested grammars on 'seen' and 'unseen' data from Karen's same domain using a probabilistic left-corner parser for PATR II unification grammars. In a similar way we have induced for our navigation agent a probabilistic grammar from a corpus of user utterances that have been obtained from several scenarios presented to (potential) visitors from the theatre. This grammar is a start. It allows the design of a primitive system and it allows bootstrapping this system from the original corpus and from corpora obtained from logging the interactions between visitors and the navigation agent. Clearly, this approach still requires the integration of speech recognition technology with natural language specification and understanding. For that reason it may be useful to investigate the generation of finite state probabilistic (unification) grammars from corpora of utterances.

In our current web-based system we have speech recognition on the server side. This requires the recording of commands on the client side and a robust transporting of the audio files. It does not require users to install speech recognition software or to download a speech recognition module as part of the virtual world from the server. Users do need however audio-software, which is usually available anyway. For speech recognition we use Speech Pearl, commercial speech understanding software from Philips. Recognition is based on keyword spotting. A next version of the software will allow a finite state specification of the user's input for speech recognition. For text-to-speech synthesis we use a Dutch text-to-speech system [4]. It runs on top of the MBROLA diphone synthesizer. Lexical resources of *Van Dale Dictionaries* have been used to obtain phonetic descriptions of the words.

5 Facial Displays and Speech Generation

5.1 Introduction

We developed a virtual face in a 3D-design environment. The 3D data has been converted to VRML-data that can be used for real-time viewing of the virtual face on WWW. We are researching various kinds of faces to determine which can be best used for our applications. Some are rather realistic and some are more in a cartoon-style (cf. Figure 3). The face is the interface between the users of the virtual theatre and the theatre information system. A dialogue window is shown when users approach the information-desk while they are navigating in the virtual theatre. Users can type their questions in a user's window. Karen answers the user in spoken language and by showing her response in a window. In addition,

Figure 3: Cartoon face for karen

a list of performances can be presented on the screen. While listening and answering Karen's face may change, reflecting her attitude towards the user and the contents of the dialogue.

How do we control the responses of the system, the prosody and the artificial face? The central module of our dialogue system is the *dialogue manager*. It maintains two data-structures: a representation of the *context* and a representation of the *plan*, the current domain-related action that the system is trying to accomplish. Based on the context, the plan and a representation of the latest user utterance, the dialogue manager selects a certain response action. Planning and action selection are based on a set of principles, the *dialogue rules*. A response action is a combination of basic domain related actions, such as database queries, and dialogue acts to convey the results of the query. Dialogue acts describe the intended meaning of an utterance or gesture. The *response generation* module selects a way to express it. It determines the utterance-structure, wording, and prosody of each system response. In addition, it also controls the orientation and expression of the face, the eyes, and the coordination of sounds and lip movement.

5.2 Prosodic and Facial Features

5.2.1 Prosodic Features and Templates

In the design of utterance generation by the information agent a list of annotated utterance templates is used. The response generation module uses a set of parameters to control the templates. Templates contain gaps to be filled with *information items*: attribute-value pairs labeled with syntactic, lexical and phonetic features. An appropriate template for a given dialogue act is selected by the following parameters: *utterance type*, *body* of the template, *given* information, *wanted* and *new* information. The utterance type and body determine the word-order and main intonation contour. The given, wanted and new slots, as well as special features, affect the actual wording and prosody. Templates respect rules of accenting and de-accenting. As a rule, information that is assumed to be given in the dialogue is de-accented, expressed as a pronoun, or even left out. Given information is repeated whenever the system is not confident it was recognized correctly, e.g., by a speech recognition module. Such verification prompts are distinguished by a rising intonation. Information that is to be presented as new, is accented. Quoted expressions, like artist names or titles of performances, are set apart from the rest of the utterance. For reading the longer texts and reviews that describe the content of performances, the system assumes a 'reading voice'.

Our text-to-speech system operates at three levels: the grapheme level, the phoneme level and a low-level representation of phones where the length and pitch of sounds is represented. It has phonetic descriptions of the usual Dutch words as they appear in lexical resources. Other prosodic information is derived by heuristic rules. It is possible to manipulate prosody by adding punctuation at the grapheme level, by adding prosodic annotations at the phoneme level or by directly manipulating the phone level.

5.2.2 Facial Features and Templates

Apart from the lips, the virtual face has a number of dynamic control parameters (Figure 4). The *eyes* can gaze at a certain direction. This can be used to direct attention towards an area. The *eyelids* may be opened and closed, for blinking. The *eyebrows* can be lifted to indicate surprise or lowered for distress. The shape of the *mouth* can be manipulated into a smile or and angry expression. The *color* of the face can be deepened, to suggest a blush that indicates shyness or embarrassment. The *orientation* of the head can be manipulated, leaning forward and backward or tilting left and right. This may produce important facial gestures like nodding and shaking one's head. It can also be used to indicate attention; leaning forward means being interested, leaning backward means loosing interest. In general the character is not still. The head will wiggle a bit and its eyes will wonder. This is called *idle behavior*. Many existing 'talking heads' look artificial because of their stillness. Moreover, not moving can also be taken as a sign. For instance, a fixed stare can indicate a misunderstanding in the dialogue. The *frequency* of idle movements is an indicator of the liveness of the character; it serves as a type of volume, to the existing emotion. So, many random movements of the head, combined with smiles and attentive eyes, indicate a very happy personality; stillness, a neutral mouth shape and looking away, indicate a withdrawn and unhappy personality. But an angry face, combined with a blush and a lot of movement, indicate increased anger. Jerky movements with wondering eyes indicate nervousness. Since our agent is supposed to be professionally friendly, she will be generally smiling and will have a moderate movement frequency.

Feature	Manipulation	Meaning
Eyes	Gaze direction	Idle behavior, attention, indexing
Eyebrows	Lift, lower	Surprise, distress, angry
Lips	Form visemes	Talk
	Stretch, round	Smile, laugh, neutral, angry, kiss
Mouth shape	Stretch, round	Smile, neutral, angry
Color	Blush	Shyness, embarrassment
Head	Orientation	Nodding, shaking head, attention
	Idle behavior	Neutral
	Movement frequency	Emotional 'volume'
Shoulders	Shrug	Indifference

Figure 4: Facial features

Each of these basic features can be combined into facial *gestures* that can be used to signal something. Gestures like nodding, shaking and shrugging can be used separately, but often utterances are combined with gestures or utterance related facial expressions. The timing of the gesture or the expression must be aligned with the utterance. We use the following general heuristic for alignment of gestures.

Like any event, an utterance and a gesture have an *entry* and an *exit* point. Moreover, an utterance can be broken down into phrases; each phrase has a so called *intonation center*, the moment where the pitch contour is highest. Since pitch accents are related to informativeness, we can assume that the accent lands on the most prominent expression. Usually the accent lands towards the end of an utterance. Similarly, each gesture has a *culmination point*. For instance for pointing, the moment that the index finger is fully extended. The visual animator extrapolates a nice curve from the entry point to the culmination and again to the exit point. Our current working hypothesis is that gestures synchronize with utterances, or precede them. So we link the gesture's entry and exit points to the entry and exit points of the utterance and make sure that the culmination point occurs before or on the intonation center.

5.3 Facial Behaviour

Facial behaviour is linked to personality and to the conversation that is being held. Facial behavior may also be related to the task an agent has to perform. For a chosen face there are different types of behavior that need to be modeled.

Firstly, permanent features like the facial expression, gazing direction and general movement characteristics, both when speaking and when idle. These can be controlled by two parameters: *mood* and *attention*. The *mood* parameter indicates the general attitude of the personality in the conversation. It is a state, that extends over a longer period. Is the agent happy, sad, angry or uncertain? The *attention* parameter controls the eyes and gazing direction. We believe that one of the benefits of a talking face is that turn taking and attention management in dialogues will be made easier. The gazing direction of the eyes and the head position are crucial for this. In section 6 we will discuss the possibility to include appropriate gaze behaviour in our human-like agents for regulating a conversational process. Usually mood and attention are fixed for a given personality. Temporary changes in emotion and attention, may result from previous utterances or from the general conversation. For instance, anger at an insult, or increased interest after a misunderstanding.

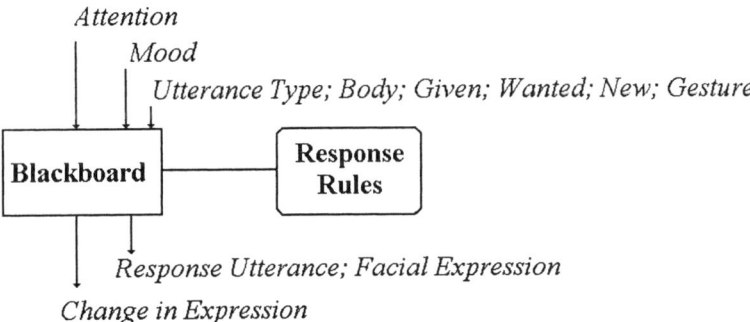

Figure 5: Blackboard architecture

Secondly, utterance related attitudes. Since we cannot monitor the user's utterances in real-time, at the moment this is limited to system utterances only. Think of smiling at a joke, raising eyebrows at a question or a pointing gesture at an indexical. Conventional gestures can be modeled as a special instance of response actions. Nodding or shrugging are coded like any other utterance synchronized with a gesture, except that they can be silent. Utterance related features are controlled by the existing utterance parameters, extended with a new parameter, *gesture*, that labels one or more facial movements to be synchronized with the utterance template. Because we know all utterance templates in advance, the synchronization can be manually adjusted if needed. The extend of the gesture and its final shape also depend on the general emotional state and attention level.

To control the many features of facial behavior we need a blackboard architecture as depicted in Figure 5. The reason for this architecture is that the parameters influence each other. Combinations of input parameters trigger a rule that produces a response action, or a more permanent change of expression.

We hope to introduce some variation in the exact choice of movement. Variation is important. For instance, it is natural to combine "yes" with a nod, but when every affirmative is combined with the same nod it looks mechanical. Another example is the raising of eyebrows. In an early version of the system the eyebrows were directly controlled by pitch level. Thus, the agent would nicely express uncertainty on a question, which has a rising intonation. But of course, pitch is also used for accenting. So the agent wrongly expressed surprise at expressions that were stressed. Synchronizing the apparently random movements with fixed speech from templates is difficult. We have chosen to align the culmination points of the movement with the intonational centers of the phrases uttered. But the exact frequency and extent of the movements will be randomly distributed, partly based on mood and attention.

5.4 Importance of Nonverbal Behavior

Nonverbal behavior as described in the previous sections is important. It provides feedback for the conversation partner of the embodied agent. This feedback helps to obtain a more smooth conversation and exchange of information. Cassell et al [2] have compared different kinds of nonverbal feedback in some experiments. They distinguished:

- content-only feedback: answering questions and executing commands related to the topic of the conversation
- envelope feedback: nonverbal behaviours related to the process of conversation, e.g., glances towards and away from the partner, manual gestures, head movements, etc.
- emotional feedback: facial displays that reference a particular emotion (smiles, look of puzzlement, etc.)

In the experiments they compared three feedback behaviours: content-only feedback (speech and actions), envelope feedback added to content feedback, and emotional feedback added to content feedback. Some questionnaires were presented to users in which they could give their opinion on life-likeness and ease of interaction. Moreover, there were measurements on the relative number of utterances, hesitations and expressions of frustrations. It could be concluded that the envelope feedback is much more important than the emotional feedback. In fact, it is argued that the possibility to visualize such behaviour is the strongest argument for using embodied agents.

6 Future and More Sophisticated Embodied Agents

6.1 Introduction

There is no need to visualize all agents in a virtual environment, especially if the environment is designed in an agent-oriented way. In that case a window appearing on the screen may be an agent, a door knob in a virtual environment may be an agent, but also a non-visible dialogue manager or database management system can be designed as an agent. We think it is useful to have a visualization of the agents that have explicit dialogues with the visitors of our environment. Maybe not necessarily during the dialogue, but, for example, to initiate a dialogue or to refer to a previous dialogue. It is clear that Karen needs to be represented as a person in the virtual world. Moreover, she needs an animated talking face and text-to-speech synthesis. It is not really clear whether the navigation agent requires a human face and body. However, if this agent is supposed to walk in front of us, it needs some visualization. Especially if we want agents that are visible for the visitor, agents that walk, agents that show how to do certain things and agents that interact with users in a natural way, then we need rather natural visualization of movements of agents (including movements of body, legs, arms, fingers, etc.) and animation of facial expressions, all in accordance with the tasks that the agent has to perform and the interaction with the visitor that is required. Generation of natural language and the corresponding synthesis of speech and facial expressions has been discussed in the previous sections. What remains is the interaction between visitors and visualized agent such that the agent's movements in the virtual environment are natural, given the task the visitor wants it to perform.

6.2 Embodied Agents and Gaze Behaviour

In our environment we have different human-like agents. Some of them are represented as communicative humanoids, more or less naturally visualized avatars standing or moving around in the virtual world and allowing interaction with visitors of the environment. In a browser which allows the visualization of multiple users, other visitors become visible as avatars. We want any visitor to be able to communicate with agents and other visitors, whether visualized or not, in his or her view. That means we can have conversations between theatre agents, between visitors, and between visitors and agents.

If a visitor, whether visualized or not, has a visual conversation partner on the screen, gaze behaviour of this partner can help in making the conversation more natural. This behaviour can be used to better convey the message (just as face expressions, head movements, gestures and intonation) and to have more natural turn-taking in dialogue or actions. Clearly, gaze behaviour of a conversational partner on the screen (whether human or artificial) should be oriented towards the human partner looking at the screen or towards its representation (maybe just as a hand or a 3D face) on the screen.

In general we want our embodied agent to look at the user when it speaks to the user or when the user is looking or speaking to the agent. However, the agent can temporarily look away from the user when starting a turn, it can look away to draw attention to something else on the same screen (another agent or user representation, object, menu or window) on the same screen or a situation not on its screen, but somewhere in the user's environment. It may also follow the user's gaze in order to show interest and to anticipate a next topic that may arise in the conversation.

Similarly, in a neutral situation we assume that a user looks at an embodied conversational partner. The user may look away from an agent to start the turn, to follow the agent's gaze or to take an independent action to look to another agent, object or window on the same screen, to look at another screen or object in the environment, or to use the keyboard or perform another task related to the conversation. The user may also decide to start looking or to look back at a particular agent. This can be taken as a sign that the user wants to address this particular agent or, in the process of a conversation, wants to give the turn, so the agent can start speaking.

We conclude that there are many reasons to model gaze behaviour of our agents, whether they represent humans or theatre employees and whether they talk to each other or to a human looking at the screen. However, as may allready have become clear from above, there are so many factors that determine gaze behaviour (including the relation between the conversational partners, the number of partners, noise in the environment, tasks that have to be performed at the same time, mood and personality of the partners, etc.) that we can not expect to be able to model all of them. Best investigated is the relation between gaze behaviour and turn-taking in a conversation. In summary: often a speaker looks away from a hearer when she starts an utterance and she looks towards he hearer when ending an utterance.[2] Cassell et al. [1,3] have investigated the relation between propositional content of utterances and gaze behaviour. This is interesting from the point of view that we can make further steps towards generating the behaviour of agents (syntactic, semantic and pragmatic content of utterances, intonation, face expressions, gaze behaviour, head and body movements) from a representation of (the history of) previous interactions, the representation of believes, desires and intentions and the representation of some personality characteristics. In the tradition of Halliday, Cassell et al distinguish between the 'theme' and the 'rheme' of an utterance. The 'theme' can be considered as the topic of the conversation. Hence, the part of the utterance that links the conversation to previous parts is the

---

[2] This gaze behaviour is only one of many cues that speakers employ. There may be gestures that indicate the end of a turn, the pitch of the utterance may fall towards the end and some specific phrases ('don't you think so?') may give the floor to an other conversational participant.

'theme', it does not really add something new to the conversation. The 'rheme' of an utterance adds something new to the conversation. That is, something which is new information for the hearer. From an empirical analysis of experimental data they conclude that:

- starting the thematic part of an utterance is frequently accompanied by gaze behaviour that looks away from the hearer, while beginning of the rheme is usually accompanied by gaze behaviour that looks towards the hearer; and
- if the start of the thematic part coincides with the start of the turn, the speaker always looks away from the hearer; in cases where the rheme coincides with the end of the turn, the speaker always looks towards the listener.

Obviously, the behaviour of computer-generated agents, their movements, face expressions, intonation of speech and the gaze behaviour described above can be controlled by our algorithms and programs in the virtual environment. In particular when the agent's contributions to the dialogue are generated from templates the 'theme' and 'rheme' of the generated utterance are known and, assuming the utterance is not too short, gaze behaviour can be adjusted to the roles the different phrases in the utterance have.[3]

In an attempt to profit as much as possible from the user's input, whether given deliberately or not, to the (agents on the) screen, it is useful to see whether the gaze behaviour of the user or visitor can not only give information about the way the conversation should proceed, but also how certain input from other modalities (speech, keyboard, mouse) should be interpreted. Obviously, this is part of our philosophy how to deal with imperfect interaction technology, i.e., the integration of different interaction modalities makes it possible to reduce the number of possible interpretations of whatever modality and to better return a system or system agent response that helps the user to continue his task and exploration. In particular we want an agent to know that the visitor is addressing him or her. That is, the agent should detect that the visitor is looking at him or her rather than looking at an other agent (window, menu, etc.) visible at an other position on the screen.

Presently, we are doing experiments with an eyetracker system. Such a system makes it possible to determine where a person is looking at. In particular, it is possible to determine to which avatar a person is looking. This allows management of multi-user conversations in a virtual environment, where each user knows when and which other users are looking at him or her. This leaves to a certain degree open how the user is represented in the environment, but at least user gaze directional information can be conveyed. Hence, visitors of our environment can address different task-oriented agents in such a way that speech recognition and language understanding are tuned to the particular task of the

---

[3] Notice that in our database with theatre information we also have reviews of performances which can be rather long and which can be read to the visitor by Karen.

agent; therefore quality of recognition and understanding can increase, since the agent may assume that words come from a particular domain and that language use is more or less restricted to this domain. That is, we can restrict lexicon and language model to the utterances that are reasonable considering the agent. Obviously, we should try to visualize agents in such a way that it is clear from their appearance what they're responsible for and what a visitor can ask them. An attempt should be made to ensure that any agent is able to determine that she isn't the right agent to answer a visitor's questions and therefore should direct the visitor to an other task-oriented agent or to an agent having global knowledge of the task-oriented knowledge of the other agents in the virtual environment.

6.3 Embodied Agents that Sell, Advise, Buy, ...

In the case of Karen, it will be clear that her boss would like her to sell as many tickets as possible. A theatre director will have certain preferences in choosing performances for the theatre, but once they have been chosen the aim is to be sold out every night. How can we program Karen (and maybe also the navigation agent) such that the behaviour towards a visitor increases the chance to reach this 'private' goal?

In the future, webpages and virtual environments will become inhabited by sales agents, hosts, guides, etcetera that offer information and answer questions (and try to sell products) in a way similar to Karen or our navigation agent, but also in a way that allows building up relationships with users through social "chit-chat" about family, work or sports. That is, these agents should have specific knowledge about a certain domain, a domain which may range from detailed knowledge about Shakespeare, performances in a next season and cars in a showroom to drinks that are available in a bar. But, in addition to that, they have some superficial global knowledge that allows them to give socially appropriate answers to keep up a conversation about topics they hardly know of and they have some special hobbyhorses and some specific knowledge with which they can steer a conversation and which makes them 'believable' to the user.

One of our concerns in the near future will be the introduction of such properties in our present agents. Maybe Karen shouldn't be allowed to give too personal opinions about performances and artists, but some more human-like conversational behaviour should be considered. Moreover, if a user really wants to know or to exchange opinions about performances and artists she should be able to communicate with other visitors of the environment or be able to address agents employed by the theatre who can share their specific knowledge (embedded in some kind of social conversation) with the visitor.

Agents that allow human-like conversation have been designed, both in research and in commercial environments. An example of a virtual web agent is Jennifer James, designed by *Extempo Systems Inc*. She sells cars in a virtual auto show room, 24 hours a day. Jennifer has a history in racing, which is useful in talking with customers. She wants to know about a customer's background and preferences. Questions are asked in a friendly and sometimes ironic way. Jennifer knows about cars and racing, but questions or comments on family or country music can be dealt with in a believable way. Jennifer has been visualized as a smiling saleswomen dressed in a red and white jump suite. Jennifer has personality, customers feel confident to talk with her, they give information about themselves and a company that employs Jennifer will build relationships of affection, trust and loyalty with its customers. The information that is elicited from a customer is stored in marketing and customer databases.

Jennifer makes a charming attempt to understand natural language and to interact with objects in the scene. The customer uses the keyboard to communicate with Jennifer. Her body and face are real-time animated, where animations attempt to be consistent with role and personality, and with the events of the dialogue. Jennifer talks back using speech synthesis. More examples of agents that display human and social qualities have been introduced. Of course, we can go back to 'conversational agents' such as Weizenbaum's Eliza and Colby's Parry (both from the sixties) or Julia, a chatterbox character of the early nineties. However, these characters have not been given the task to inform visitors, to give information, to guide and help or to purposely try to seduce potential buyers to buy commercial products, to visit sites and events and to take part in activities.

7 MM-Retrieval in the VR Environment

We investigate how to store, index and retrieve theatre-related multi-media objects (text, pictures, audio, video) in such a way that a visitor of our virtual theatre environment can address Karen in such a way that she is not only able to inform the user about artists, performances, dates and available seats, but also knows how to provide the user with pictures, audio and video fragments (previews) of performances. In general, if a visitor asks about a particular performance, it is not difficult to let Karen present the (available) associated multi-media information. However, if the visitor's questions are about or address associated information rather than explicitly making references to a particular performance, then our theatre agents have to start reasoning with the information that is available in the database and information that is available in a knowledge representation scheme in which general knowledge about the theatre domain has been stored.

At this moment no explicit, comprehensive and suitable ontology of the theatre information and booking domain has been designed. Karen knows about this domain, but her knowledge is hidden in the linguistic (syntactic, semantic) and the dialogue management (pragmatic) knowledge of the present system. This knowledge nevertheless allows a better mapping on the information that is available in the theatre database. However, other agents have no access to this implicit knowledge. Similarly, our navigation agent knows about the geography of the theatre and how users generally ask questions about the theatre. However, knowledge about the geography has been represented as a list of coordinates of particular locations and possible ways to refer to them. Hence, the user has some freedom how to address these locations. However, whatever synonyms are allowed, the list is predefined and only the navigation agent knows how to access this information. Our speech recognition software, that is, the number of phrases it can recognize reasonably correctly in normal circumstances, does very much determine the navigation's agent intelligence.

How to handle a user's question about artists and performances, where the user doesn't use names that are available in the theatre database? That is, can we provide Karen with general knowledge about artists and performances which goes beyond the information that is available in the database? , and, if it turns out that the user doesn't know about Karen's knowledge and such knowledge is not complete, can we nevertheless map the user's question to a database query?

As an example, suppose someone asks: "Well, what's her name, Michelle, eh, I know she received an Oscar last week. Are there any movies with her showing this week?" At this moment we think that it is reasonable to expect from our system that it starts searching WWW with a question that has as keywords: 'Oscar' and 'Michelle'. The intelligent theatre search engine should deliver the name of the actress and this should be sufficient to search the database for movies showing this week (or maybe later this month) with this particular actress in one of the roles. A knowledge representation network with the main concepts and their relationships in the world of theatre (opera, musicals, plays, music, cabaret, etc.) and theatre actors (directors, performers, technicians, reviewers, etc.) has to be designed in order to connect information available on WWW and information available in local databases through a process of reasoning.

Textual information is important, since it is widely available. Newspapers, magazines and WWW pages contain reviews about performances, movies and performers. However, there is no need to confine ourselves to textual information. We are involved in national and European projects on indexing and retrieval of multi-media information, including pictures, captions, audio, video, subtitles and (transcripts of) spoken movie dialogues. Integrating this research on multimedia indexing and retrieval and natural language (spoken) dialogue systems embedded in environments where users can 'say what they see' (ranging from handheld devices to immersive virtual reality environments) is a topic high on our research agenda. We think it is fruitful to consider WWW as an extension of our theatre databases and therefore we need an intelligent agent that knows how to filter and present the results of a WWW search to a naive user looking for some information that may help in having a nice evening.

8 Presence and Believable Agents

'Presence', as defined by Lombard and Ditton [10], is the perceptual illusion of nonmediation, that is, 'a person fails to perceive or acknowledge the existence of a medium in his/her communication environment and responds as he/she would if the medium were not there.' They mention that this illusion can occur in two distinct ways:

- The medium can appear to be invisible, with the medium user and the medium content sharing the same physical environment; and
- The medium can appear to be transformed into a social entity.

From the previous sections it may have become clear that rather naturally emerging topics of our interest are closely related to the issue of presence. The environment that is offered and the locations that can be visited look familiar, the functions of several objects and what to do with them is clear from their appearance and the multimodality approach allows a variety of user input and the production of different sensory outputs. The agents in the environment are assumed to be friendly and cooperative and the embedding of talking faces and moving avatars in the environment will increase the tendency to treat agents as social actors, rather than as software modules.

We have looked at possibilities to increase a user's commitment to the system (like we would in similar systems, e.g., for electronic commerce) with the aim to obtain co-operative behavior. One obvious reason which makes us loose a user is when use of imperfect technology is not sufficiently backed up by context (including other modalities) which seduces the user to a certain interaction behavior and helps to disambiguate the users utterances.

In our world mediated entities (our agents in the virtual world) play the role of social actors, they increase the feeling of presence and they help to increase the user's commitment to the system. A more technical reason to have agents as social actors is that they can influence the interaction behavior of the users in such a way that it remains restricted to the task and/or domain knowledge, therefore hiding shortcomings of imperfect interaction technology, in particular speech and language technology. With other words, in order to increase the quality of the web-based information and transaction services we are offering, it seems to be useful to exploit the possibility to increase the role of social actors in our environment.

It should be clear from previous sections that in our environment we decided to explore the possibility to increase the user's feeling of presence in order to increase his or hers commitment to the system. We certainly do not want to advocate such an approach in general. In our application we think this is a useful approach. We don't think this should be the general approach to the design of computer interfaces. Is the interface serving a leisure activity, is it task-oriented or is it assumed to change the user?

9 Conclusions

In this chapter we reported about on-going and future research on interactions in virtual worlds. We intend to continue with the interaction between experimenting with the virtual environment (adding agents and interaction modalities) and theoretic research on multi-modality, formal modeling, natural language and dialogue management. In particular the integration of visitors in an agent platform that models a uniform verbal and non-verbal behavior is required in order to be able to maintain and extend the virtual environment.

Our approach to designing a virtual environment for an interest community is bottom-up. At this moment the system has two embodied agents with different tasks and limited interaction between them. Moreover, the agents do not employ a model of a user or of user groups. In general, when we talk about interface agents we mean software agents with a user model, that is, a user model programmed in the agent by the user, provided as a knowledge base by a knowledge engineer or obtained and maintained by a learning procedure from the user and customized according to his preferences and habits and to the history of interaction with the system. In this way we have agents that make personalized suggestions (e.g. about articles, performances, etc.) based on social filtering (look at others who seem to have similar preferences) or content filtering (detect patterns, e.g. keywords) of the items that turn out to be of interest to the visitor. In this way, visitors can be provided with personal assistants ('butlers') that know about the visitors' preferences, that can exchange information with other personal assistants and that can search for and filter information that is of interest for the visitor. In particular we hope to integrate research that comes available from the European projects *Magic Lounge* (which aims at developing tools that allow intelligent communication services in virtual meeting places; these tools include shared white boards and chat environments, but also tools that record and store interaction events such that it becomes possible to browse earlier interactions and inspect individual contributions) and *PERSONA* (a project about navigation in two- and three-dimensional interfaces; in this project the concept of social navigation is explored, that is, navigation that exploits the possibility to talk to other users and to agents for obtaining information, including the following of their trails in the information space).

Contrary to many other virtual environments the public version of the environment is in use by the general audience. The environment contains actual information about the theatres, it is accessible on WWW and in addition it is part of an exposition in a technology activity centre where visitors can get explanation about the environment.

References

1. Cassell, J., Pelachaud, C., Badler, N., Steedman, M., Achorn, B., Becket, T., Douville, B., Prevost, S. & M. Stone. Animated Conversation: Rule-Based Generation of Facial Expression, Gesture and Spoken Intonation for Multiple Conversational Agents. Proceedings of SIGGRAPH '94, 1994.
2. Cassell, J. & K.R. Thórisson. The power of a nod and a glance: envelope vs. Emotional feedback in animated conversational agents. Applied Artificial Intelligence, to appear.
3. Cassell, J., O.E. Torres, S. Prevost. Turn taking vs. Discourse Structure: How Best to Model Multimodal Conversation. In: Wilks (ed.), Machine Conversations. The Hague: Kluwer, to appear.
4. Dirksen, A. and Menert, L. Fluent Dutch text-to-speech. Technical manual, Fluency Speech Technology/OTS Utrecht, 1997.
5. Doest, H. ter. Towards Probabilistic Unification-Based Parsing. Ph.D. Thesis, University Twente, February 1999.
6. Friedman, B. (ed.). Human Values and the Design of Computer Technology. CSLI Publications, Cambridge University Press, 1997.
7. Hulstijn, J. & A. van Hessen. Utterance Generation for Transaction Dialogues. Proceedings 5th International Conf. Spoken Language Processing (ICSLP), Vol. 4, Sydney, Australia, 1998, 1143-1146.
8. Jönsson, A. Dialogue Management for Natural Language Interfaces. PhD thesis, Linköping University, 1993.
9. Lie, D., J. Hulstijn, A. Nijholt, R. op den Akker. A Transformational Approach to NL Understanding in Dialogue Systems. Proceedings NLP and Industrial Applications, Moncton, New Brunswick, August 1998, 163-168.
10. Lombard, M. & T. Ditton. At the heart of it all: The concept of presence. Journal of Mediated Communication 3, Nr.2, September 1997.
11. Nass, C., B. Reeves & G. Leshner. Technology and roles: A tale of two TVs. Journal of Communication 46 (2), 121-128.
12. Sproull, L., M. Subramani, S. Kiesler, J. Walker & K. Waters. When the interface is a face. In [6], 163-190.

Chapter 9. Virtual BioBots

Michael Paulin and Rachel Berquist

Department of Zoology and Centre for Neuroscience, University of Otago,
Dunedin, New Zealand
mike.paulin@stonebow.otago.ac.nz

Abstract. Virtual biobots are computer models that look and behave like animals. They work by simulating the biomechanics and nervous systems of real animals in virtual environments that simulate the physics of the real world. We have developed anatomical, neurophysiological and software methods for building virtual biobots. Virtual biobots have direct applications in biology teaching and video animation, as well as research applications in biology and engineering. Our virtual biobot builder technology may underpin the development of novel biomimetic technologies by facilitating the development of complex software models of biological systems.

Keywords. Virtual reality, neural model, robotics, animation.

1 Introduction

Virtual biobots are software robots that look and behave like animals. They are not animations but autonomous agents living in virtual worlds. We use virtual biobots as research tools in investigating the neural basis of animal behaviour. Our goal is to build virtual biobots that not only look and behave like real animals, but that do this by simulating the biomechanics and brains of real animals.

In this chapter we will explain how and why we use virtual biobots, how we create them and what they do. We will speculate on how virtual biobots may evolve and their potential applications in the next millennium. The most obvious immediate application of virtual biobots outside academia is in video animation. Virtual biobots will be employed as virtual actors, taking verbal directions in natural language: "Mouse, come out from behind the chair and run out the door!" Perhaps more significant outputs of our virtual biobot project will not be biobots themselves, but the biobot builder technology and the various tools and techniques developed for making biobots. This emerging technology makes it progressively easier to realistically model and simulate complex biological structures. It may underpin the development of a wide range of technologies based on mimicking biological systems.

2 Computational Neuroethology

Our virtual biobots have evolved within a research programme in computational neuroethology. That is an attempt to understand and simulate the neural basis of animal behaviour, by building computer models of neurons, neural networks and behaviour. As zoologists we take a broad interdisciplinary approach. Development and evolution are fundamental aspects of our approach.

A major problem that we must confront is the complexity of brain structure. There are about a hundred billion neurons (10^{11}), and a million billion connections (10^{15}) in a human brain. Large numbers of neurons in widely distributed parts of it are activated during even simple behaviours, such as blinking or chewing. Because of this complexity it is not possible to map circuitry underlying any meaningful fragment of vertebrate behaviour by examining the circuitry during the behaviour.

Fortunately it is not necessary to specify the detailed structure of a neural circuit in order to simulate it realistically. This must be true, because there are only about three billion (10^9) DNA base pairs in the human genome, which specify the structure of 80-100,000 genes. This is not enough to specify how all of your neurons are interconnected, but is enough to build a brain that works. In fact only some of these genes influence brain development. They do this by specifying general rules and constraints on cell division, migration and connections. Detailed brain structure is determined by dynamical processes involving the organism and its environment, operating under these genetic influences (Deacon, 1990, 1998; Raff, 1996).

The relationship between structure and function in neural systems can be clarified by considering an analogy with elementary geometry. Suppose an operation T translates a point a to b. This can be expressed in coordinate-free operator notation as b = Ta. Now if we want to actually implement this transformation, for example by moving a dot on a computer screen using the MATLAB computer language, we must introduce a coordinate system and write b = T*a. The MATLAB variables represent sets of numbers in the computer's memory. T is a matrix and a and b are vectors. The vectors can be expressed as visible points on the screen using MATLAB instructions. Suppose we choose a new coordinate system, related to the first via a transformation matrix S such that a point with coordinates x in the original coordinate system has coordinates xx=S*x in the new one. Then we can implement the same transformation in the new coordinate system using the MATLAB instruction bb=TT*aa, where TT= STS^{-1}. The numbers are different, but the result is the same. What you see on the computer screen is a point moving from a to b. It is easy to check, either by doing a little algebra or empirically using MATLAB, that TT has the same eigenvectors and eigenvalues as T. It is these global properties *not individual matrix entries* that determine what geometric operation is performed. Infinitely many matrices related via the similarity transformation TT= STS^{-1} will do the same thing. This result extends to matrix-vector analysis of networks and dynamical systems.

Input-output behaviour of a network can be expressed in terms of geometric transformations, in which matrix entries represent strengths of connections between components. There are infinitely many different networks that have a particular behaviour.

The implication for brain construction is that connectivity patterns and individual synaptic strengths don't matter. A neural network can maintain its function while individual synaptic weights wander around apparently at random. Conversely, it is pointless to attempt to relate changes in synaptic strengths at individual synapses to changes in the function of a neural network, unless you can measure all of the synaptic weights in the network. It is certain that no two brains the same, it is likely that no single brain is the same from moment to moment. Brain building is not about specifying complex structure, it is about dynamically maintaining global properties of a complex structure within a satisfactory range. Developmental genetics, and the mismatch in the information content of genes and brains, tells us that this can be done in a way that requires many orders of magnitude less information than is present in the structure itself. This is why we consider functional neuroanatomy to be a part of developmental biology, and think that computational neuroethology should be less about building neural models and more about getting neural models to build themselves.

Development is a key to understanding and modelling biological structures. This is how real bodies and brains are built. Brains are so complex that it seems unlikely that there is any practical means of building genuinely brain-like devices except by creating self-assembling models. These models must mimic real developmental mechanisms at least in principle. Until we understand those principles, and as a way of developing that understanding, we can proceed by mimicking particular developmental processes.

Vertebrate nervous system structure is to some extent induced from the periphery. That is, neural structures that develop in your brain for perceiving and behaving depend on what equipment – sense organs, limbs and so on – your brain is connected to as it develops (Deacon, 1990). These developmental processes are genetically canalized, restricting particular species to a limited range of possibilities. Thus genetically engineering human photoreceptors to be sensitive to ultraviolet light may be sufficient to produce babies able to see in the ultraviolet range. It is less clear what would happen to the developing visual system if photoreceptors were engineered to respond to, say, commercial radio transmissions.

We use models of development to construct some details of our biobots. For example, peripheral nerve endings sprout and seek out their targets. This means that it is not necessary for us to specify certain fine details by taking data from anatomical images. We plan to incrementally increase the self-organizing capabilities of our biobot builder software and extend it by using evolutionary programming methods. Evolution works by creating new mixtures of successful developmental programmes and by tweaking parameters of these programmes - genes - at random. The former mechanism is more important in creating diversity, which is useful for long-term survival and adaptation and why sex is so popular. The significance for virtual biobots is that if we can create software that automates the construction of, say, a dogfish nervous system, then it may be a depressingly small technological challenge, and an even smaller conceptual challenge, to create a virtual human brain. Indeed, if we had such technology there would be no need to set our sights so low.

Genes have been selected by evolution to ensure that developmental processes under normal conditions lead to adaptive structures and mechanisms, i.e. useful stuff. Adaptations include mechanisms that allow tissues to adaptively modify their structure in response to internal or external events. These changes occur over appropriate timescales. For example, you may blink if an object approaches your eye, or your skin may harden where it is regularly abraded. Thus biological structure is determined by dynamical processes operating over a wide range of timescales, with evolution and development at one end of the continuum, behaviour at the other, and learning and memory somewhere in between.

Small changes in developmental rules may lead to large changes in adult structure, but basic materials and mechanisms are conserved. Because evolution is constrained to create new structures by modifying old ones with a restricted range of materials, complex structures may sometimes evolve to perform simple tasks. For example, a class of neurons might evolve families of ion channel proteins and use mixtures of different channels to achieve certain membrane properties, either because no single type of channel protein could produce the required properties, or because there is no pathway along which such a protein could evolve. In such a case we might realistically model the neuron at a biophysical level using only a single type of channel. The real mechanisms are implicit in the model and could be made explicit if required. It may not actually be possible or reasonable to reduce families of ion channels to a single channel in a model. Our point is just that simplification and abstraction are not incompatible with realism in biological models. This point seems to elude many computational neural modellers, enamoured with the rapid expansion of computer power. Although we do not presently know what kinds of simplifications are possible, in general it may be possible to produce greatly simplified yet realistic models of brain circuitry because the structure of computer models is not constrained in the same way that the structure of real brains is.

There are more than 50,000 species of vertebrates. As with other aspects of vertebrate anatomy, brains have a common basic design that varies systematically in ways that are related to shared ancestry and to the sensory, motor and behavioural characteristics of each species. By comparing and contrasting the brains of different species, and identifying characteristics of brains that are associated with particular capacities or behaviours, it is possible to deduce something about the functional roles of different brain structures.

For example, the bottlenose dolphin, *Tursiops*, has a very much larger cerebellum than humans do. We might ask why this is the case. Early studies on humans indicated that the cerebellum contributes to the agility of movements.

Dolphins are indeed agile. Comparison across species shows that large cerebellums are associated with agile movements, but that there are some exceptions to this rule. In particular some species of fish that sense electric fields have greatly enlarged cerebellums. The common factor across all species is cerebellar involvement in tracking and predicting trajectories of moving objects, including movements of the animal itself. Motor agility requires an ability to track and predict movements, and therefore depends on the cerebellum. Thus the strange cerebellum of strange fish is a key to discovering the function of this part of the vertebrate brain (Paulin, 1993; 1997).

The diversity of vertebrate brain structure also means that there are many different examples to choose from in studying any particular brain structure, sensory system or behaviour. Some animals have specialisations that provide opportunities for particular investigations. For example, barn owls have a hypertrophied auditory system associated with their acute ability to locate sounds. Investigations of this system have uncovered general principles and mechanisms that all vertebrates use to compute sound source location.

Ultimately virtual biobots will be grown from virtual eggs. To make progress towards that goal we are developing software and related methods that progressively automate the biobot-building process. In doing so we are confronting an important problem that has hovered on the edge of mainstream biology for much of this century (Darcy Thompson, 1961; Stewart, 1997). This is the relationship between genes and form. The public can be forgiven for thinking that genes create forms, since some geneticists appear to believe this. However, the attempt to build virtual biobots makes it clear that identifying genes, gene products and expression patterns does not provide any useful information about how to construct virtual animals. We need to simulate the physical developmental processes that give rise to biological form.

Knowing that the emergence of forms is correlated with gene expression is unhelpful. For example, handedness in humans is presaged by left-right asymmetries in gene expression (Ramsdell and Yost, 1998), but saying that this is the genetic basis of handedness merely begs the question of why the genes are expressed this way. Asymmetric gene expression is presumably part of a distributed process that undergoes a spontaneous break in spatial symmetry. Analogously, a piece of toast balanced on its edge will, sooner or later, fall left or right without being explicitly pushed. We may observe that it usually falls buttered-side down, but it would be a mistake to say that butter causes the toast to fall even if the butter or something in the buttering process biases the fall in that direction. A vertical piece of toast in a gravitational field falls over because a vertical piece of toast is unstable in a gravitational field. The butter may bias the outcome, or it may be no more than a convenient way to label the direction of the fall. It does not instruct the toast to fall.

Our virtual biobot builder methods already involve virtual genes – parameters of developmental simulations – but the main problem in biobot building is not in selecting parameters but in creating models of developmental processes that are capable of generating biomimetic forms under *some* parameterization. Evolutionary programming methods will become genuinely useful for biobot development *inter alia* when they are properly integrated with developmental or self-organizing software methods.

In the case of a simple geometric transformation implemented using vectors and matrices, we saw that it is not helpful to know matrix entries ('connection strengths') unless you know all of them. Only when you know all of them is it possible to relate these local parameters to global parameters that specify the geometric transformation – the function of the matrix. But there is more to it than this, namely that in this special example we have a general solution to a class of geometric problems (a nonsingular nxn matrix reversibly translates a point in n-space) and we know how to analyse or synthesize a structure that performs a particular operation in this class. We can also do this in more sophistocated contexts, such as control systems or electronic circuit design. It is instructive to read the comments of Rudolf Kalman, a leading figure in the development of modern matrix-vector based or state space control theory. The following quote comes after Kalman has given an example of how algebraic methods may be used to design the 'simplest control system of interest' within the state space framework:

> *Even in this simple case, the 'canonical wiring diagram' ... is too complex to be intuitively understandable. Here the superiority of modern control theory is especially striking; we obtained the wiring diagram without any preconceived notion of what it might look like when we were finished.*
>
> *It is tempting to speculate about the implications of this result for problems in biology, especially the theory of the brain in higher animals. It could be true that it is hopeless to try to understand brain functions solely on the basis of anatomy (wiring diagrams). Perhaps the problem will become relatively transparent only after we have developed a theory (vaguely analogous to the present one) powerful enough to give us the main features of the anatomy.*
>
> R.E. Kalman (1969) p66.

If simple artificial control systems are too complex to be understood by their designers without mathematical theory that relates structure to function, it does seem doubtful that nervous systems might be understood without similar theory. Current methods in computational neuroscience resemble the quantitative methods used with limited success in engineering control theory until the 1950s, which Kalman refers to as 'a very primitive kind of applied mathematics'. It is no coincidence that aerospace technology literally took off at about the time that Kalman published new methods for tracking trajectories of dynamical systems in state space in the mid 1950s. State space theory can not be applied directly to model neural circuitry because the operators and operands of the theory do not correspond to the operators and operands of neural circuits, i.e. neurons and spikes. This correspondence does occur in artificial systems because efforts are made to ensure that electronic circuit components have algebraically tractable properties. In particular, op-amps are designed to be linear. But we can't redesign brain components, and so we have to reformulate state space theory to provide the required calculus (Paulin, 1998). Virtual biobots will provide a test-bed for neuromorphic models that use state-space methods to ensure that circuits are capable of performing tasks required of them in the real world.

We are a long way off being able to create biobots entirely using self-organizing software. In a sense this is fortunate because it is obvious that the computer power required to simulate development of whole organisms will be massive by today's standards. Our strategy is to proceed in parallel with progress in hardware and software development to do what is possible with what we have, and trust that that progress will enable us to incrementally transform our dreams into virtual reality.

2   The Desktop Dogfish Builder

We build biobots using *Desktop Dogfish Builder*, an application written in MATLAB v5.3. MATLAB (www.mathworks.com) is a numerical analysis package with object-oriented graphics. The ability to attach computations to graphical objects, including computations that alter the properties of objects, makes it possible for these objects to interact with each other and with users. MATLAB objects can interact with software objects in other applications running concurrently, or with real objects via a hardware interface. Thus a MATLAB virtual brain could operate a real biobot.

Desktop Dogfish Builder was named because it was initially developed to create a model of the spiny dogfish *Squalus acanthias*. We digitize plastinated thick sections of a dogfish and import these into the *Desktop Dogfish Builder*. An expert user, familiar with the anatomy of the dogfish, identifies anatomical structures by pointing and clicking on these images. The *Builder* assembles these isolated points into three-dimensional images based on information that the user provides about the shape of the structures. For example, if the user specifies that a certain set of points lie on the boundary of a white convex structure named 'EYE', then the *Builder* draws a solid white surface which is the convex hull of this set and attaches the tag 'EYE' to it. The tag is an invisible label that can be used to find and modify the virtual eye.

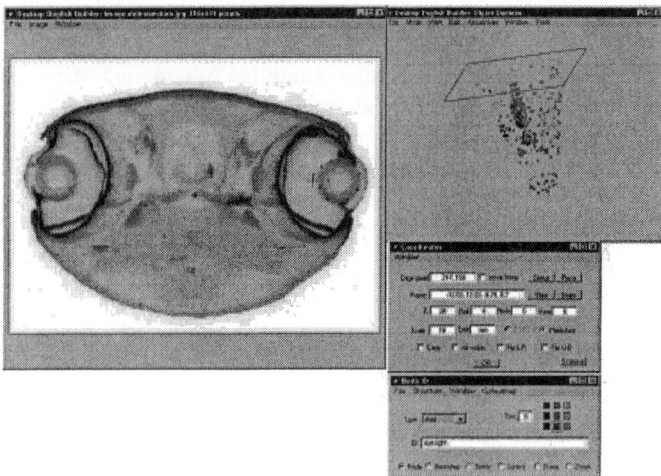

Figure 1: DDB user interface. The user has labelled the eye in the image of a plastinated tissue section on the left. The corresponding points appear in three dimensions in the node window, upper right. The other windows allow the user to specify coordinates and properties of the structure.

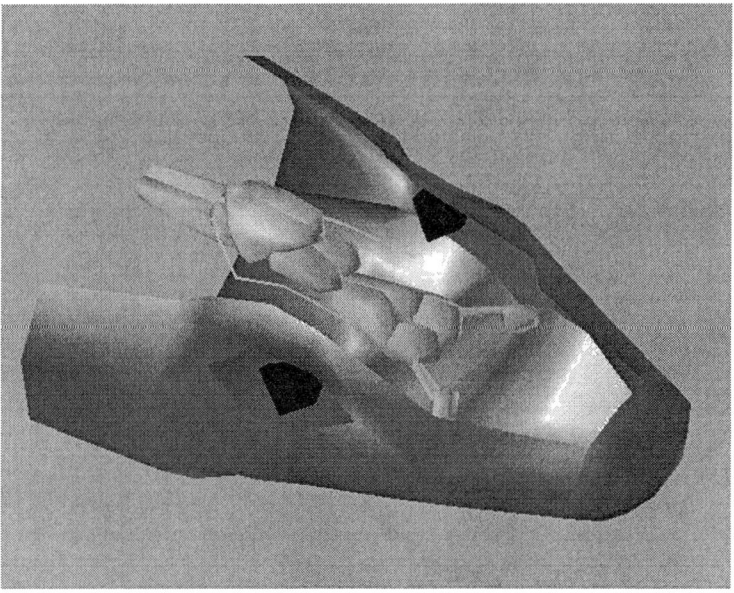

Figure 2: Partially assembled virtual dogfish. The cranium and the skin over the top of the head have been edited out to reveal the brain.

The next step is to attach functions to the virtual biobot components. *Desktop Dogfish* has no moving parts yet, but does have computational models of cranial sensory nerves that respond to virtual stimuli. We obtain these models by applying system identification methods to real spike train data (Paulin, 1993b). These models accurately mimic firing patterns of real neurons responding to arbitrary stimulus waveforms within physically realistic bounds. We have identified models of electrosensory neurons in dogfish and vestibular sensory neurons in bullfrogs (Paulin, 1996; Paulin and Hoffman, 1998). We simulate responses of other cranial sensory neurons using models of the same form adjusted to give plausible firing patterns consistent with literature reports.

Naturalistic virtual stimuli are generated by a virtual crab living in *Desktop Dogfish's* virtual world. It scuttles along the sea floor and swims through the water column, autonomously or following instructions from the user. The crab has very simple anatomy – it is a polygon with triangular legs – but it demonstrates how MATLAB can be used to animate virtual biobots. The user specifies what kind of stimuli the crab generates. This can be any combination of electrical, hydromechanical, olfactory or visual signals. The cues detected by the dogfish are computed from simple physical models. For example, in virtual electroreception we treat the crab as an electrical dipole and compute electric field strengths across the *Desktop Dogfish's* electroreceptor organs.

Teaching Applications

The initial application of the *Desktop Dogfish* will be teaching at undergraduate level. Dissection of the spiny dogfish is widely used to teach introductory vertebrate neuroanatomy. *Desktop Dogfish* will not only provide a three-dimensional dissection guide, but will enhance these lessons by allowing students to interactively explore the functions of the dogfish's cranial nerves. Instead of providing students with a table of cranial nerves and their functions we will require the students to discover these functions and make up their own tables by experimenting with the *Desktop Dogfish* and the virtual crab.

A variety of supplementary teaching material can be attached to the *Desktop Dogfish*. For example, a student might point a virtual probe at a structure and be taken to a web page that summarizes what is known about the structure and provides links to other pages that describe it in detail and discuss recent research. Clicking on the virtual crab could lead to web-based tutorials and information about the physical properties of the marine environment relevant to predation and predator avoidance. This may be an effective way to introduce biology majors to aspects of biology that many of them have difficulty with. Biology is a branch of physics - the physics of complex systems (Kauffman, 1993, 1995). That biology presently appears to be relatively nontechnical in comparison to physics or engineering is simply a reflection of how little we presently know about the physics of complex systems. The *Desktop Dogfish* will help to prepare a new generation of biology students by allowing them to explore basic physics in an interesting biological context.

In future we hope to extend the *Desktop Dogfish* into a strategy game that introduces students to the complex physics of vertebrate brains and behaviour. We would hope to redirect some of the energy and enthusiasm currrently being expended on computer games into learning something about the real world. Narrative, context and personal involvement are powerful mediators of learning. It is too easy, especially in a subject as broad and as complex as neurobiology, for instruction to focus on content, and it is too hard for students to assimilate complex factual material. We envisage students taking on the role of predator or prey, and learning to survive in the virtual ocean by working out how the opposition perceives, plans and behaves.

At some point it will become possible for students to play against each other in a virtual ocean containing many predators and prey, using interfaces that allow them to experience and not merely observe their avatar's perceptions. Will they also discover what it is like to be a shark? Perhaps consciousness is 'merely' the ability of an agent to simulate and predict its own behavior in interactions with the world including other agents (Mandler, 1992; Vandervert, 1997; Paulin, 1997). In that case, shark-consciousness is not only real but humans may experience it through this computer technology. Conversely, realistic humanoid virtual biobots may one day provide computers with the opportunity to experience what it is like to be human. Biology, physics, mathematics and computing have come long way in the last 50 years, and a lot could happen in another 1000. Besides, we are not worried about facing criticism if our speculations seem niaeve at the end of the third millenium.

Research Applications

We are using *Desktop Dogfish* to address a problem in modelling neural mechanisms that allow sharks and their relatives to detect tiny electrical signals generated by prey. Sharks can detect electric fields as small as a few nanovolts per centimeter (Kalmijn, 1982). This is remarkable enough, but even more so when you consider that the organs which the shark uses to detect electric fields are located on the surface of an electrical noise source – the shark's own head – which generates signals having spectral characteristics that are similar to those produced by the prey. No artificial noise-suppression technology performs as well.

Neurophysiological measurements have shown that nerves leading from electroreceptor organs to the shark's brain respond predominantly to variations in the shark's own electric field. Responses to externally applied fields are barely discernable against this level of self-induced sensory input, even when the applied field strength is well above the threshold for behavioural responses by the shark. The primary neurons from the electroreceptors connect to secondary neurons that transmit electrosensory information to higher levels in the brain. Remarkably, many of these secondary neurons do not respond at all to self-generated electric fields, but are exquisitely sensitive to external fields such as those generated by prey (Montgomery and Bodznick, 1999)

There are several mechanisms consistent with the neuroanatomy of the electrosensory hindbrain in elasmobranchs that might contribute to the shark's remarkable ability to strip out unwanted noise as a signal enters its brain (Montgomery and Bodznick, 1994; Nelson and Paulin, 1995; Paulin et al., 1998). In order to test models of what goes on in the shark's brain, we need to be able to realistically simulate the spatial and temporal patterns of sensory input to the brain. *Desktop Dogfish* will allow us to provide existing computational models of filtering in the shark hindbrain with 'real' data to filter. We suspect that none of them will be able to replicate the observed behaviour of secondary neurons under realistic conditions. Thus *Desktop Dogfish* will clarify the problem posed by shark's ability to attend only to relevant stimuli, and provide a platform for attempts to solve it by building realistic models.

This application of *Desktop Dogfish* illustrates how virtual biobots may be used as research tools generally. A great deal of early progress was made in physics by studying simple systems in isolation. In neuroethology there are no simple systems. Animals are distributed nonlinear time-varying dynamical systems, and we are specifically interested in how these systems perceive and behave in complex unstructured stochastic environments. While it is generally a good idea to simplify, it is important to retain aspects of biomechanics and physics that are important to the neural system's function. The difficulty of course is in determining what those are. This is where virtual biobots may help.

The brainstem structure that removes unwanted noise from shark electrosensory inputs has the same neural organization as cerebellar cortex. There is overwhelming evidence that the cerebellum has a role in dynamical control of movements. Not in generating movements *per se*, but in producing movements that accurately attain goals when actuators and targets have stochastic nonlinear nonstationary dynamics. For example, the cerebellum would seem to significantly improve a shark's ability to strike accurately at prey that is fleeing and attempting to evade. Comparative evidence strongly suggests that the specific role of the cerebellar cortex is an aspect of trajectory prediction rather than trajectory control *per se* (Paulin, 1993). Current models of cerebellar control of movement and adaptive filtering by cerebellar-like circuits assume linearity and stationarity in the dynamics of the controlled or observed processes. These assumptions are wrong. This is no big deal, because real-world systems are never perfectly linear or stationary yet it is often possible for engineers to build effective controllers or filters by assuming that they are. *Desktop Dogfish* will enable us to discover what assumptions are reasonable for modelling movement control and sensory filtering in real dogfish. More importantly, it will provide a means for engineers and biologists to communicate effectively about adaptive sensory filtering and movement control problems, and the contribution of cerebellar(-like) neural circuits.

More generally, virtual biobots will provide a framework for developing new methods and devices in signal processing, control, artificial intelligence and robotics, by making it possible to build models of complex biological systems that have capabilities beyond those of current technology. For example, adaptive filters based on the shark electrosensory hindbrain may provide improved hearing aids, wider cell-phone coverage and the ability to conduct normal conversations and attend to relevant signals in high-noise environments.

Further Applications

The most obvious first application of virtual biobots, beyond their use as teaching and research tools, is as virtual actors for video animation. Physics-based simulations are already used for certain aspects of video animation, such as autonomous bouncing balls, and ambitious attempts to extend this are under way (e.g. www.mathengine.com). Our work, focussing on specialised animal behaviours that are relatively simple biomechanically and relatively well understood, will complement these attempts.

It is unlikely that virtual biobots with neuromorphic control systems will be applied in real-time applications such as video games in the near future, because of the massive computer power that will be required to simulate even vaguely brain-like control systems with realistic biomechanics and environments. However, hybrid systems that use combinations of neuromorphic, neural-network (connectionist), fuzzy logic and other 'soft computing' methods a.k.a. 'kluges' will be useful for both commercial and research applications. Even with sophisticated computer software (e.g. www.kinetix.com) video animation is a specialized and time-consuming task. The ability to automate even small aspects of the task will be valuable in video animation and video game applications. These hybrid virtual biobots will also be useful for researchers, who will be able to focus on particular biological problems in a whole-organism context.

Finally, once we learn to evolve and grow virtual biobots by mimicking biological structures and processes, there is no reason to be constrained by biology. Instead of looking to biology to solve technological problems we can allow self-organizing systems to develop outside the constraints of biological evolution, and build new kinds of information technology. As Steven Vogel points out, historical claims to have developed new technology by copying nature generally turn out to be misguided or overblown. Innovative engineering requires finding new ways to put together what you have as much as gathering and tabulating new data and mimicking existing systems. It would be a mistake to focus entirely on 'biomimetics' and not pay attention to current engineering theory and technologies. Ultimately virtual biobots may be most important as a means to an end – providing new technology for information processsing systems engineering – and not just an end in themselves. In the meantime, they are fun to build and play with.

Acknowledgements

Supported by the University of Otago Division of Sciences; Committee for the Advancement of Learning and Teaching; New Zealand Dental Research Foundation. Thanks to Russell Barnett of the Department of Anatomy and Structural Biology for providing plastinated tissue sections.

References

1. Deacon T.W. (1990) Rethinking mammalian brain evolution. American Zoologist 30(3):629-705.
2. Deacon, T.W. (1998) The Symbolic Species: The Co-evolution of language and the brain. W.W. Norton and Co.
3. Kalman, R.E. (1969) Elementary control theory from the modern point of view. In: Kalman, Falb and Arbib: Mathematical System TheoryMcGraw-Hill, New York. pp25-66.
4. Kalmijn, A.J. (1982) Electro-perception in sharks and rays. Nature 212:1232-1233.
5. Kauffman, S.A. (1993) The Origins of Order. Oxford University Press.
6. Kauffman, S.A. (1995) At Home in the Universe. Viking, New York.
7. Mandler, J.M. (1992) How to build a baby. Psychological Review 99(4): 587-604.
8. Montgomery J.C. and Bodznick D. (1999) Signals and noise in the elasmobranch electrosenspry system. Journal of Experimental Biology 202: 1349-1355.
9. Montgomery J.C. and Bodznick D. (1994) An adaptive filter that suppresses self-induced noise in the electrosensory and lateral line mechanosensory systems of fish. Neuroscience Letters 174: 145-148.
10. Nelson M.E. and Paulin M.G. (1995) Neural simulations of adaptive reafference suppression in the elasmobranch electrosensory system. J. Comp. Physiol. A. 177: 723-736.
11. Paulin, M.G. (1993) The role of the cerebellum in motor control and perception. Brain, Behaviour and Evolution. 41(1):39-50.
12. Paulin, M.G. (1993b) A method for constructing data-based models of spiking neurons using a dynamic linear – static nonlinear cascade. Biol. Cybern. 69:67-76.
13. Paulin M.G. (1996) System identification of spiking neurons. In: Advances in Processing and Pattern Analysis of Biological Signals. I. Gath and G. Inbar (eds). Plenum, New York. pp183-194.
14. Paulin, M.G. (1997) Neural representations of moving systems. International Review of Neurobiology. 41:515-533.
15. Paulin, M.G. (1997b) The evolutionary origin of image schemas. New Ideas in Psychology 15(2):133-136.
16. Paulin M.G. and Hoffman L.R. (1998) Modelling the firing pattern of bullfrog vestibular neurons responding to naturalistic stimuli. Neurocomputing26: 223-228.
17. Paulin, M.G., Senn, W., Yarom, Y., Meiri, H. and Cohen, D. (1998) A model of how rapid changes in local input resistance of shark electrosensory neurons could enable the detection of small signals. In: Computational Neuroscience. J. Bower (ed) Plenum, New Yor. pp239-244.

18. Raff, R.A. (1996) The Shape of Life. University of Chicago Press.
19. Ramsdell A.F. and Yost, H.J. (1998) Molecular mechanisms of vertebrate left-right development. Trends in Genetics 14(11):459-465.
20. Stewart, I. Life's Other Secret (1997). John Wiley and Sons Inc.
21. Thompson, D.W. (1961) On Growth and Form. (New edition edited by J.T. Bonner) Cambridge University Press.
22. Vandervert, L. (1997) The evolution of Mandler's conceptual primitives (image-schemas) as neural mechanisms for space-time simulation structures. New Ideas in Psychology 15(2):105-123.
23. Vogel, S. (1998) Cat's paws and catapaults: Mechanical Worlds of Nature and People. W.W. Norton and Co.

Part III

New Connectionist Computational Paradigms: Brain-Like Computing and Quantum Neural Networks

Chapter 10. Future Directions for Neural Networks and Intelligent Systems from the Brain Imaging Research

John G. Taylor

Department of Mathematics, King's College, Strand, London WC2R2LS, UK
john.g.taylor@kcl.ac.uk

Abstract. An overview is given of recent results coming from non-invasive brain imaging (PET, fMRI, EEG & MEG), and how these relate to, and illuminate, the underpinning neural networks. The main techniques are briefly surveyed and data analysis techniquest that are presently being used are reviewed. The results of the experiments are then summarised. The most important recent technique used in analysing PET and fMRI, that of structural modelling, is briefly described, results arising from it presented, and the problems this approach presents in bridging the gap to the underlying neural networks of the brain described. New neural networks approaches are summarised which are arising from these and related results, especially associated with internal models. The relevance of these for indicating future directions for the development of intelligent systems concludes the article.

Keywords. Brain imaging, structural models, neural networks, working memory, internal models, intelligent systems

1 Introduction

There is increasing information becoming available from functional brain imaging on how the brain solves a range of tasks. The specific brain modules active while human subjects solve various cognitive tasks are now being uncovered [1],[2]. The data show that there are networks of modules active during task solution, with a certain amount of overlap between networks used to solve different problems. This use of non-invasive imaging to give a new 'window' on the brain has aroused enormous interest in the neuroscience community, and more generally begun to shed light on the way the brain solves hard tasks. The nature of the networks used to support intelligent processing is now being explored by various means, especially in terms of the creation of internal models of the environment. This indicates a clear direction for the future development of intelligent machines, a feature to be considered at the end of the paper.

There are many problems to be faced in interpreting functional brain images (those obtained whilst the subject is performing a particular task). Although the experimental paradigms used to obtain the functional brain imaging data already contain partial descriptions of the functions being performed by the areas detected, the picture is still clouded. The overall nature of a task being performed while subjects are being imaged can involve a number of more primitive sub-tasks which themselves have to be used in the determination of the underlying functions being performed by the separate modules. This means that there can be several interpretations of the roles for these modules, and only through the convergence of a number of experimental paradigms will it be possible to disentangle the separate primitive functions.

It is the hope that use of the latest techniques of analysis of resulting data, as well as the development of new techniques stretching the machines to their very limits, will allow solution to these problems of ambiguity, and a truly global model of the brain will result. The paper starts with a review of brain imaging techniques and data analysis. A survey of results obtained by analysis of PET and fMRI data, termed structural modelling is then given. This approach has the potential to lead to a global processing model of the brain. As part of that, the connection between structural models and the underlying neural networks of the brain is then explored. Several recent brain imaging results are then considered which indicate new neural network architectures and processing styles. The manner in which the brain uses internal models of the environment, including its own active sensors and effectors as well as of its inputs, is then explored, leading to a general program of analysis of the global processing by the brain. This leads to a possible approach to building intelligent systems, which is discussed briefly in the final section.

2 A Survey of Brain-Imaging Machines

2.1 PET & fMRI

These machines investigate the underlying neural activity in the brain indirectly, the first (the acronym PET= positron emission tomography) by measuring the 2 photons emitted in the positron annihilation process in the radio-active decay of a suitable radio-nuclide such as $H_2\ ^{15}O$ injected into a subject at the start of an experiment, the second (fMRI = functional magnetic resonance imaging) that of the uneven distribution of nuclear spins (effectively that of the proton) when a subject is in a strong magnetic field (usually of 1.5 Tesla, although a 12 T machine is being built especially for human brain imaging studies). The PET measurement allows determination of regions of largest blood flow, corresponding to the largest 2-photon count. The fMRI measurement is termed that of BOLD (blood oxygen-level dependent). This signal stems from the observation that during changes in neuronal activity there are local changes in the amount of oxygen in tissue, which alters the amount of oxygen carried by haemoglobin, thereby disturbing the local magnetic polarisability.

Spatially these two types of machines have a few millimetres accuracy across the whole brain. However temporally they are far less effective. PET measurements need to be summed over about 60-80 seconds, limiting the temporal accuracy considerably. fMRI is far more sensitive to time, with differences in the time of activation of various regions being measurable down to a second or so by the 'single event' measurement approach. This has already produced the discovery of dissociations in the time domain between posterior and anterior cortical sites in working memory tasks [3].

That regions of increased blood flow correspond exactly to those of increased neural activity, and these also identify with the source of the BOLD signal, is still the subject of considerable dispute. The siting of the BOLD signal in the neurally most active region was demonstrated recently [4] by a beautiful study of the positioning of the rat whisker barrel cortex from both 7T fMRI measurement and by direct electrode penetration. The present situation was summarised effectively in [5], with a number of hypotheses discussed as to the sources of the blood flow and BOLD signals. I will assume that these signals are all giving the same information (at the scale of size we are considering).

Many cognitive studies have been performed using PET; there are now as many using fMRI, some duplicating the PET measurements. These results show very clear localisation of function and the involvement of networks of cortical and subcortical sites in normal functioning of the brain during the solution of tasks. At the same time there has been considerable improvement in our understanding of brain activity in various mental diseases, such as schizophrenia, Altzheimer's and Parkinson's diseases. There have also been studies of patients with brain damage, to discover how the perturbed brain can still solve tasks, albeit slowly and inefficiently in many cases.

2.2 MEG & EEG

The magnetic field around the head due to neural activity, although very low, is measurable by sensitive devices, such as SQUIDs (superconducting quantum interference devices). Starting from single coils to measure the magnetic field at a very coarse level, MEG (MEG = magneto-encephalography) measurements are now being made with sophisticated whole-head devices using 148 [6] or even 250 measuring coils [7]. Such systems lead to ever greater spatial sensitivity, although they have a number of problems before they can be fully exploited. In particular it is first necessary to solve the inverse problem, that of uncovering the underlying current sources producing the magnetic field. This is non-trivial, and has caused MEG not to be as far advanced in brain imaging as PET and fMRI systems. However that situation is now changing. There is good reason to bring MEG up to the same standard of data-read-out simplicity as PET and fMRI since, although it does not have the same spatial sensitivity as the other two it has far better temporal sensitivity - down to a millisecond. Thus MEG fills in the temporal gap on the knowledge gained by PET and fMRI. This is also done by EEG (EEG = electro-encephalography), which is being consistently used by numbers of groups in partnership with PET or fMRI so as to determine the detailed time course of activation of known sites already implicated in a task by the other machines [2].

3 Data Analysis

3.1 Statistical Parameter Maps (SPMs)

The data arising from a given experiment consists of a set of data points, composed of a time series of activations collected during an experiment, for a given position (termed a pixel) in the head. The data is reasonably immediate for PET and fMRI, although there must be careful preprocessing performed in order to remove movement artefacts and to relate the site being measured to its co-ordinates in the head according to some standard atlas. All the machines now use a static MRI measurement of the brain of a subject as a template, to which the data being measured are referred. Some analyses use a more detailed warping of the brain of a given subject to that of a standard brain, as given especially by the standard brain atlas arrived at by the anatomical analysis of a number of human brains [8]. However this can introduce distortions so a time-consuming but more accurate method is to identify similar regions from the brains of different people by common landmarks, and then compare (or average) the activity at the similar regions decided on in this manner over the group of subjects to remove noise. The use of independent component analysis (ICA) has recently proved of value in data cleaning and especially in the removal of artefacts such as effects of breathing and heart beat.

The data at a given pixel is a set of activation levels. These have been obtained from a number of measurements taken under a set of known conditions. For example, I am involved in studying the motion after-effect (MAE) by fMRI [9]. This occurs due to adaptation to movement in one direction, and arises, for example, if you look at a waterfall for a period of about 20 or so seconds and then turn your gaze to the side. The static rock face then seems to move upwards for about 10 seconds afterwards. Our measurements were taken using an 'on-off' paradigm, with a set of moving bars being observed by a subject during 10 measurements (each lasting 3 seconds) , then 10 with static bars, then 10 with the bars moving up and down, then another 10 with static bars. The MAE occurs at the end of a period of movement in one direction only, so the purpose of the study was to determine the change of BOLD signal after the one-way movement, in comparison to the two-way movement. Regions were searched for with a suitably high correlation of their activation with the 'box-car' function, equal to +1 during movement and just afterwards and -1 during other static periods and the up-and-down movement. This correlation at each pixel in the head leads to the statistical parameter map of the heading of this sub-section. Significance levels can then be attached to the value at any one point in terms of the difference of that value as compared to that for a control condition in which there is no condition of interest being applied. This leads to the standard t-test and to maps of t- or z-parameters throughout the brain. More sophisticated analysis can then be performed to detect regions of interest (a number of adjacent pixels with significant z-scores) and their significance. There are now a number of software packages available to perform such analysis, and they have been recently been compared [10]. Fast data analysis techniques are now available so that on-line results can be obtained and thereby allow for optimisation of paradigms being employed [11].

3.2 The Inverse Problem

There is a hard inverse problem - to determine the underlying current distribution causing the observed field strengths - for both EEG and MEG. That for the former is more difficult due to the conduction currents that 'wash out' the localisation of deep sources. This does not occur for MEG, but there is still the problem that certain currents, such as radial ones in a purely spherical head, are totally 'silent', leading to no external magnetic field. Modulo this problem, the standard approach to uncovering the underlying neural generators has been to assume that there is a limited set of current dipoles whose parameters (orientation, strength and position) can be varied to optimise the mean square reconstruction error (MSE) of the measured data. Such approaches are limited, especially when fast temporal effects are being investigated. More recently magnetic field tomography (MFT) has been developed to give a distributed source representation of the measurements [12].

The use of MFT and similar continuous source distribution systems is now becoming more commonplace in MEG, so that high-quality data are now becoming available for an increasing number of cognitive tasks similar to those from fMRI and PET but with far greater temporal sensitivity.

3.3 Structural Modelling

Initially the results of PET and fMRI studies have uncovered a set of active brain sites involved in a given task. This is usually termed a 'network', although there is no evidence from the given data that a network is involved but only an isolated set of regions. It is possible to evaluate the correlation coefficients between these areas, either across subjects (as in PET) or for a given subject (in fMRI). There is great interest in using these correlation coefficients to determine the strength of interactions between the different active areas, and so uncover the network involved. Such a method involves what is called 'structural modelling', in which a linear relation between active areas is assumed and the path strengths (the linear coefficients in the relation) are determined from the correlation matrix. In terms of the interacting regions of figure 1:

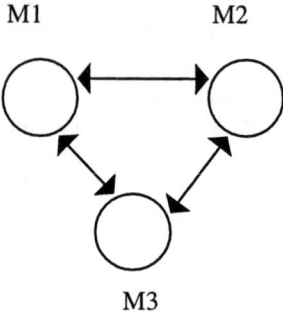

Figure 1. A set of three interacting modules in the brain, as a representative of a simple structural model. The task of structural modelling is to determine the path strengths with which each module affects the others in terms of the cross-correlation matrix between the activities of the modules.

Corresponding activities z_i (i =1,2,3) of the modules satisfy a set of linear equations

$$z_i = \Sigma\, a_{ij}\, z_j + \eta_i$$

where the variables η_I are assumed to be independent random noise variables. It is then possible to calculate the coefficients a_{ij} from the correlation coefficients $C(i,j)$ between the variables z. This can be extended to comparing different models by a chi-squared test so as to allow for model significance testing to be achieved.

3.4 Bridging the Gap

An important question is as to how to bridge the gap between the brain imaging data and the underlying neural network activity. In particular the interpretation of the structural model parameters will then become clearer. One way to achieve this is by looking at simplified neural equations which underpin the brain activations. We can describe this activity by coupled neural equations, using simplified neurons:

$$\tau\, dU(i, t)/dt = - U(i,t) + \sum_{neurons, j} C(i,j)\, f[U(j, t-t(i,j))]$$

where U(i , t) is the vector of membrane potential of the neurons in module i at time t, C(i ,j) is the matrix of connection strengths between the modules, and t(i,j) the time delay (assumed common for all pairs of neurons in the two modules).

There are various ways we can reduce these coupled equations to structural model equations. One is by the mean field approximation <U> = u**1** (where **1** is the vector with components equal to 1 everywhere) so that the particular label of a neuron in a given module is averaged over. The resulting averaged equations are:

$$\tau \, du(i,t)/dt = -u(i,t) + \sum_j c(i,j) \, f[u(j, t-t(i,j))]$$

where c(i ,j) is the mean connection strength between the modules i and j. In the limit of $\tau = 0$ and for linearly responding neurons with no time delay there results the earlier structural equations:

$$u(i,t) = \sum_j C(i,j) \, u(j,t)$$

These are the equations now being used in PET and fMRI. The path strengths are thus to be interpreted as the average connection weights between the modules. Moreover the path strengths can be carried across instruments, so as to be able to build back up to the original neural networks. This involves also inserting the relevant time delays as well as the connection strengths to relate to MEG data.

The situation cannot be as simple as the above picture assumes. Connection strengths depend on the paradigm being used, as well as the response modality (such as using a mouse versus verbal report versus internal memorisation). Thus it is necessary to be more precise about the interacting modules and their action on the inputs they receive from other modules; this will not be a trivial random linear map but involve projections onto certain subspaces determined by the inputs and outputs. The above reduction approach can be extended to such aspects, using more detailed descriptions of the modules [14].

4 Results of Static Activation Studies

There are many psychological paradigms used to investigate cognition, and numbers of these have been used in conjunction with PET or fMRI machines. These paradigms overlap in subtle ways that it is difficult at times to 'see the wood for the trees'. To bring in some order we will simplify the case by considering a set of categories of cognitive tasks. We do that initially along lines suggested by Cabeza and Nyberg [15], who decomposed cognitive tasks into the categories of:

- attention (selective/sustained)

- perception (of object/face/space/top-down)
- language (word listening/word reading/word production)
- working memory (phonological loop/visuospatial sketchpad)
- memory (semantic memory encoding & retrieval/episodic memory encoding & retrieval)
- priming
- procedural memory (conditioning/skill learning)

So far the data indicate that there are sets of modules involved in the various cognitive tasks. Can we uncover from them any underlying functionality of each of the areas concerned? The answer is a partial 'yes'. The initial and final low-level stages of the processing appear more transparent to analysis than those involved with later and higher level processing. Also more study has been devoted to primary processing areas. Yet even at the lowest entry level the problem of detailed functionality of different regions is still complex, with about 30 areas involved in vision alone, and another 7 or 8 concerned with audition. Similar complexity is being discerned in motor response, with the primary motor area being found to divide into at least two separate subcomponents. However the main difficulty with the results of activated areas is that they are obtained by subtraction of a control condition, so that areas activated in common will disappear in the process. That can be avoided by using structural modelling introduced in the previous section. We will therefore turn to survey the still small but increasing sets of data now being analysed by that approach.

5 Structural Models of Particular Processes

5.1 Early Spatial and Object Visual Processing

This has been investigated in a series of papers concerned with the dorsal versus ventral routes of visual processing as related to spatial and object processing respectively. The object experiment used simultaneous matching of a target face to two simultaneously presented faces. The spatial task used a target dot located in a square with one side a double line to be compared to a simultaneously presented pair of similar squares containing dots; the matching test stimulus had a dot in the same position as the test stimulus in relation to the double line. The results of these researches [17], [18] showed that there are indeed two pathways for spatial and object processing respectively, the former following the ventral pathway from the occipital lobe down to the temporal lobe, the latter the dorsal route from occipital up to the parietal lobe, as shown in figure 2.

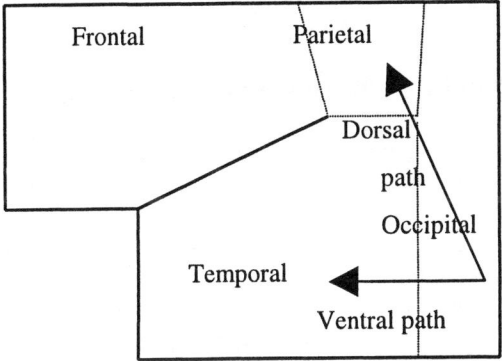

Figure 2. A schematic of the cortex, showing its subdivision into the frontal, parietal, temporal and occipital lobes, and the vantral and dorsal routes for visual input to further cortex from early visual processing.

5.2 Face Matching

A careful study was performed as to how the paths involved change with increase of the time delay between the presentation of one face and the pair to which the original target face must be matched [19]. The delays ranged from 0 to 21 seconds, with intermediate values of 1, 6, 11 and 16 seconds. The lengthening of the time duration for holding in mind the target face caused interesting changes in the activated regions and their associated pathways. In particular the dominant interactions during perceptual matching (with no time-delay) are in the ventral visual pathway extending into the frontal cortex and involving the parahippocampal gyrus (GH) bilaterally, with cingulate (in the middle of the brain). The former of these regions is known to be crucial for laying down long-term memory, the latter for higher-order executive decision-making. This activity distribution changed somewhat as the delays increased, first involving more frontal interactions with GH in the right hemisphere and then becoming more bilaterally symmetric and including cingulate-GH interactions, until for the longest delays the left hemisphere was observed to have more interactions with GH and anterior occipital areas on the left, although still with considerable feedback from right prefrontal areas. These changes were interpreted [19] as arising from different coding strategies. At the shortest time delays the target face can be held by a subject in an 'iconic' or image-like representation, most likely supported by right prefrontal cortices. As the time delay becomes larger there is increasing difficulty of using such an encoding scheme, and the need for storage of more detailed facial features and their verbal encoding. That will need increasing use of the left hemisphere. However there is also a greater need for sustained attention as the duration gets longer, so requiring a greater use of right hemisphere mechanisms.

5.3 Memory Encoding and Retrieval

There is considerable investigation of this important process both for encoding and retrieval stages. In earlier imaging experiments there had been a lot of trouble in detecting hippocampal regions in either hemisphere during memory processing. The use of unsubtracted activations leads to clear involvement of these regions in both stages, as is clear from the careful study of Krause et al using PET [20]. This has resulted in a structural model for both the encoding and retrieval stages. There is considerable complexity in these structures. However it can be teased out somewhat by the use of various quantifications, such as by my introduction of the notion of the total 'traffic' carried by any particular site [21] (defined as the number of activated paths entering or leaving a given area). This allows us to conclude:

- there is a difference between the total traffic for encoding and retrieval between the two hemispheres, where in encoding the L/R ratio for the total traffic in all modules is 51/40 while for retrieval it has reversed to 46/50;
- the anterior to posterior cortex ratio is heavily weighted to the posterior in retrieval, the anterior/posterior traffic ratios for encoding being 32/38 and for retrieval 28/54;
- the sites of maximum traffic are in encoding left precuneus (in the middle of the brain at the top of the parietal lobes) with a traffic of 11, and right cingulate (with 8) and in retrieval the right precuneus (with 11) and its left companion (with 9). By comparison the hippocampal regions have traffic of (5 for R, 6 for L) in encoding and (8 for L, 6 for R) in retrieval.

We conclude that the first of these result is in agreement with the HERA model of Tulving and colleagues [22]: there is stronger use of the left hemisphere in encoding while this changes to the right hemisphere in retrieval. As noted by Tulving et al [22] there is considerable support for this asymmetry from a number of PET experiments. The structural models show that in both conditions there is more bilateral prefrontal involvement.

The anterior to posterior difference is not one which has been noted in the context of the paired-associate encoding and recall tasks. However there are results for posterior/anterior asymmetry associated with the complexity of the working memory tasks, as in the n-back task [23] (where a subject has to indicate when a particular stimulus of a sequence of them has occurred n times previously in the sequence). When n becomes greater than 2 (so for delay times in the task longer than 20 seconds) it has been reported that anterior sites become activated and there is an associated depression in posterior sites which were previously active for the case of n=0 and 1. The difference between the anterior and posterior traffic in our case indicates that the processing load is considerably lower in the retrieval part of the task than in the encoding component. This is consistent with expectations: the hard work in the task involves setting up the relations between the words as part of the encoding process. Once these relations are in place then there can be a reasonable level of automaticity.

5.4 Hearing Voices in Schizophrenia

A study has been reported by Bullmore and colleagues [24], in which subjects had to decide, for a set of 12 words presented sequentially, which of them referred to living entities, which to non-living ones. Subjects were asked to rehearse their reply subvocally. This was compared to a baseline condition in which subjects looked at a featureless isoluminant screen. A path analysis of the data showed a network involving extrastriate cortex ExCx, posterior superior temporal gyrus STG (Wernicke's area), dorsolateral prefrontal cortex DLPFC, inferior frontal gyrus IFG (Ba 44/45 L) and supplementary motor area SMA, where B, L and M denote bilateral, left-sided or mesial respectively.

The path analysis for normal subjects showed the flow pattern

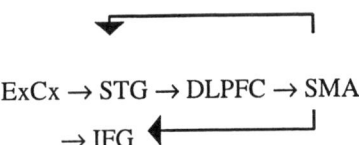

whereas for schizophrenic subjects there was no such clear path model in which there was SMA feedback to IFG and STG. This fits well with the model of Frith [25] of lack of control of feedback from the voice production region by SMA to STG in schizophrenics. They do not know they are producing internal speech and think they are hearing voices speaking to them. Awareness of this process could thus be in STG.

6 New Paradigms for Neural Networks?

The results now pouring in from brain imaging, as well as from single cell and deficit studies, lead to suggestions of new paradigms for neural networks. In brief, some of these are:

recurrent multi-modular networks for temporal sequence processing, based on cartoon versions of the frontal lobes, with populations of excitatory and inhibitory cells similar to those observed in cortex and basal ganglia. These are able to model various forms of temporal sequence learning [26] and delayed tasks and deficits observed in patients [27].

Attention is now recognised as sited in a small network of modules: parietal and prefrontal lobes and anterior cingulate. This supports the 'central representation' [28], with central control modules (with global competition) coupled to those with semantic content, and extends feature integration..

Hierarchically coded modules, with overall control being taken by a module with (oscillatory) coupling to the whole range of the hierarchy

However the above set of paradigms do not take account of the input→output nature of the brain, which has developed across species so as to be increasingly effective in granting its possessor better responses for survival. One of the most important of these is, through increased encephalisation, the ability of the brain to construct internal models of its environment. This includes its own internal milieu, leading to ever better control systems for response in the face of threats. Such internal models are now being probed by a variety of techniques that are increasingly using brain imaging to site the internal models. It is these models which will be elaborated on next before being used to give a more general view of intelligent brain processing. From this picture it will be possible to obtain a view of intelligent processing in a more general framework than up to now.

6.1 Spatial Arm Movement Estimation

There are many paradigms involving motor actions (posture control, eye movements of various sorts,..) which have led researchers to suppose that internal models of the motor effectors were involved. One important case study involved estimation of how far a subject has moved their hand in the dark during a time varying from one trial to another [29]. The accuracy of this assessment was tested by the subject giving an estimate of this position by moving a marker using the other hand. It was found that the mean error (obtained by averaging over several trials) increased as the duration of time for the movement increased till a maximum value was reached and then remained about constant. The amount of spread in this error had a similar behaviour.

To explain this result it was assumed that an internal model of the hand position had been already created in the brain, so that it could be updated by feedback from expected sensory signals arising in the arm from the change of its position as compared to further actual feedback from the proprioceptive sensors in the arm muscles activated during the movement. There would also be visual feedback, although in this paradigm (carried out in the dark) this would not be present (but will be used in the following paradigm). The total model of this system is shown in figure 3.

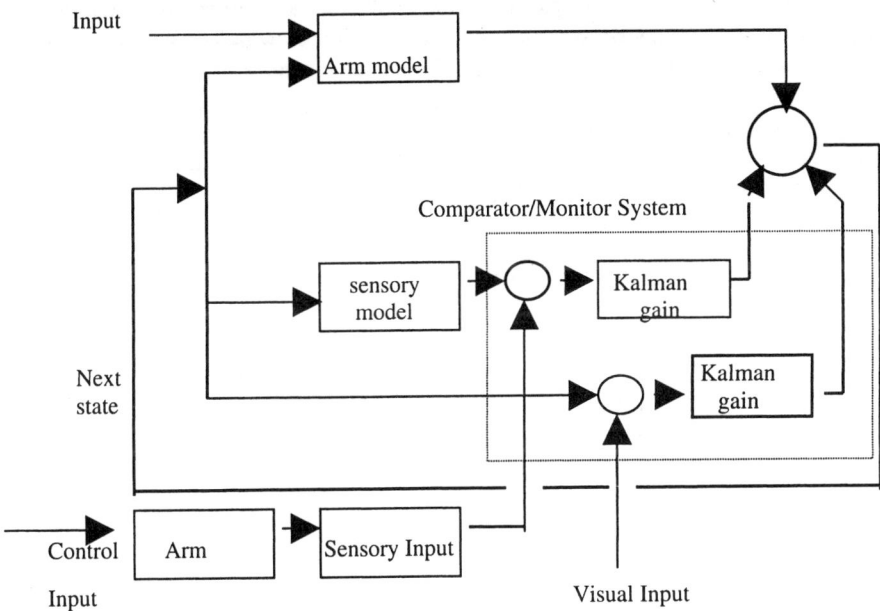

Figure 3. Internal arm control model. The arm model in the upper box updates the position of the arm by a linear differential equation, using the external control signal (from elsewhere in the brain). Feedback from sensory and visual input is used to give correction terms (the 'Kalman gain' terms) which are added to the next state value arising from the linear arm model, to give an estimate of the next arm position. This is then used for further updating.

The over-all system of figure 3 uses a linear differential equation for obtaining an initial update of the position $x(t)$ of the arm at time t:

$$d^2 x/dt^2 = -(b/m)dx/dt + F \qquad (1)$$

where m is the weight of the arm, F the force being applied to it by the muscles and b a constant. The first term on the right hand side of the equation is a counter-force arising from the inertia of the arm muscles. The gain term is of form

$$K(t)[y - Cx(t)] \qquad (2)$$

where C is the scaling factor modifying the position to give the expected position from proprioceptive feedback signals up the arm to the brain. The gain factor $K(t)$ is time dependent, with time dependence obtainable from a well-defined set of equations (termed the Kalman filter equations) [30]. The results of using this model to determine the time development of the mean value of $x(t)$ and of the variance of $x(t)$ were in close agreement with the observed curves from these quantities for a set of subjects [29].

This result is strong, but indirect support, for the existence of internal models of motor action in the brain. In particular they indicate the presence of a monitor system, enclosed in the broken-line rectangle in fig 2, which assesses errors between predicted values of feedback and the actual feedback experienced. Both this and the internal arm model, however, have not been directly observed in the brain. To develop this 'internal model' or so-called 'observer' approach [30] further other evidence will now be considered which helps to site the internal model and monitor system more precisely.

6.2 Accuracy of Arm Movement

The above model can be applied to another set of results, now with comparison having been made with subjects with certain deficits. The paradigm being considered is that of the movement of a pencil held in the hand from a given start position to a marked square on a piece of paper. If the side of the marked square to be moved to is denoted by W and A is the amplitude of the total arm movement from the start position to the centre of the marked square, then the time T taken to make the movement to the marked square increase as the width W of the square is reduced according to what is known as Fitt's law [31]:

$T = a + b\ln[A/W]$

where ln is the natural logarithm, a and b are constants with b being positive.

An extension of this paradigm has been used by asking subjects to imagine they were making the movement, not actually moving their arm at all, and stating when they have reached, in their mind's eye, the required position in the target square. The results for normals is that there imagined movement times still obey Fitt's law, with the same constants a and b above. However there are two sorts of deficits which cause drastic alterations to this result [31]:

a) for patients with motor cortex damage, Fitt's law still describes the relation between movement time and width of the target square, but now the time for both the actual and imagined movement (still closely the same for each value of the width W of the target square) is considerably slower for the hand on the opposite side to the cortical damage as compared to that on the same side as the damage (where there is a cross-over of control from side of the brain to side of the body). In other words the movement is slowed - noted by subjects as feeling as they are moving the hand through 'glue' - but there is no difference between imagined and actual movement. Thus the site for 'imagining' has not been damaged.

b) for patients with parietal damage this relation between actual and imagined movement is broken. There is still slower actual movement of the opposite hand to the damaged cortex as compared to that on the same side as the damage. However there is now no dependence on the width of the target square for making the imagined movement with the hand on the opposite side to the damage. Fitt's law has been broken in this case. The site of imaginary location of the hand as it makes it movement is no longer available.

We conclude that

1) the comparator/monitor system used in estimating the position of the hand in relation to the square is contained in the parietal lobe. Loss of this region leads to no ability to assess the error made in making the imagined movement.

2) the internal arm model itself (of equation 1 above) is sited in the motor cortex. Its degradation leads to an increase of the friction term b in the above model, as reported in the experience of patients with such damage.

This approach through observer-based control theory needs to be extended, or blended in, to other brain processes. In particular it is needed to relate the internal models described above to the results of coupled networks of active modules observed by structural models and recounted in section 5 above. This extension is difficult to achieve since there is still little deep understanding of the underlying functionality of the various regions being observed. However such a blending will be started here, commencing with a small extension to the case of fading of sensations in a patient with parietal damage.

6.3 The Fading Weight

The nature of continued reinforcement of activity allowing the continued experience of an input has been demonstrated in a remarkable fashion by a case of fading of the experience of a weight placed on the hand of a patient who had suffered severe loss of parietal cortex due to a tumour in that region [32]. When a weight was placed on her hand the sensation of it faded, in spite of her directed attention to it, in a time proportional to the mass of the object placed in her hand: a 50 gram weight faded in 5 seconds, a 100 gram one in 8 seconds, a 150 gram weight in 10 seconds, and so on.

It was suggested that such fading occurs due to the presence of a store for holding an estimate of the state of the hand (in this case the mass it is supporting) which has been lost by the patient with parietal lobe damage. In that case, with no such store available, the nature of the model of equation (1) indicates that the natural time constant is m/b. Thus it is to be expected, on that model, that such fading would occur (with no store) in a time proportional to the mass placed on the hand. In the presence of a store there will be active refreshment of such knowledge, so avoiding the fading of experience in such a fashion. Active refreshment is a process that again requires action, as did the arm motion in section 6.1. It is to be expected that such 'action' is in motor cortex or related frontal areas. We turn in the next sub-section to consider experimental evidence for the presence of processes in frontal lobes achieving actions on activity in posterior regions.

6.4 Rotating in the Mind

Working memory [33] is a term introduced in the 70's to describe short-term memory, as used, say, when you look up a phone number and hold it in mind until you have made the phone call. You need not make a permanent memory of the number; it fades from your mind afterwards unless you deliberately attempt to remember it for later use. Such working memory has two components: one involved a buffer store, in which items are activated by decay away in a few seconds if not refreshed. Such buffer working memory stores have been observed by brain imaging in a number of modalities, such as for words (the phonological store) for space (the visuo-spatial sketch-pad) for timing of signals (on the left hemisphere). In general these buffer sites are placed in the parietal lobes [34]. At the same time there is a more active system of refreshment in the frontal lobes, so that when you held in mind the phone number it could be rehearsed (subvocally) so as to hold it over a number of seconds if need be.

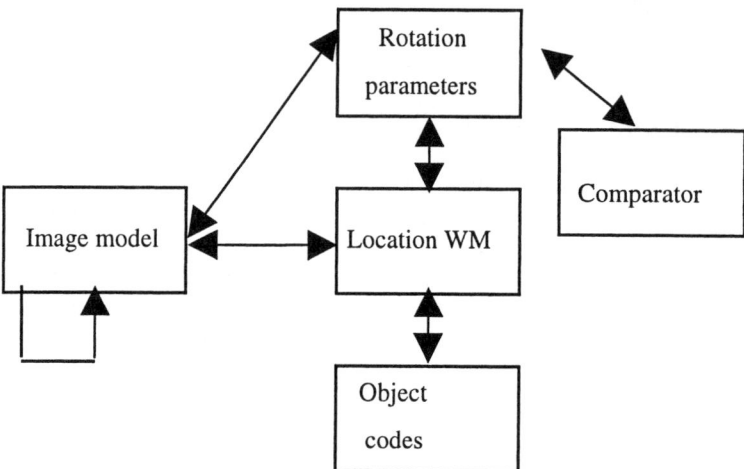

Figure 4. Internal model of processing for the internal rotation of inputs associated with comparing pictures of two identical solid objects rotated/reflected with respect to each other. The image model is used both to refresh the image of the object held in the location module, in conjunction with the object code module and the rotation parameter module, and to achieve rotation of this image by augmenting the appropriate parameter. The comparator module is used to determine when the rotation has been completed and alignment between the two objects achieved (or not).

The manner in which working memory is involved, and its sites in the brain exposed, was made clear in a beautiful fMRI brain imaging experiment of rotating images in the mind to compare two three dimensional figures, seen at different angles, as the same or mirror images [35]. It was found that the time for response increased linearly with the amount one figure had been rotated from the other. On imaging the brain, a number of regions were observed to have activity which also increased linearly with the angle required to line up the two figures with each other. This was especially clear in both parietal lobes, as well as to a lesser extent in the inferior temporal and motor cortical areas.

These results can be understood in similar terms to the internal action model of figure 3 by the model of figure 4.

The internal model works in a similar manner to that of figure 3, in that it has a module, entitled the 'image model', which has a representation of the dynamical equation for updating the position of the input image, in this case just the rotation parameter θ. The equation of motion being implemented by this model is the linear equation

$d\theta/dt = \omega$

where ω is the angular velocity used in the internal rotation process. It is very likely that ω is determined by eye movement machinery, as part of preparing saccades. The increase of the rotation parameter θ by the dynamics of frontal lobe activity (to which we turn in a moment) will be sent back to the rotation parameter module, which then acts to rotate the spatial image being held on the location working memory, being guided by the object codes module to keep the shape of the object during rotation consistent with its original shape, although now with a rigid rotation added. During this process the comparator module assesses the effects of the rotation change and brings it to s halt when there is identification between the rotated and stationary image. All the time there must be rehearsal of the held images (or just the rotated on, the static one being referred to from the outside picture). This rehearsal must also involve an activity, so will be in a frontal site involving motor acts.

From the above account the identifications can be made:

1) the object code module is contained in the temporal lobe;

2) the location working memory module is in the parietal lobe (possibly in inferior parietal lobe);

3) the rotation parameter module is also in the parietal lobe (possibly in superior parietal lobe);

4) the comparator is also in the parietal lobe (as part of the monitoring system of sub-section 6.2);

5) the active rotation is performed in a frontal site, such as the supplementary eye field, where saccades are programmed.

The above identifications are based on the identifications already made in the previous sub-section. However we can add further support for placing the active module of 5) above in the frontal lobes in terms of the general recurrent features of the frontal lobe connectivity. This is well known to be in terms of a set of recurrent neural loops involving not only the frontal cortices but also sub-cortical structures termed the basal ganglia and the thalamus. The latter is a way-station for all activity entering the cortex from outside, the former are involved in the recurrent pathway

cortex→basal ganglia→thalamus→cortex

This set of loops is in actuality far more complicated than the above simple flow diagram would indicate, and distinguishes the frontal lobes from posterior cortex, which lacks any similar recurrent system with basal ganglia as a controlling and guiding unit. Models of this frontal system indicate its power to enable the learning of temporal structure in response patterns to inputs, a quality lacking in the posterior cortex [36].

Such powers the frontal system posses indicate it is ideal to support an iterative updating of activity denoting a state of the body (or other parts of the brain). Thus the above equation of rotation parameter update can be written as the updating equation

$\theta(t+\Delta t) = \theta(t) + \omega$

which can be seen to be achieved by updating the value of θ at any time by addition of the amount ω to it. Such a process can be achieved very straightforwardly by a linear neuron coding for θ being updated cyclically by a constant input ω input every time step. A similar interpretation of recurrent updating can be given for the mode of computation of the linear internal arm model considered in sub-section 6.1.

The various components of the model of figure 4 will have activity which in general will increase as the required amount of rotation is increased. The details of such increase still have to be worked out, and will add a valuable constraint to the modeling process, as well as amplify the sites involved and their functionality.

6.5 Conclusion on Processing

A number of experimental supports have been given for a style of processing in the brain based on internal 'active' models of either body parts or other brain activities, from buffer working memory sites. The actions involved have so been modeled above by simple linear differential equations; Th. true from of these models will in general be much more complex. However that is not a problem since neural networks are able to act as universal approximators of any dynamical system. These active models, based in the frontal lobes and depending on recurrent circuits for their powers. are coupled to posterior codes for inputs. Some of these codes have a categorical form, such as of object representations. Others are of a spatial form, giving the actual spatial parameters of the inputs to be available for active grasping, reaching or other actions to be made on the inputs. Finally there are also posterior monitoring systems for detecting mismatches and making corrections to frontal actions, which for fast response are based on assessments of inputs and their changes. The expected changes then have to be matched to actual changes as measured by inputs from outside (or from proprioceptive inputs from the body). The matching process takes place in posterior sites in the cortex.

In general, then, we have a division of labour in the brain:

FRONTAL	POSTERIOR
Active, updating posterior representations	Passive, creating & holding representations, & monitoring the relation between frontal and posterior activities
internal model state updating, rotation, rehearsal	preprocessing, object representations, spatial maps
maximum time scale: many seconds	maximum time scale: 1-2 seconds

It is possible to observe that the development of the action-frontally-based models of the above sub-sections attack the problem of understanding the networks of the brain in a different fashion from those described in section 4. There the structural models had a timeless character. Now the models from this section

7 Conclusions

There are considerable advances being made in understanding the brain based on brain imaging data. This is made particularly attractive by the use of structural modelling to deduce the strengths of interconnected networks of active regions during task performance. Neural models are being developed to allow these results to be advanced, as well as indicate new paradigms for artificial neural networks. In particular the development of action-frontally-based models of the previous section attack the problem of understanding the networks of the brain in a different fashion from those described in section 4. The structural models derived there had a timeless character. Now the models from the previous section involved a strict separation of duties: the posterior sites were involved in preprocessing, holding object and spatial representations, with short term memory components holding activity on-line for no more than 1-2 seconds. Such a division of labour is an important result arising from fusing together the two ways of approaching the brain: structural models and analysis of active processing in terms of internal models. What do these results hold out for the development of intelligent systems

The deepest definition of intelligence is that it involves the manipulation of internal neural representations to achieve desired goals. The above discussion in section 6 indicated how models of some simple internal transformations may be made. Thus the more general conclusion arrived at the end of section 6 is of relevance for the development of more advance intelligent systems. Such systems need to possess the ability to take internal actions on internal models of the environment, be they either models of the effector system of the system itself, or models of the external world as held in buffer working memory sites, based on models of inputs slowly constructed during development. Thus the most important path to reach artificial intelligence is to construct artificial models of such internal actions.

References

1. Proceedings of the colloquium "Neuroimaging of Human Brain Function", Proc Natl Acad Sci USA, vol 95, February, 1998.
2. Posner M and Raichle ME (1994) Images of Mind, San Francisco: Freeman & Co
3. Ungerleider LG, Courtney SM & Haxby JVA neural system for human visual working memory Proc Natl Acad Sci USA 95, 883-890
4. Yang X, Hyder F & Shulman RG (1997) Functional MRI BOLD Signal Coincides with Electrical Activity in the Rat Whisker Barrels, NeuroReport 874-877
5. Raichle M E (1998) Behind the scenes of functional brain imaging: A historical and physiological perspective. Proc Natl Acad Sci USA 95, 765-772
6. Bti, Ltd, San Diego, CA
7. NEC, private communication from AA Ioannides
8. Talairach J & Tournoux P (1988) Co-planar steroetaxic atlas of the human brain. Stuttgart: G Thieme.
9. Schmitz N., Taylor J.G., Shah N.J., Ziemons K., Gruber O., Grosse-Ruyken M.L. & Mueller-Gaertner H-W. (1998) The Search for Awareness by the Motion After-Effect, Human Brain Mapping Conference '98, Abstract
10. Gold S, Christian B, Arndt S, Zeien G, Cizadio T, Johnoson DL, Flaum M & Andreason NC (1998) Functional MRI Statistical Software Packages: A Comparative Analysis. Human Brain Mapping 6, 73-84
11. D Gembris, S Posse, JG Taylor, S Schor et al (1998) Methodology of fast Correlation Analysis for Real-Time fMRI Experiments, submitted.
12. Ioannides AA (1995) in Quantitative & Topological EEG and MEG Analysis, Jena: Druckhaus-Mayer GmbH
13. Taylor JG, Ioannides AA, Mueller-Gaertner H-W (1999) Mathematical Analysis of Lead Field Expansions, IEEE Trans on Medical Imaging.
14. Taylor JG, Krause BJ, Shah NJ, Horwitz B & Mueller-Gaertner H-W (1998) On the Relation Between Brain Images and Brain Neural Networks, submitted.

15. Cabeza R & Nyburg L (1997) Imaging Cognition. J Cog Neuroscience 9, 1-26
16. Gabrieli JDE, Poldrack RA & Desmond JE (1998)The role of left prefrontal cortex in memory Proc Natl Acad Sci USA, 95, 906-13
17. McIntosh AR, Grady CL, Ungerleider LG, Haxby JV, Rapoport SI & Horwitz B (1994) Network analysis of cortical visual pathways mapped with PET, J Neurosci 14, 655-666
18. Horwitz B, McIntosh AR, Haxby JV, Furey M, Salerno JA, Schapiro MB, Rapoport SI & Grdy CL (1995) Network analysis of PET-mapped visual pathways in Alzheimer type dementia, NeuroReport 6, 2287-2292
19. McIntosh AR, Grady CL, Haxby J, Ungerleider LG & Horwitz B (1996) Changes in Limbic and Prefrontal Functional Interactions in a Working Memory Task for Faces. Cerebral Cortex 6, 571-584
20. Krause BJ, Horwitz B, Taylor JG, Schmidt D, Mottaghy F, Halsband U, Herzog H, Tellman L & Mueller-Gaertner H-W (1999) Network Analysis in Episodic Encoding and Retrieval of Word Pair Associates: A PET Study. Eur J Neuroscience 4/99 (in press).
21. Taylor JG, Krause BJ, Horwitz B & Mueller-Gaertner H-W (1998) Modeling Memory-Based Tasks, in preparation
22. Tulving E, Kapur S, Craik FIM, Moscovitch M & Houles S (1995) Hemispheric encoding/ retrieval asymmetry in epsodic memory: Positron emission tomography findings. Proc Natl Acad Sci USA 91, 2016-20.
23. Cohen JD, Perlstein WM, Braver TS, Nystrom LE, Noll DC, Jonides J& Smith EE (1997) Temporal dynamics of brain activation during a working memory task. Nature 386, 604-608
24. Bullmore E, Horwitz B, Morris RG, Curtis VA, McGuire PK, Sharma T, Williams SCR, Murray RM & Brammer MJ (1998) Causally connected cortical networks for language in functional (MR) brain images of normal and schizophrenic subjects. submitted to Neuron
25. Frith CD (1992) The Cognitive Neuropsychology of Schizophrenia. Hove UK: L Erlbaum Assoc.
26. Taylor JG & Taylor N (1998) Hard wired models of working memory and temporal sequence storage and generation. Neural Networks (to appear).
27. Monchi O & Taylor JG (1998) A hard-wired model of coupled frontal working memories for various tasks. Information Sciences (in press).
28. Taylor JG (1999) 'The Central Representation', Proc IJCNN99
29. Wolpert DM, Ghahramani Z & Jordan M (1996) An Internal Model for Sensorimotor Integration. Science 269:1880-2
30. Jacobs OLR (1993) Introduction to Control Theory (2nd ed). Oxford: Oxford University Press.
31. Sirigu A, Duhamel J-R, Cohen L, Pillon B, DuBois B & Agid Y (1996) The Mental Representation of Hand Movements After Parietal Cortex Damage. Science 273:1564-8.
32. Wolpert DM, Goodbody SJ & Husain M (1998) Maintaining internal representations: the role of the human superior parietal lobe. Nature neuroscience 1:529-534
33. Baddeley A (1986) Working Memory. Oxford: Oxford University Press.
34. Salmon E, Van der Linden, Collette F, Delfiore G, Maquet P, Degueldre C, Luxen A & Franck G (1996) Regional brain activity during working memory tasks. Brain 119:1617-1625

35. Carpenter PA, Just MA, Keller TA, Eddy W & Thulborn K (1999) Graded Functional Activation in the Visuospatial System with the Amount of Task Demand. J Cognitive Neuroscience 11:9-24.
36. Taylor NT & Taylor JG (1999) Temproal Sequence Storage in a Model of the Frontal Lobes. Neural Networks (to appear)

Chapter 11. Quantum Neural Networks

Alexandr A. Ezhov[1] and Dan Ventura[2]

[1] Department of Mathematics, Troitsk Institute of Innovation and Fusion Research
142092 Troitsk, Moscow Region, Russia
alexandr.ezhov@usa.net

[2] Applied Research Laboratory, The Pennsylvania State University
University Park, PA 16802-5018 USA

Abstract. *This chapter outlines the research, development and perspectives of quantum neural networks – a burgeoning new field which integrates classical neurocomputing with quantum computation [1]. It is argued that the study of quantum neural networks may give us both new understanding of brain function as well as unprecedented possibilities in creating new systems for information processing, including solving classically intractable problems, associative memory with exponential capacity and possibly overcoming the limitations posed by the Church-Turing thesis.*

Keywords. *Quantum neural networks, associative memory, entanglement, many universes interpretation*

1 Why Quantum Neural Networks?

There are two main reasons to discuss quantum neural networks. One has its origin in arguments for the essential role which quantum processes play in the living brain. For example, Roger Penrose has argued that a new physics binding quantum phenomena with general relativity can explain such mental abilities as *understanding, awareness* and *consciousness* [2]. However, this approach advocates the study of intracellular structures, such as microtubules rather than that of the networks of neurons themselves [3]. A second motivation is the possibility that the field of classical artificial neural networks can be generalized to the quantum domain by eclectic combination of that field with the promising new field of quantum computing [4]. Both considerations suggest new understanding of mind and brain function as well as new unprecedented abilities in information processing. Here we consider quantum neural networks as the next natural step in the evolution of neurocomputing systems, focusing our attention on artificial rather than biological systems. We outline different approaches to the realization of quantum distributed processing and argue that, as in the case of quantum computing [5], Everett's many universes interpretation of quantum mechanics [6] can be used as a general framework for producing quantum analogs of well-known classical artificial neural networks. We also outline some perspectives on quantum neurocomputers in the next century.

2 Neural Networks: Toward Quantum Analogs

There are many different approaches to what we can call *quantum neural networks*. Many researchers use their own analogies in establishing a connection between quantum mechanics and neural networks. The main concepts of these two fields may be considered as follows [7-8]:

Table 1. Main concepts of quantum mechanics and neural networks

Quantum mechanics	Neural Networks
wave function	neuron
Superposition (coherence)	interconnections (weights)
Measurement (decoherence)	evolution to attractor
Entanglement	learning rule
unitary transformations	gain function (transformation)

One should be careful *not* to consider corresponding concepts in the two columns as analogical – in the table above their order is arbitrary. Indeed, the establishment of such correspondences is a major challenge in the development of a model of quantum neural networks.

To date, quantum ideas have been proposed for the effective realization of classical – rather than neural – computation. The concept of quantum computation may arguably be traced back to the pioneering work of Richard Feynman [1], who examined the role quantum effects would play in the development of future hardware. As hardware speeds continue to increase, hardware scales correspondingly continue to decrease and at some point in the not too distant future, Feynman realized, gates and wires may consist of only a few atoms, and quantum effects will then play a major role in hardware implementation[1]. Feynman concluded that

---

[1] It is somewhat remarkable that in 1982 just as Richard Feynman published his first paper on quantum computation, John Hopfield proposed his model of neural content-addressable memory [9], which attracted many physicists to the field of artificial neural networks.

such quantum devices can have significant advantages over classical computational mediums. In 1985 David Deutsch formalized the foundations of quantum computation [5].

3 Some Quantum Concepts

Quantum computation is based upon physical principles from the theory of quantum mechanics (QM), which is in many ways counterintuitive. Yet it has provided us with perhaps the most accurate physical theory (in terms of predicting experimental results) ever devised by science. The theory is well-established and is covered in its basic form by many textbooks (see for example [10]). Several necessary ideas that form the basis for the study of quantum computation are briefly reviewed here.

Linear superposition is closely related to the familiar mathematical principle of linear combination of vectors. Quantum systems are described by a wave function ψ that exists in a Hilbert space. The Hilbert space has a set of states, $|\phi_i\rangle$, that form a basis, and the system is described by a quantum state $|\psi\rangle$,

$$|\psi\rangle = \sum_i c_i |\phi_i\rangle$$

$|\psi\rangle$ is said to be in a linear superposition of the basis states $|\phi_i\rangle$, and in the general case, the coefficients c_i may be complex. Use is made here of the Dirac bracket notation, where the ket $|\cdot\rangle$ is analogous to a column vector, and the bra $\langle\cdot|$ is analogous to the complex conjugate transpose of the ket. In quantum mechanics the Hilbert space and its basis have a physical interpretation, and this leads directly to perhaps the most counterintuitive aspect of the theory. The counter intuition is this -- at the microscopic or quantum level, the state of the system is described by the wave function ψ, that is, as a linear superposition of all basis states (i.e. in some sense the system is in all basis states at once). However, at the macroscopic or classical level the system can be in only a single basis state. For example, at the quantum level an electron can be in a superposition of many different energies; however, in the classical realm this obviously cannot be.

Coherence and decoherence are closely related to the idea of linear superposition. A quantum system is said to be coherent if it is in a linear superposition of its basis states. A result of quantum mechanics is that if a system that is in a linear superposition of states interacts in any way with its environment, the superposition is destroyed. This loss of coherence is called decoherence and is governed by the wave function ψ. The coefficients c_i are called probability amplitudes, and $|c_i|^2$ gives the probability of $|\psi\rangle$ collapsing into state $|\phi_i\rangle$ if it decoheres. Note that the wave function ψ describes a real physical system that must collapse to exactly one basis state. Therefore, the probabilities governed by the amplitudes c_i must sum to unity. This necessary constraint is expressed as the unitarity condition

$$\sum_i |c_i|^2 = 1$$

In the Dirac notation, the probability that a quantum state $|\psi\rangle$ will collapse into an eigenstate $|\phi_i\rangle$ is written $|\langle\phi_i|\psi\rangle|^2$ and is analogous to the dot product (projection) of two vectors. Consider, for example, a discrete physical variable called spin. The simplest spin system is a two-state system, called a spin-1/2 system, whose basis states are usually represented as $|\uparrow\rangle$ (spin up) and $|\downarrow\rangle$ (spin down). In this simple system the wave function ψ is a distribution over two values (up and down) and a coherent state $|\psi\rangle$ is a linear superposition of $|\uparrow\rangle$ and $|\downarrow\rangle$. One such state might be

$$|\psi\rangle = \frac{2}{\sqrt{5}}|\uparrow\rangle + \frac{1}{\sqrt{5}}|\downarrow\rangle$$

As long as the system maintains its quantum coherence it cannot be said to be either spin up or spin down. It is in some sense both at once. Classically, of course, it must be one or the other, and when this system decoheres the result is, for example, the $|\uparrow\rangle$ state with probability

$$|\langle\uparrow|\psi\rangle|^2 = \left(\frac{2}{\sqrt{5}}\right)^2 = 0.8$$

A simple two-state quantum system, such as the spin-1/2 system just introduced, is used as the basic unit of quantum computation. Such a system is referred to as a quantum bit or *qubit*, and renaming the two states $|0\rangle$ and $|1\rangle$ it is easy to see why this is so.

Operators on a Hilbert space describe how one wave function is changed into another. Here they will be denoted by a capital letter with a hat, such as $\hat{A}$, and they may be represented as matrices acting on vectors. Using operators, an eigenvalue equation can be written $\hat{A}|\phi_i\rangle = a_i|\phi_i\rangle$, where a_i is the eigenvalue. The solutions $|\phi_i\rangle$ to such an equation are called eigenstates and can be used to construct the basis of a Hilbert space as discussed previously. In the quantum formalism, all properties are represented as operators whose eigenstates are the basis for the Hilbert space associated with that property and whose eigenvalues are the quantum allowed values for that property. It is important to note that operators in quantum mechanics must be linear operators and further that they must be unitary so that $\hat{A}^\dagger \hat{A} = \hat{A}\hat{A}^\dagger = \hat{I}$, $\hat{I}$ is the identity operator, and $\hat{A}^\dagger$ is the complex conjugate transpose, or adjoint, of $\hat{A}$.

Interference is a familiar wave phenomenon. Wave peaks that are in phase interfere constructively (magnify each other's amplitude) while those that are out of phase interfere destructively (decrease or eliminate each other's amplitude). This is a phenomenon common to all kinds of wave mechanics from water waves to optics. The well-known double slit experiment demonstrates empirically that at the quantum level interference also applies to the probability waves of quantum mechanics. As a simple example, suppose that the wave function described above is represented in vector form as

$$|\psi\rangle = \frac{1}{\sqrt{5}}\begin{pmatrix} 2 \\ 1 \end{pmatrix}$$

and suppose that it is operated upon by an operator $\hat{O}$ described by the following matrix,

$$\hat{O} = \frac{1}{\sqrt{2}}\begin{bmatrix} 1 & 1 \\ 1 & -1 \end{bmatrix}$$

The result is

$$\hat{O}|\psi\rangle = \frac{1}{\sqrt{2}}\begin{bmatrix} 1 & 1 \\ 1 & -1 \end{bmatrix}\frac{1}{\sqrt{5}}\begin{pmatrix} 2 \\ 1 \end{pmatrix} = \frac{1}{\sqrt{10}}\begin{pmatrix} 3 \\ 1 \end{pmatrix}$$

and therefore now

$$|\psi\rangle = \frac{3}{\sqrt{10}}|\uparrow\rangle + \frac{1}{\sqrt{10}}|\downarrow\rangle$$

Notice that the amplitude of the $|\uparrow\rangle$ state has increased while the amplitude of the $|\downarrow\rangle$ state has decreased. This is due to the wave function interfering with itself through the action of the operator -- the different parts of the wave function interfere constructively or destructively according to their relative phases just like any other kind of wave.

Entanglement is the potential for quantum states to exhibit correlations that cannot be accounted for classically. From a computational standpoint, entanglement seems intuitive enough -- it is simply the fact that correlations can exist between different qubits -- for example if one qubit is in the $|1\rangle$ state, another will be in the $|1\rangle$ state. However, from a physical standpoint, entanglement is little understood. The questions of what exactly it is and how it works are still not resolved. What makes it so powerful (and so little understood) is the fact that since quantum states exist as superpositions, these correlations exist in superposition as well. When the superposition is destroyed, the proper correlation is somehow communicated between the qubits, and it is this "communication" that is the crux of entanglement. Mathematically, entanglement may be described using the *density matrix* formalism. The density matrix ρ_ψ of a quantum state $|\psi\rangle$ is defined as

$$\rho_\psi = |\psi\rangle\langle\psi|$$

For example, the quantum state

$$|\xi\rangle = \frac{1}{\sqrt{2}}|00\rangle + \frac{1}{\sqrt{2}}|01\rangle$$

appears in vector form as

$$|\xi\rangle = \frac{1}{\sqrt{2}}\begin{pmatrix}1\\1\\0\\0\end{pmatrix}$$

and it may also be represented as the density matrix

$$\rho_\xi = |\xi\rangle\langle\xi| = \frac{1}{2}\begin{pmatrix}1 & 1 & 0 & 0\\1 & 1 & 0 & 0\\0 & 0 & 0 & 0\\0 & 0 & 0 & 0\end{pmatrix}$$

while the state

$$|\psi\rangle = \frac{1}{\sqrt{2}}|00\rangle + \frac{1}{\sqrt{2}}|11\rangle$$

is represented as

$$\rho_\psi = |\psi\rangle\langle\psi| = \frac{1}{2}\begin{pmatrix}1 & 0 & 0 & 1\\0 & 0 & 0 & 0\\0 & 0 & 0 & 0\\1 & 0 & 0 & 1\end{pmatrix}$$

and the state

$$|\zeta\rangle = \frac{1}{\sqrt{3}}|00\rangle + \frac{1}{\sqrt{3}}|01\rangle + \frac{1}{\sqrt{3}}|11\rangle$$

is represented as

$$\rho_\zeta = |\zeta\rangle\langle\zeta| = \frac{1}{3}\begin{pmatrix}1 & 1 & 0 & 1\\1 & 1 & 0 & 1\\0 & 0 & 0 & 0\\1 & 1 & 0 & 1\end{pmatrix}$$

where the matrices and vectors are indexed by the state labels 00, ..., 11. Now, notice that ρ_ξ can be factorized as

$$\rho_\xi = \frac{1}{\sqrt{2}}\left(\begin{pmatrix} 1 & 0 \\ 0 & 0 \end{pmatrix} \otimes \begin{pmatrix} 1 & 1 \\ 1 & 1 \end{pmatrix} \right)$$

where $\otimes$ is the normal tensor product. On the other hand, ρ_ψ can not be factorized. States that cannot be factorized are said to be *entangled*, while those that can be factorized are not. Notice that ρ_ζ can be partially factorized two different ways, one of which is

$$\rho_\zeta = \frac{1}{\sqrt{3}}\left(\begin{pmatrix} 1 & 1 \\ 1 & 1 \end{pmatrix} \otimes \begin{pmatrix} 0 & 0 \\ 0 & 1 \end{pmatrix} \otimes \begin{pmatrix} 1 & 1 & 0 & 1 \\ 1 & 0 & 0 & 0 \\ 0 & 0 & 0 & 0 \\ 1 & 0 & 0 & 0 \end{pmatrix} \right)$$

(the other contains the factorization of ρ_ξ and a different remainder); however, in both cases the factorization is not complete. Therefore, ρ_ζ is also entangled, but not to the same degree as ρ_ψ (because ρ_ζ can be partially factorized but ρ_ψ cannot). Thus there are different degrees of entanglement and much work has been done on better understanding and quantifying it [11-12]. It is interesting to note from a computational standpoint that quantum states that are superpositions of only basis states that are maximally far apart in terms of Hamming distance are those states with the greatest entanglement. For example, ρ_ψ is a superposition of only the states 00 and 11, which have a maximum Hamming spread, and therefore ρ_ψ is maximally entangled. Finally, it should be mentioned that while interference is a quantum property that has a classical cousin, entanglement is a completely quantum phenomenon for which there is no classical analog.

4 Interpretations of Quantum Theory

It is important to note that much of the power of classical artificial neural networks is due to their massively parallel, distributed processing of information and also due to the nonlinearity of the transformation performed by the network nodes (neurons). On the other hand, quantum mechanics offers the possibility of an even more powerful *quantum parallelism* which is expressed in the principle of superposition. This principle provides quantum computing an advantage in processing huge data sets. Though quantum computing implies parallel processing of all possible configurations of the state of a register composed of N qubits, only one result can be read after the decoherence of the quantum superposition into one of its basis states. However, entanglement provides the possibility of measuring the states of all qubits in a register whose values are interdependent. Though the mathematics of quantum mechanics is fairly well understood and accepted, the physical reality of what the theory *means* is much debated and there exist different interpretations of quantum mechanics, including:

- Copenhagen interpretation [7];
- Feynman path-integral formalism [13];
- Many universes (many-world) interpretation of Everett [6], etc.

The choice of interpretation is important in establishing different analogies between quantum physics and neurocomputing.

The field of neural networks contains several important basic ideas, which include the concept of a processing element (*neuron*), the *transformation* performed by this element (in general, input summation and nonlinear mapping of the result into an output value), the *interconnection structure* between neurons, the network *dynamics*, and the *learning rule* which governs the modification of interconnection strengths. A major dichotomization of neural networks can be realized by considering whether they are trained in a supervised or unsupervised manner. An example of the latter is the *Hopfield model* of content-addressable memory using the concept of attractor states [9].

We shall argue below that it is adequate to choose such a Hopfield network as a reference point for the consideration of neural models in general. In fact, the Hopfield model itself was proposed during a previous "invasion" of physics into the theory of artificial neural networks in 1982. What Hopfield discovered was an analogy between networks with symmetrical bonds and *spin glasses*.

While quantum mechanics is a linear theory, neurocomputing is very dependent upon nonlinear approaches to data processing. At first glance, this appears to complicate the establishment of a correspondence between the two fields. However there are different ways to overcome this difficulty.

As mentioned earlier, evolutionary operators in quantum mechanics must be unitary, and certain aspects of any quantum computation must be considered as evolutionary. For example, storing patterns in a quantum system demands evolutionary processes since the system must maintain a coherent superposition that represents the stored patterns. On the other hand, other aspects of quantum computation preclude unitarity (and thus linearity) altogether. In particular, decoherence is a non-unitary process.

In the *Copenhagen interpretation*, non-unitary operators do exist in quantum mechanics and in nature. For example, any observation of a quantum system can be thought of as an operator that is neither evolutionary nor unitary. In fact, the Copenhagen school of thought suggests that this non-evolutionary behavior of quantum systems is just as critical to our understanding of quantum mechanics as is their evolutionary behavior. Now, since recalling a pattern from a quantum system would require the decoherence and collapse of the system at some point (at the very latest when the system is observed), it can be argued that pattern recall may be considered as a non-unitary process. In which case, the use of unitary operators becomes unnecessary. Since the decoherence and collapse of a quantum wave function is non-unitary and since pattern recall in a quantum system requires decoherence and collapse at some point, why not make explicit use of this non-unitarity, in the pattern recall process? This decoherence of a quantum state can be considered as the analog of the evolution of a neural network state to an attractor basin. This analogy has been mentioned in the work of Perus [14][2].

As a second approach to reconciling the linearity of quantum mechanics with the nonlinearity inherent in artificial neural networks, consider the *Feynman interpretation* of quantum mechanics, which is based on the use of path integrals. The probability of an event is expressed by the formula,

$$|\psi(t)\rangle = \sum_{all\ paths} e^{-\frac{i}{\hbar}\int_0^t [\frac{m\dot{x}^2(\tau)}{2} - V(x(\tau))]d\tau}$$

Here nonlinearity can be due both to the nonlinear form of the potential V(x) and also to the operation of the exponent. This fact has been used in approaches to modelling quantum neural networks by Elizabeth Behrman and coworkers [16-17] and Ben Goertzel [18] (some analogies used for the development of quantum neural networks are summarized in Table 2). Behrman et al. first developed a *temporal* model of a quantum neural network which utilizes a quantum dot molecule coupled to a substrate lattice through optical phonons [16]. In this model temporal evolution of the system resembles the equations for virtual neurons and the timeline discretization points for the Feynman path integral serve as these virtual neurons. The concept of neurons used here is rather artificial, and in fact the number of neurons depends on the parameters of the temporal discretization scheme, rather than on the number of quantum particles involved. Recently, this group working at Wichita State University proposed a *spatial* model for a quantum neural network based on the use of a spatial array of quantum dot molecules. It was shown that any logical gate, including a purely quantum one – phase shift – can be performed using these systems [17]. Note that another approach to quantum neural networks used by Ron Chrisley from the University of Sussex [19-20], considers the positions of slits in an interference experiment (similar to Young's double-slit experiment) as representing neuron state values while the positions of other slits encode the values of the network weights. Obviously, there is a high diversity of possible approaches to the construction of a model of quantum neural networks.

But we shall try to argue that as in the case of quantum computing the most consistent way to obtain a general model seems to be *Everett's many universes interpretation* of quantum mechanics. Everett's approach suggests that decoherence or collapse of the wave function is an *illusion*, and that actually the wave function obeys the Schrödinger equation at all times. Rather than causing the wave function to collapse, the effect of the measurement is to split the observer

---

[2] However, in some sense, the formalism described by him is much more similar to the concept of the synergetical computer proposed by Hermann Haken [15].

into a number of copies, each copy observing just one of the possible results of a measurement, unaware of the other possible outcomes. It follows that there exist many, mutually unobservable but equally real universes, each corresponding to a single possible outcome of the measurement [21].

Using this metatheory of quantum mechanics as a starting point, we can combine the field of artificial neural networks with that of quantum computation in a natural way. For our purposes it is sufficient to consider the application of neural approaches, in their simplest forms, to pattern recognition. We shall then see how a concept of quantum neural networks naturally emerges from the theory of neurocomputing.

Table 2. Quantum analogies used for different concepts of artificial neural networks

Model	Neuron	Connections	Transformation	Network	Dynamics
Perus	quantum	Green function	linear	temporal	collapse as convergence to attractor
Chrisley	classical (slit position)	classical (slit position)	nonlinear through superposition	multilayer	non-superpositional
Behrman et al.	time slice, quantum	interactions through phonons	nonlinear through potential energy and exponent function	temporal and spatial	Feynman path integral
Goertzel	classical	quantum	nonlinear	classical	Feynman path-integral
Menneer and Narayanan	classical	classical	nonlinear	single-item networks in many universes	classical
Ventura	qubit	entanglement	linear and nonlinear	single-item modules in many universes	unitary and non-unitary transformations

5 How Pattern Recognition Leads Us to Quantum Neural Networks

One simple approach to pattern recognition can be termed a template-based method, in which examples of different pattern classes are stored separately as multiple templates. A presented stimulus can then be recognized (classified) according to the class of the template most similar to the input stimulus. To be efficient this process should be performed in parallel since the number of stored templates can be prohibitive to sequential processing. In general, this scheme is characterized by rather low performance due to a lack of generalization and also due to the need to guarantee invariant recognition.

Neural networks provide the ability to use only *one* system to store *multiple data* belonging to different classes and to classify the presented stimulus in a parallel, distributed manner [22]. Thus, the problem of parallelism is naturally solved in this approach. Further, the capability of approximating arbitrarily complex functions makes neural networks very effective for creating classification systems.

It is often desirable to use multimodular systems consisting of so-called single-class neural networks [23-25]. In this scheme, a network is trained using only examples of patterns belonging to a single class, and a different network is trained for each class to be recognized. Classification is performed by presenting the input stimulus to each of the different modules, comparing their outputs and using some criterion to choose a winning module. The problem of parallelism arises again in this approach (though, not nearly as acutely as in the case of a template-based method), but it has many advantages associated with the usefulness of spurious memories for generalization [26]. Various types of neural systems can be used as the basis for such a multimodular recognition scheme, including auto-associative perceptrons, but what is especially pertinent to this discussion is the fact that Hopfield networks seem to be especially good candidates for this role.

It is well known that any state of the Hopfield network is either a stable attractor or evolves to some such attractor. It is usual to interpret attractors as memorized patterns, or sometimes as spurious memories, while non-stable states can be considered as corrupted versions of memories containing enough partial information to retrieve the memorized pattern stored as the nearest stable state. Numerous studies have been performed to investigate the properties of content-addressable memories which can be implemented by the Hopfield model and its various derivatives [27]. The main drawback of such memories is their limited capacity. However, using a probabilistic interpretation of the network state energy – the functional which governs state dynamics – it can be argued that the Hopfield network is best suited for the extraction of the locally most plausible version of a single prototype, for which all stored patterns can be considered as corrupted versions [28]. This approach can be also thought of as implementing the use of distributed templates in the sense that all representatives of a given class are compared with an external stimulus in a parallel, distributed manner. What is even more interesting, this approach opens the door to the development of quantum neural networks by suggesting a further generalization of the idea of class-specific neural networks. Namely, we can generalize the idea to its extreme, considering a system of separate networks each trained with *only a single pattern*! This, in turn, brings us naturally to the many universes approach to quantum mechanics.

6 Many Universes Approach

In memorizing a set of patterns, why not use a set of many Hopfield networks, each of which stores a single pattern? In the classical Hopfield model we typically use only one network to store many patterns, and we sum all pattern correlations in order to build the network's Hebbian interconnections as follows.

$$T_{ij} = \sum_s \sigma_i^s \sigma_j^s, \quad T_{ii} = 0 \quad i,j = 1,\ldots,N$$

This summation causes multiple problems if we want to consider a network as a passive memorization system. The interference of the different patterns leads to a loss of the stability for some memories (producing, instead, a spurious memory) and, as a result, to a rather restricted memory capacity, which grows at best only linearly with the number of neurons [27].

If, on the other hand, we simply generate multiple Hopfield networks which store only one pattern each, we lose any parallelism in processing the information. But what about a quantum approach? Imagine, that we can store all patterns as the quantum superposition.

$$|\psi\rangle = \sum_s |\sigma_1^s \sigma_2^s \ldots \sigma_N^s\rangle$$

In this case, each of the patterns can be considered as existing in a separate universe. Moreover, the interaction of such a superposition with the environment is performed in parallel, and further, this parallelism has a quantum nature. It has in fact been shown in, given a set of patterns, how such a superposition may be created [29], and each of the basis states in the superposition will play the role of a single memory state independent of the number of them that exist in the superposition. In theory, then, a *quantum associative memory can have exponential memory capacity!* (See [30].) It should also be mentioned here that although spurious states can arise in such a quantum memory, these spurious state are not the result of an interference of memories as in the classical case but instead arise for a completely different reason in the retrieval phase and therefore do not directly influence stored patterns.

Let us imagine that instead of the various memory states existing in parallel universes, we have single-memory, Hopfield-type networks existing in these universes. In the classical Hopfield network, the existence of symmetric, Hebbian connections guarantees the stability of a unique stored pattern; similarly, in a quantum analog of the Hopfield network the integrity of a stored pattern (basis state) is due to entanglement [31]. This property characterizes multi-particle systems and is the basis of all known quantum algorithms. Now we can consider quantum associative memory as a realization of the extreme condition of using many Hopfield networks, each storing a single pattern in parallel quantum universes!

Continuing this line of reasoning, we can further imagine more complex neural structures existing in such parallel worlds. Such an idea has been explored by Menneer and Narayanan, who consider a set of multilayer perceptrons, each trained on only one pattern that are combined into a quantum network whose weights are superpositions of the weights of all perceptrons existing in parallel universes [32].

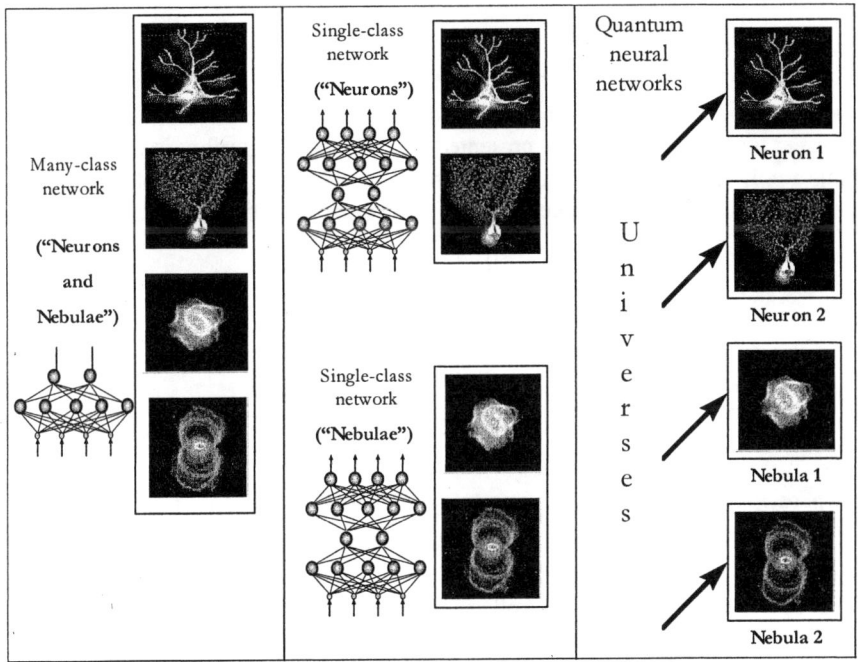

Fig. 1. Many-class networks are trained using the examples from different classes (here "Neurons" and "Nebulae" together) – left; A set of modular single-class neural networks use for training only the objects belonging to one class (two networks for two classes: "Neurons" and "Nebulae" separately) – center; Quantum neural networks may be trained using only pattern each! (four networks for four examples in many universes) – right.

Moreover, they also consider the many universes approach to quantum neural networks as methodologically correct and cognitively plausible. Indeed, fast learning of the networks in separate universes avoids the objection to neural network models being adequate accounts of mind because multiple presentations of patterns is implausible for human learning [32].

7 Quantum Associative Memory

One of the most promising approaches to quantum neurocomputing is the quantum associative memory, of which one approach is described in [33-35]. The task of pattern association can be broken down into two major components: memorization and recall. The memorization step consists of storing patterns in the memory while the recall step entails pattern completion or pattern association based on partial and/or noisy input.

Memorization

An efficient quantum algorithm for constructing a coherent state over n qubits to represent a set of m patterns is presented in [29]. The algorithm is implemented using a polynomial number (in the length and number of patterns) of elementary operations on one, two, or three qubits. The key operator in this process is

$$\hat{S}^p = \begin{bmatrix} 1 & 0 & 0 & 0 \\ 0 & 1 & 0 & 0 \\ 0 & 0 & \sqrt{\frac{p-1}{p}} & \frac{-1}{\sqrt{p}} \\ 0 & 0 & \frac{1}{\sqrt{p}} & \sqrt{\frac{p-1}{p}} \end{bmatrix}$$

where $m \geq p \geq 1$. This is actually a set of operators that are conditional transforms – there is a different $\hat{S}^p$ operator associated with each pattern to be stored. The algorithm also makes use of various versions of some standard quantum computational operators such as the Controlled-Not and Fredkin gates. Now given a set P of m binary patterns of length n, the quantum algorithm for storing the patterns requires a set of $2n+1$ qubits, the first n of which actually store the patterns and can be thought of as n neurons in a quantum associative memory. The remaining $n+1$ qubits are ancillary qubits used for bookkeeping and are restored to the state $|\overline{0}\rangle$ after every storage iteration. Each iteration through the algorithm makes use of a different $\hat{S}^p$ operator and results in another pattern being incorporated into the quantum system. The result is a coherent superposition of states that correspond to the patterns, with the amplitudes of the states in the superposition all being equal. The algorithm requires O(mn) steps to encode the m patterns as a quantum superposition over n quantum neurons. This is optimal in the sense that just reading each instance once cannot be done any faster than O(mn).

Recall – completion

The recall capability of the quantum associative memory can be implemented using the quantum search algorithm due to Grover [36]. This algorithm has been traditionally considered as implementing a search for an item in an unsorted (quantum) database of N items, and it performs this task in O($\sqrt{N}$) time, a feat that is impossible classically. In the quantum computational setting, finding an item in the database means measuring the system and having the system collapse to the basis state which corresponds to the item in the database for which we are searching. Now, we can equally well consider the algorithm as accomplishing the task of pattern completion in a quantum associative memory. The basic idea of Grover's algorithm is to invert the phase of the desired basis state and then to invert all the basis states about the average amplitude of all the states. Repetition of this process produces an increase in the amplitude of the desired basis state to near unity followed by a corresponding decrease in the amplitude of the desired state back to its original magnitude. The process has a period of $\frac{\pi}{4}\sqrt{N}$ and thus after O($\sqrt{N}$) operations, the system may be observed in the desired state with near certainty. Define

$$\hat{I}_\phi = \text{identity matrix except for } i_{\phi\phi} = -1$$

which inverts the phase of the basis state ϕ,

$$\hat{W} = \frac{1}{\sqrt{2}}\begin{bmatrix} 1 & 1 \\ 1 & -1 \end{bmatrix}$$

which is often called the Walsh transform, and

$$\hat{G} = -\hat{W}\hat{I}_0\hat{W}$$

which effects the inversion about average. Now to perform the search on a quantum database of size N, begin with the system in the $|\overline{0}\rangle$ state and apply the $\hat{W}$ operator. This initializes all the possible states to have the same amplitude. Finally, apply the operator $\hat{G}\hat{I}_\tau$ (recall that operators are applied right to left), where τ is the state being sought, $\frac{\pi}{4}\sqrt{N}$ times and observe the system.

Combining the algorithms

A quantum associative memory can now be implemented by combining the two algorithms just discussed. Define $\hat{P}$ as an operator that implements the algorithm for memorizing patterns. Then the operation of the memory can be described as follows. Memorizing a set of patterns is simply

$$|\psi\rangle = \hat{P}|\overline{0}\rangle$$

with $|\psi\rangle$ being a quantum superposition of appropriate basis states, one for each pattern. Now, suppose we know $n-k$ bits of a pattern and wish to recall the entire pattern. We can use a modification of Grover's algorithm to complete the pattern, producing one of the stored patterns that matches on the $n-k$ bits that we know. Thus, with $2n+1$ neurons (qubits) the quantum associative memory can store up to 2^n patterns in O(mn) time and requires O($\sqrt{2^n}$) time to recall a pattern. This last bound is somewhat slower than desirable and may be improved with a non-unitary recall mechanism. In fact, Grover's search algorithm has been proven to be optimal in the number of steps required when unitarity is required. Thus, we have another motivation for non-unitary processes in quantum neural computation.

Recall – association

Of course, in general, a quantum memory should not only be able to complete patterns but also to correct them. In other words, given a noisy stimulus, the memory should produce the pattern most similar to that input. This can be accomplished with further modification of the basic quantum memory model we have been discussing. This modification involves the use of distributed queries and is presented in detail in [37]. Briefly, a distributed query is a distribution of the form

$$|b^p\rangle = \sum_{x=0}^{2^d-1} b_x^p |x\rangle$$

over the amplitudes of all possible states in the memory. The index p marks one of these states, $|p\rangle$, which is the center of the distribution (real-valued amplitudes are distributed such that the maximal value occurs at this center, and the amplitudes of the other basis states decrease monotonically with Hamming distance from the center state). This leads to the introduction of spurious memories into the recall process; however, counter to intuition the presence of these spurious memories may actually facilitate memory recall [37]. Table 3 summarizes the analogies used in developing a quantum associative memory.

It should be noted that the "neuron" in the first row of the Table 3 is strictly artificial and should not be considered as a model of its biological analog. Really, as stated by Penrose "...it is hard to see how one could usefully consider a quantum superposition consisting of one neuron firing, and simultaneously not firing" [2]. There are many other arguments against attributing any biological meaning to this scheme, so we should consider it only in the context of the development of *artificial* quantum associative memory.

Table 3. Corresponding concepts from the domains of classical neural networks and quantum associative memory

Classical neural networks		Quantum associative memory	
Neuronal State	$x_i \in \{0,1\}$	Qubit	$\|x\rangle = a\|0\rangle + b\|1\rangle$
Connections	$\{w_{ij}\}_{ij=1}^{p-1}$	Entanglement	$\|x_0 x_1 ... x_{p-1}\rangle$
Learning rule	$\sum_{s=1}^{p} x_i^s x_j^s$	Superposition of entangled states	$\sum_{s=1}^{p} a_s \|x_0^s ... x_{p-1}^s\rangle$
Winner search	$n = \max_i \arg(f_i)$	Unitary transformation	$U: \psi \to \psi'$
Output result	n	Decoherence	$\sum_{s=1} a_s \|x^s\rangle \Rightarrow \|x^k\rangle$

8 Implementation of QNN

How can quantum neural networks be implemented as real physical devices? First, let us mention briefly some of the difficulties we might face in the development of a physical realization of quantum neural networks.

Coherence

One of the most difficult problems in the development of any quantum computational system is the maintainence the system's coherence until the computation is complete [38]. This loss of coherence (decoherence) is due to the interaction of the quantum system with its environment. In quantum cryptography this problem may be resolved using error-correcting codes [38]. What about quantum neural networks? It has been suggested that if fact these systems may be implemented before ordinary quantum computers will be realized because of significantly lower demands on the number of qubits necessary to represent network nodes and also because of the relatively low number of state transformations required during data processing in order to perform useful computation [35, 39]. Another approach to the problem of decoherence in quantum parallel distributed processing proposed by Chrisley excludes the use of superpositional states at all and suggests the use of quantum systems for implementing standard neural paradigms, i.e. multilayer neural systems trained with backpropagation learning [20]. This model, however, takes no advantage of the use of quantum parallelism. A more promising approach to the implementation of quantum associative memory based on the use of Grover's algorithm is provided by bulk spin resonance computation (see below).

Connections

The high density of interconnections between processing elements is a major difficulty in the implementation of small-scale integration of computational systems. In ordinary neurocomputers these connections are made via wires. In (non-superpositional) quantum neurocomputers they are made via forces. In the quantum associative memory model discussed here, these connections are due to the entanglement of qubits.

Physical systems

Now we can outline what kind of physical systems might be used to develop real quantum neural networks and how these systems address the problems listed above.

- **Nuclear Magnetic Resonance.** A promising approach to the implementation of quantum associative memory based on the use of Grover's algorithm is provided by *bulk spin resonance computation*. This technique can be performed using Nuclear Magnetic Resonance systems for which coherence times on the order of thousands of seconds have been observed. Experimental verification of such an implementation has been done by Gershenfeld and Chuang [40] (among others), who used NMR techniques and a solution of chloroform ($CHCl_3$) molecules for the implementation of Grover's search on a system consisting of two qubits – the first qubit is decribed by the spin of the nucleus of the isotope C^{13}, while second one is described by the spin of the proton (hydrogen nucleus). Rather interestingly, this approach to quantum computation utilizes not a single quantum system but rather the statistical average of many copies of such a system (a collection of molecules). It is precisely for this reason that the maintenance of system coherence times is considerably greater than for true quantum implementations. Further, this technology is relatively mature, and in fact coherent computation on seven qubits using NMR has recently been demonstrated by Knill, et al. [41]. This technology is most promising in the short term, and good progress in this direction is possible in the early 21st century.

- **Quantum dots.** These quantum systems basically consist of a single electron trapped inside a cage of atoms. These electrons can be influenced by short laser pulses. Limitations to these systems which must be overcome include 1) short decoherence times due to the fact that the existence of the electron in its excited state lasts about a microsecond, and the required duration of a laser pulse is around a nanosecond; 2) the necessity of developing a technology to build computers from quantum dots of very small scale (10 atoms across); 3) the necessity of developing special lasers capable of selectively influencing different groups of quantum dots with different wavelengths of light. The use of quantum dots as the basis for the implementation of QNN is being investigated by Behrman and co-workers [16-17].

- **Other systems.** There are many other physical systems which are now being considered as possible candidates for the implementation of quantum computers (and therefore possibly quantum neurocomputers). These include various schemes of *cavity QED* (quantum electrodynamics of atoms in optical cavities), *ion traps*, *SQUIDs* (superconducting quantum interference devices), etc. Each has its own advantages and shortcomings with regard to decoherence times, speed, possibility of miniaturization, etc. More information about these technologies can be found in [4, 31].

9 Can QNN Outperform Classical Neural Networks?

It is now known that quantum computing gives us unprecedented possibilities in solving problems beyond the abilities of classical computers. For example Shor's algorithm gives a polynomial solution (on a quantum computer) for the problem of prime factorization, which is believed to be classically intractable [42]. Also, as previously mentioned, Grover's algorithm provides super-classical performance in searching an unsorted database.

What of quantum neural networks? Will they give us some advantages unattainable by either traditional von Neumann computation or classical artificial neural networks? Compared to the latter, quantum neural networks will probably have the following advantages:

- exponential memory capacity [30];
- higher performance for lower number of hidden neurons [39];
- faster learning [32];
- elimination of catastrophic forgetting due to the absence of pattern interference [32];
- single layer network solution of linearly inseparable problems [32];

- absence of wires [17];
- processing speed (10^{10} bits/s) [17];
- small scale (10^{11} neurons/mm^3) [17];
- higher stability and reliability [39];

These potential advantages of quantum neural networks are indeed compelling motivation for their development. However, the more remote future possibilities of QNN may be even more exciting.

10 Frontiers of QNN

It is generally believed that the right hemisphere is responsible for spatial orientation, intuition, semantics etc., while the left hemisphere is responsible for temporal processing, logical thinking and syntax. Given this view, it is very natural to consider that neurocomputers can be thought of as imitating our right brain function while von Neumann computers can be thought as mimicing the functionality of our left brain. Penrose characterizes these two types of computation as *bottom-up* and *top-down* respectively. Nevertheless, he argues that higher brain functions such as consciousness cannot be modelled using just these types of computation. The ideas discussed in this chapter introduce the possibility of combining the unique computational abilities of classical neural networks and quantum computation, thus producing a computational paradigm of incredible potential. However, we make no effort here to relate any of these concepts to biological systems; in fact, much of what we have discussed is most likely very different from biological neural information processing. Therefore it seems unlikely that quantum neural networks, at least in the context discussed here, could be considered a candidate for the basis of consciousness. However, Perus has suggested that neural networks can be a "macroscopic replica of quantum processing structures". If so, they "could be an interface between the macro-world of man's environment and the micro-world of his non-local consciousness" [43]. Thus, it is not out of the realm of possibility that future models of quantum neural networks may afterall provide significant insight into the workings of the mind and brain.

There are some proponents for the idea that QNN may be developed that have abilities beyond the restrictions imposed by the Church-Turing thesis. Simply put, according to this thesis, all existing computers are equivalent in computational power to the Universal Turing Machine. Moreover, all *algorithmic* processes we can perform in our mind can be realized on this machine and *vice versa*. No existing neurocomputers, nor any quantum computers theorized to date can escape the bounds imposed by the Church-Turing thesis. But what about quantum neural networks? Dan Cutting has posed the query, "*Would quantum neural networks be subject to the decidability constraints of the Church-Turing thesis?*" [39]. For existing models of QNN the answer seems surely to be "no", but some speculative physical systems (wormholes, for example) are discussed as possible candidates for the basis of QNN that could exceed these bounds [39]. This is a very intriguing question, and it is a challenge for the future to try to develop a theory of quantum neural networks that will give us completely new computational abilities for tackling problems that cannot now be solved even in principle. In the process we shall certainly be examining the concept of computation in a very different light and in so doing will be likely to make discoveries that to this point have been overlooked.

Acknowledgements

We are grateful to Professor Nikola Kasabov for his invitation to prepare this chapter. We also acknowledge useful discussions with Mitja Perus, Tony Martinez, Ron Chrisley, Dan Cutting, Elizabeth Behrman, and Subhash Kak on various aspects of quantum neural computation.

References

1. Feynman, R. (1986) Quantum mechanical computers. Foundations of Physics, vol. 16, pp.507- 531.
2. Penrose, R. (1994) Shadows of the Mind. A search for the missing science of consciousness. Oxford University Press, New York, Oxford.
3. Hameroff, S. and Rasmussen, S. (1990) Microtubule Automata: Sub-Neural Information Processing in Biological Neural Networks. In: Theoretical Aspects of Neurocomputing, M. Novak and E. Pelikan (Eds.), World Scientific, Singapore, pp.3-12.
4. Brooks, M. (Ed.) (1999) Quantum computing and communications, Springer-Verlag, Berlin/Heidelberg.
5. 5. Deutsch, D. (1985) Quantum theory, the Church-Turing principle and the universal quantum computer, Proceedings of the Royal Society of London, A400, pp.97-117.
6. Everett, H. (1957) "Relative state" formulation of quantum mechanics. Review of modern physics, vol.29, pp.454-462.
7. Dirac, P.A.M. (1958) The principles of quantum mechanics. Oxford, Claredon Press.
8. Domany, E., van Hemmen, J.L., and Schulten, K. (Eds.) (1992) Models of neural networks, Springer-Verlag. Berlin, Heidelberg, New York.
9. Hopfield, J.J. (1982) Neural networks and physical systems with emergent collective computational abilities, Proceedings of the National Academy of Sciences USA, vol.79, pp.2554-2558.
10. Feynman, R.P., Leighton, R.B., and Sands, M. (1965) The Feynman Lectures on Physics, vol. 3, Addison-Wesley Publishing Company, Massachusetts.

11. Vedral, V., Plenio, M.B., Rippin, M.A., and Knight, P.L. (1997) Quantifying Entanglement. Physical Review Letters, vol. 78 no. 12, pp. 2275-2279.
12. Jozsa, R. (1997) Entanglement and Quantum Computation. Geometric Issues in the Foundations of Science, S.Hugget, L.Mason, K.P. Tod, T.Tsou and N.M.J. Woodhouse (Eds.), Oxford University Press.
13. Feynman, R.P. and Hibbs, A.R. (1965) Quantum Mechanics and Path Integrals. McGraw-Hill, New-York.
14. Perus, M. (1996) Neuro-Quantum parallelism in brain-mind and computers, Informatica, vol. 20, pp.173-183.
15. Haken, H. (1991) Synergetic computers for pattern recognition, and their control by attention parameter. In Neurocomputers and Attention II: connectionism and neurocomputers, V.I. Kryukov and A. Holden (Eds.), Manchester University Press, UK, pp 551-556.
16. Behrman, E.C., Niemel, J., Steck, J.E., and Skinner, S.R. (1996) A quantum dot neural network. Proceedings of the 4th Workshop on Physics of Computation, Boston, pp.22-24, November.
17. Behrman, E.C., Steck, J.E., and Skinner, S.R. (1999) A spatial quantum neural computer., Proceedings of the International Joint Conference on Neural Networks, to appear.
18. Goertzel, B. Quantum Neural Networks. http://goertzel/org/ben/quantnet.html
19. Chrisley, R.L. (1995) Quantum learning. In Pylkkänen, P., and Pylkkö, P. (Eds.) New directions in cognitive science: Proceedings of the international symposium, Saariselka, 4-9 August, Lapland, Finland, pp.77-89, Helsinki, Finnish Association of Artificial Intelligence
20. Chrisley, R.L. (1997) Learning in Non-superpositional Quantum Neurocomputers, In Pylkkänen, P., and Pylkkö, P. (Eds.) Brain, Mind and Physics. IOS Press, Amsterdam, pp 126-139.
21. Deutsch, D. (1997) The fabric of reality. Alen Lane: The Penguin Press.
22. Bishop, C.H. (1995) Neural networks for pattern recognition, Clarendon Press, Oxford.
23. Cotrell, G.W., Munro, P., and Zipser D. (1985) "Learning internal representation from gray-scale images: An example of extensional programming", Proceedings of the Ninth Annual Conference of the Cognitive Science Society, Irvine, CS.
24. Gasquel, J.-D., Moobed, B., and Weinfeld, M. (1994) "An internal mechanism for detecting parasite attractors in a Hopfield network", Neural Computation, vol.6, pp.902-915.
25. Schwenk, H., and Milgram, M. (1994) Structured diabolo-networks for hand-written character recognition. International Conference on Artificial Neural Networks, 2, Sorrento, Italy, pp.985-988.
26. Ezhov, A.A., and Vvedensky, V.L. (1996) Object generation with neural networks (when spurious memories are useful), Neural Networks, vol. 9, pp.1491-1495.
27. Müller, B., Reinhardt, J., and Strickland, M.T. (1995) Neural Networks, Springer-Verlag, Berlin, Heidelberg.
28. Ezhov, A.A., Kalambet, Yu.A., and Knizhnikova, L.A. (1990) "Neural networks: general properties and particular applications". In: Neural Networks: Theory and Architectures. V.I. Kryukov and A. Holden (Eds.) , Manchester University Press, Manchester, UK, pp.39-47.

29. Ventura, D. and Martinez, T. (1999) Initializing the amplitude distribution of a quantum state", submitted to Foundations of Physics Letters.
30. Ventura, D. and Martinez, T. (1998) Quantum associative memory with exponential capacity, Proceedings of the International Joint Conference on Neural Networks, pp.509-513.
31. Milburn, G.J. (1998) The Feynman Processor, Perseus Books, Reading MA.
32. Menneer, T. and Narayanan, A. (1995) Quantum-inspired neural networks. Technical report R329, Department of Computer Science, University of Exeter, UK
33. Ventura, D. and Martinez, T. (1999) A quantum associative memory based on Grover's algorithm. Proceedings of the International Conference on Artificial Neural Networks and Genetic Algorithms, pp.22-27.
34. Ventura, D. (1998) Artificial associative memory using quantum processes. Proceedings of the International Conference on Computational Intelligence and Neuroscience, vol.2, pp.218-221.
35. Ventura, D. and Martinez, T.(1999) Quantum associative memory. Information Sciences, in press.
36. Grover, L.K. (1996) A fast quantum mechanical algorithm for database search. Proceedings of the 28th Annual ACM Symposium on the Theory of Computation, pp.212-219.
37. Ezhov, A.A.,Nifanova, A.V., and Ventura, D. (1999) Quantum Associative Memory with Distributed Queries, in preparation.
38. Gruska, J. (1999) Quantum computing, McGraw-Hill, UK.
39. Cutting, D.(1999) Would quantum neural networks be subject to the decidability constraints of the Church-Turing thesis? Neural Network World, N.1-2, pp.163-168
40. Gershenfeld, N.A. and Chuang, I.L. (1996) Bulk Spin Resonance Quantum Computation. Science, 257 (January 17), p.350.
41. Knill, E. , Laflamme, R., Martinez, R. and Tseng, C.-H. (1999) A Cat-State Benchmark on a Seven Bit Quantum Computer, Los Alamos pre-print archive, , quant-ph/9908051
42. Shor, P.W. (1997) Polynomial-time algorithm for prime factorization and discrete lpgarithms on a quantum computer, SIAM Journal on Computing, vol.26, pp.1484-1509.
43. Perus, M. (1997) Neural networks, quantum systems and consciousness. Science Tribune, Article - May. http:// www.tribunes.com/tribune/art97/peru1.htm

Chapter 12. Suprathreshold Stochastic Resonance in a Neuronal Network Model: a Possible Strategy for Sensory Coding

Nigel G. Stocks[1] and Riccardo Mannella[2]

[1] School of Engineering, University of Warwick, Coventry CV4 7AL, England
[2] Dipartimento di Fisica, Università di Pisa and INFM UdR Pisa,
  Via Buonarroti 2, 56100 Pisa, Italy

Abstract. *The possible mechanism to explain the dynamics of the transduction in sensory neurons is investigated. We consider a parallel array of noisy FitzHugh-Nagumo model neurons, subject to a common input signal. The information transmission of the signal through the array is studied as a function of the internal noise intensity. The threshold of each neuron is set suprathreshold with respect to the input signal. A form of stochastic resonance, termed suprathreshold stochastic resonance (SSR), which has recently been observed in a network of threshold devices [8] is also found to occur in the FHN array. It is demonstrated that significant information gain, over and above that attainable in a single FHN element, can be achieved via the SSR effect. These information gains are still achievable under the assumption that the thresholds are fully adjustable.*

1 Introduction

Despite its relatively simple dynamics, the exact strategy by which sensory neurons encode stimulus information is still not clear. Much of the problem lies in the fact that neurons are really black boxes, and we should infer what happens inside by probing them from outside. Here, we approach the problem of coding by sensory neurons from a modelling perspective: we will present a minimal nonlinear system, we will then maximise the amount of information flowing in the system (which is, arguably, what ideal sensory neurons should do), and we will show that the proposed mechanism does better than the other mechanisms known so far. Furthermore, the mechanism is robust against changes in the model parameters. The paper is organised as follows: first, we review some background material, and introduce some terminology; then we describe the nonlinear phenomenon that we believe may be at the core of the signal transduction mechanism (the Suprathreshold Stochastic Resonance); we will apply it to a parallel array of FitzHugh-Nagumo neurons; finally, conclusions will be drawn.

2 Background

The phenomenon of stochastic resonance (SR) has been much in the news recently, partly on account of its wide occurrence in many areas of science [1]. SR is commonly said to occur when the response of a nonlinear system to a weak signal is enhanced by an increase in the ambient or the internal noise intensity; a stronger definition requires that the signal/noise *ratio* (SNR) should also increase. The usual observation is that the SNR increases with increasing noise intensity, passes through a maximum, and then decreases again. Thus, the general behaviour is somewhat similar to a conventional resonance curve, except the noise intensity instead of frequency is used as the tuning variable. One of the reasons for the interest in SR is that in some cases (for instance, in biological systems) there is an intrinsic noise which cannot be removed: hence, it makes sense to ask whether the noise itself could be "tuned" to yield an optimal SNR. The original work on SR concentrated mostly on bistable systems, and on periodic input signals. Trying to apply SR to biological systems, however, we are forced to deal with aperiodic (broadband) signals.

One of the first applications of SR to neural networks is the work by Collins and coworkers [2]. There, for the first time the term of Aperiodic Stochastic Resonance (ASR) was introduced. It had been known for some time that SR could be enhanced by arrays of coupled resonators: it was therefore natural to carry the idea over, and to try to use a similar approach to understand and explain the behaviour of a neural network. Much in the spirit of signal analysis, the typical approach used to characterise the behaviour of such networks has been Linear Response Theory (LRT) [3]. Given an input signal $s(t)$ and the response of the network $r(t)$, LRT [4] implies that

$$r(t) = \int_{-\infty}^{t} \chi(t-t')s(t')dt'. \qquad (1)$$

Clearly, if the network is transmitting information between input and output, we would expect some kind of correlation between $s(t)$ and $r(t)$: also, we would expect that the better the transmission properties of the network are, the larger this correlation should be. A quantity which has been used to quantify information transmission is the cross-correlation coefficient

$$C_1 = \frac{\langle (s(t) - \langle s(t) \rangle)(r(t) - \langle r(t) \rangle) \rangle}{\sqrt{\langle (s(t) - \langle s(t) \rangle)^2 \rangle}\sqrt{\langle (r(t) - \langle r(t) \rangle)^2 \rangle}} \qquad (2)$$

where $\langle ... \rangle$ means time averages. It is immaterial in the expression above that $s(t)$ should be periodic or aperiodic, hence (2) it is a definition which could be used both in SR or in ASR.

Collins and coworkers [2] considered a parallel array of FitzHugh-Nagumo (FHN) model neurons. Each neuron evolved under the influence of a common input signal but independent internal noise. The overall response was obtained by adding the spike trains of each neuron at a common summing point. A single neuron remains in its quiescent (zero) state unless the input signal exceeds some level (threshold), in which case the neuron fires. Concentrating on so called "sub-threshold" signals (signals which are below threshold, in the absence of noise), one of the main results by Collins et al. has been that, using C_1 as an indicator, and in the presence of broadband signals, the transmission of the network improves as the number N of elements is increased, approaching unity (note that it does not reach one, as we will comment further down). Furthermore, whereas for small N, C_1 shows a maximum as a function of the noise intensity very much like in classical SR, when N is increased, this maximum broadens into a plateau and the dependence of the maximal $C_1(N)$ on noise intensity becomes weak. This leads to the conjecture that as $N \to \infty$ we observe the so called "Stochastic resonance without tuning" (to the noise intensity).

It has been however argued in the literature [5], that similar performances (i.e. C_1 almost order of unity) could be achieved even in a single FHN element: the idea is to change the threshold of the single element, so that the neuron is in a firing state even in the absence of any signal. Biasing the element in a supra-threshold regime and furthermore adding some refractory period in the spike-detection mechanism (basically, disregarding spikes within some short time after a firing event), the single FHN element can perform as well as a network of subthreshold FHN elements. This, of course, casts doubts as to the reason why neurons would set their thresholds to be subthreshold to the input stimulus. Indeed, it has been argued that, for broadband signals, optimisation of the information flow should be achieved via adjusting the threshold level rather than by tuning to noise [6]. Optimising the threshold level certainly results in much better signal transmission than could be hoped to be achieved tuning to noise (i.e using classical SR). In the work described below a possible solution to this problem, using the notion of suprathreshold stochastic resonance (SSR), is forwarded.

Finally, a side conclusion of [5] was that the mechanism behind SR without tuning is really the phenomenon of noise induced linearisation (NIL) [7]: as the noise is increased, the relation between output and input becomes more and more linear and hence C_1 increases. Of course, the behaviour of C_1 is a balance between linearisation (which is independent of N) and the ability of the network to ensemble average out the noise. For sufficiently large noise C_1 must decreases due to increased noise at the output. This will be discussed further in Sect. 4.

3 A Threshold Network

Recent work on a network of threshold elements [8] modelled as Heaviside functions (i.e., without internal dynamics) seems to suggest that a new form of SR, termed suprathreshold SR (SSR), can extend the dynamic range of multilevel systems to larger, suprathreshold, signal strengths. The output $y_i(t)$ of each element in the network is given by the response function

$$y_i(t) = \begin{cases} 1 \text{ if } s(t) + \xi_i(t) > \theta_i \\ 0 \text{ otherwise} \end{cases} \qquad (3)$$

where $s(t)$ is the signal (the same for all elements) and $\xi_i(t)$, with common intensity σ_ξ, is the noise at the input of the element i-th. The quantity θ_i is the threshold level for the i-th element. The response of the network $r(t)$ is obtained adding the output of the various elements, and clearly it is numerically equal to the number, n, of elements which are triggered at any instant of time.

Dynamically, the main difference between this network and the FHN network is the implicit inclusion of a refractory period in the FHN model. However, the statistics, and hence response, of these types of networks are governed by the setting of the threshold levels and not by the exact nature of the response of each individual device. For these

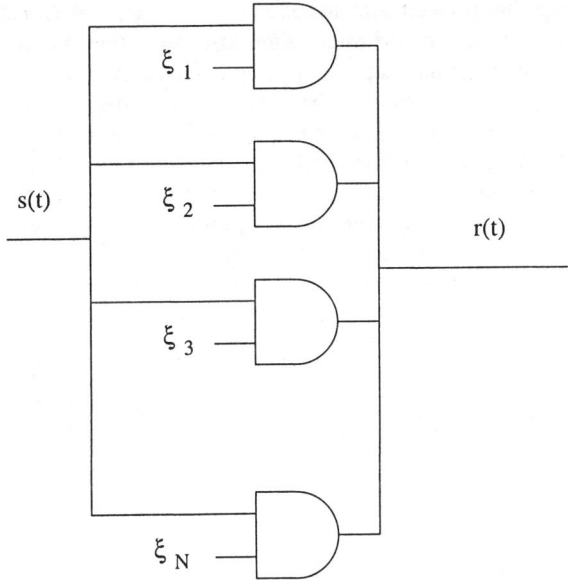

Fig. 1. A schematic diagramme of the network

reasons, the network of threshold devices is expected to capture the main qualitative features of the FHN network but at the same time be more amenable to exact theoretical analysis.

One of the problems with using C_1 (see (2)) as a measure is that it really only makes sense if the dynamics are linear. Although it may be argued that, following [5], linearization occurs in the system, at perhaps large enough noise amplitudes, it is however far from satisfactory to use a linear indicator in systems where, a priori, we should expect to find a nonlinear response. In general, an information theoretic measure is more advantageous than a linear signal processing technique, such as cross-correlation, because it is robust against a deterministic nonlinear deformation of the signal. Consequently, it is possible to quantify the dynamics of these networks, and hence SR, largely independently of the linearity of the response.

In [8] a more general indicator of the amount of information transmitted through the network was used, the so called *average mutual information* (AMI), which is an information theoretic measure widely used in signal analysis. The AMI (or transmitted) information, for the threshold network (which in information theory is regarded as a semi-continuous channel) can be written [9]

$$\text{AMI} = H(r) - H(r|s) = -\sum_{n=0}^{N} P_r(n) \log_2 P_r(n)$$
$$- \left(-\int_{-\infty}^{\infty} ds \sum_{n=0}^{N} P(n|s) P_s(s) \log_2 P(n|s) \right) \quad (4)$$

where $H(r)$ is the information content (or entropy) of $r(t)$ and $H(r|s)$ can be interpreted as the amount of encoded information lost in the transmission of the signal. $P_r(n)$ is the probability of the output $r(t)$ being numerically equal to n and $P(n|s)$ is the conditional probability density of the output being in state n given knowledge of the signal value, s. The logarithms are taken to base 2 so the AMI is measured in bits. If all information is lost in transmission $H(r|s) = H(r)$ (which occurs as $\sigma_\xi \rightarrow \infty$) and hence AMI=0. Alternatively, if all encoded information is transmitted ($\sigma_\xi = 0$) $H(r|s) = 0$ and AMI = $H(r)$. Given it is straightforward to show [9] that for any non zero σ_ξ, $H(r|s) < H(r)$, it would seem to follow that maximum information transfer occurs when there is no internal noise. However, this is not necessarily the case, because internal noise also serves to increase $H(r)$. Consequently, the maximisation of AMI by internal noise is a balance between possible additional information generated by the noise and the increased loss in information transmitted through the network with increasing σ_ξ. It is this ability of noise to maximise the transmitted information that is termed SR. We should add that the AMI has been used in the context of neuronal dynamics, among others, in [6,10].

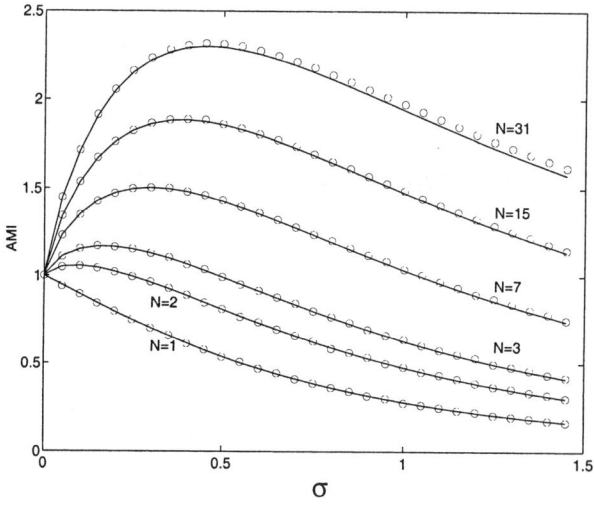

Fig. 2. Plot of transmitted information against $\sigma = \sigma_\xi/\sigma_s$ for various N and all $\theta_i = 0$. The quantity σ_s is the standard deviation of the broadband input signal. The data points are the results of a digital simulation of the network and the solid lines were obtained by numerically evaluating (5)

In addition to the subthreshold signal enhancements previously reported [2], in this model, by an appropriate choice of threshold levels, noise can also optimise the detection of suprathreshold signals. Figure 2 shows the results for all thresholds set equal to zero (the mean of the signal) and various N. In this situation the signal is strongly suprathreshold, yet SR type behaviour (i.e. a noise induced maximum) is still observed for all $N > 1$. As N increases the maximum value attained by the AMI also increases.

This type of SR was be termed suprathreshold stochastic resonance (SSR). In contrast to classical SR, SSR is observed when the threshold levels are packed about the mean of the signal with a standard deviation of less than σ_s. In other words, SSR is observed when all thresholds adapt to the DC-component level of the signal distribution. Again in contrast to classical SR, SSR effects do not diminish with increasing signal magnitude - as long as the noise is scaled accordingly. Consequently, these effects can be observed with signals of any magnitude without having to modify the threshold levels.

The mechanism giving rise to SSR is quite different to that of classical SR and is not connected to a previously reported form of suprathreshold SR[11]. In the absence of noise, all devices switch in unison and consequently the network acts like a single bit analogue to digital converter (AMI = 1bit). Finite noise results in a distribution of thresholds that gives access to more bits of information; effectively the noise is accessing more degrees of freedom of the system and hence generating information. ¿From the law of large numbers it is easy to establish that the signal response will grow $\sim N$ whilst the error scales as $\sim \sqrt{N}$, consequently the signal-noise ratio, and hence the transmitted information, improves with increasing N. The network also linearises in a similar way described in [5]. However, whilst noise induced linearisation (NIL) nearly always occurs in strongly nonlinear systems, SR, or SSR, does not.

If all thresholds are set to the same value (taken to be θ) the AMI can be calculated exactly:

$$\text{AMI} = -\sum_{n=0}^{N} P_r(n) \log_2 P'(n) - \left(-N \int_{-\infty}^{\infty} dx P_s(x)\right.$$
$$\left. (P_{1|x} \log_2 P_{1|x} + P_{0|x} \log_2 P_{0|x})\right),$$
$$P'(n) = \int_{-\infty}^{\infty} dx P_s(x) P_{1|x}^n P_{0|x}^{N-n} \quad (5)$$

where $P_r(n) = C_n^N P'(n)$, C_n^N is the Binomial coefficient. The signal distribution is taken as Gaussian,

$$P_s(x) = (1/\sqrt{2\pi\sigma_s^2}) \exp(-x^2/2\sigma_s^2),$$

and $P_{1|x} = 1/2 \,\text{Erfc}[(\theta-x)/\sqrt{2\sigma_\xi^2}]$, is the conditional probability of $y_i = 1$ for a given x (note all devices are identical) and similarly $P_{0|x} = 1 - P_{1|x}$ is the probability of a zero given a signal value x. Erfc is the complementary error function. The integrals and summation in (5) can be calculated numerically (solid lines in Fig. 2). Good agreement between simulation and theory is observed confirming the existence of the SR effect.

4 The FHN Network

We will now go on to investigate whether SSR effects do indeed exist in more realistic neural networks. The model we considered is a network of N FitzHugh-Nagumo (FHN) [12] equations, driven by signal and noise. The single system equation considered has the form (for the i-th neuron)

$$\epsilon \dot{v}_i = -v_i(v_i^2 - \frac{1}{4}) - w_i + A - b_i + S(t) + \xi_i(t)$$
$$\dot{w}_i = v_i - w_i \quad (6)$$

where the $v_i(t)$ are fast (voltage) variables and $w_i(t)$ are slow (recovery) variables. All neurons are subjected to the same aperiodic signal $S(t)$, and in general they could have different thresholds b_i. The stochastic process $\xi_i(t)$ will be taken Gaussian, with zero average and correlation given by

$$\langle \xi_i(t)\xi_j(s) \rangle = \delta_{ij} 2D\delta(t-s)$$

In the following we will also fix some of the variables to their customary values, namely $\epsilon = 0.005$ (time scale of the fast variables), $A = -0.2451$ (constant, tonic activation signal).

The integration of the relevant equations of motion was done with the Heun algorithm [13]. The output of the different neurons were added together, as depicted in Fig. 1 and like in the previous section, so the output signal is given by

$$\tilde{r}(t) = \sum_{i=1}^{N} v_i(t).$$

The first problem we have to face is how to introduce the AMI for our model. We follow the rate encoding method used in [2] and pass $\tilde{r}(t)$ through a low pass filter, according to the equation

$$\dot{r}(t) = -\frac{1}{t_p}r(t) + \tilde{r}(t)$$

to reconstruct the response $r(t)$. In the following, the quantity t_p will be loosely referred to as "pole time". We should add that a finite pole time could be thought of as having an analogy with signal reconstruction at the membrane of a summing neuron. Unlike the threshold networks studied in Sect. 2, the FHN network response has a continuous distribution and therefore, a continuous description of the AMI must be used. This is similar in definition to (4) with the summation replaced by an integral over all output states r.

The dynamics of the FHN equation is dominated by a Hopf bifurcation: in the absence of input, below/above the bifurcation point the "neuron" is in a quiescent/firing state. Simulations for the FHN network need some care: for the parameter chosen above, we generally took an integration time step order of 10^{-2} or smaller; the input signals were sequences of length equal to $2^{18} \times 10^{-2}$ seconds, obtained filtering a white Gaussian process through a low pass filter with cut-off of 0.2 rads^{-1}. Typically, we took 16 to 32 independent signal sequences (to build the signal probability $P(s)$) and 16 to 32 noise realizations for each signal sequence. The conditional probability distribution $P(r|s)$ and the response probability $P(r)$ were built, and from them the AMI computed. Typically, the one dimensional probabilities were digitized into 1024 or 2048 bins, $P(r|s)$ into an array of 1024×1024 or 2048×2048 bins. Typically, $t_p = 2$s.

First, motivated by the results on the threshold network, we set all $b_i = 0$. In this condition, the FHN equation is sitting just at the Hopf bifurcation point, but in suprathreshold with respect to the signal. The result of the simulations is shown in Fig. 3, for varying N.

It is clear that even for this model SSR is present. Similar to the results on the threshold network, the AMI peaks at finite values of the noise intensity for all $N > 1$, and its maximum value increases as N is increased. Networks with different N's will

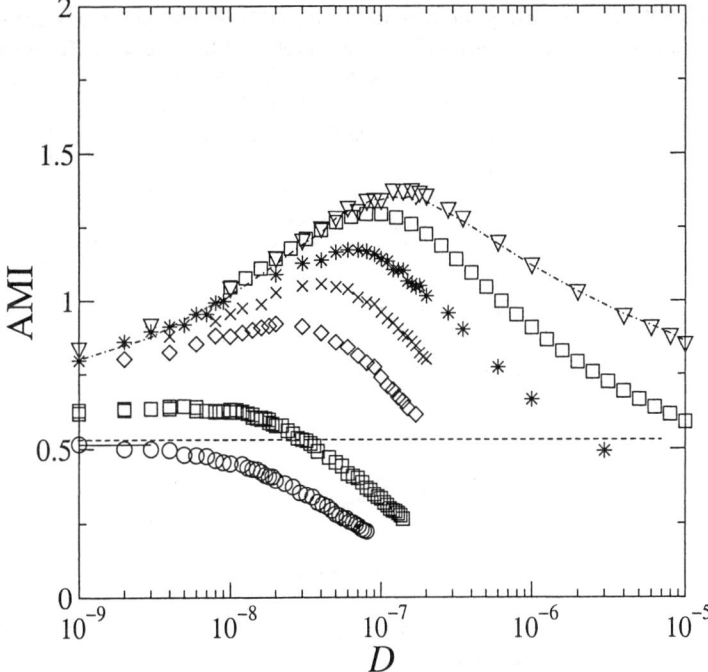

Fig. 3. AMI for a FHN model with $b_i = 0$, as function of D, for different N and $\sigma_s = .00387$. From bottom to top, $N = 1, 2, 8, 16, 32, 64, 128$. The dashed horizontal line is the maximum AMI for a single firing neuron, the short solid line is the AMI for $D = 0$ (same AMI for all N). The dot-dashed line is the expected AMI under the assumption that the network is acting as a Gaussian linear channel.

yield the same AMI for $D = 0$, and this is shown by the short solid line on the graph. Note that the AMI for $N = 1$ is a monotonic decreasing function. The maximum AMI trasmitted by a single neuron (obtained by varying b_i and setting it to its optimal value) is shown by the dashed line. Consequently, the network can significantly outperform a single neuron even with the thresholds set to a suprathreshold level.

One may wonder if NIL is achieved in our network. To understand this, we plotted on Fig. 3 with a dot-dashed line the expected AMI computed under the assumption that the network can be treated as a Gaussian linear channel for $N = 128$. This line was obtained by evaluating C_1 from the simulations, and using the equation [14]

$$\text{AMI}_G = -\frac{1}{2}\log_2(1 - C_1^2) \tag{7}$$

The agreement with the AMI obtained from the simulations is excellent at large noise intensities, but small discrepancies start to appear at lower noise values. Whilst these discrepancies appear small they are in fact significant. Linear Gaussian channel theory predicts that $C_1 \to 1$ as $D \to 0$ and hence AMI $\to \infty$, which clearly cannot occur in these networks. Hence, the discrepancies point to a breakdown of the linear assumption and the increasing importance of nonlinearity.

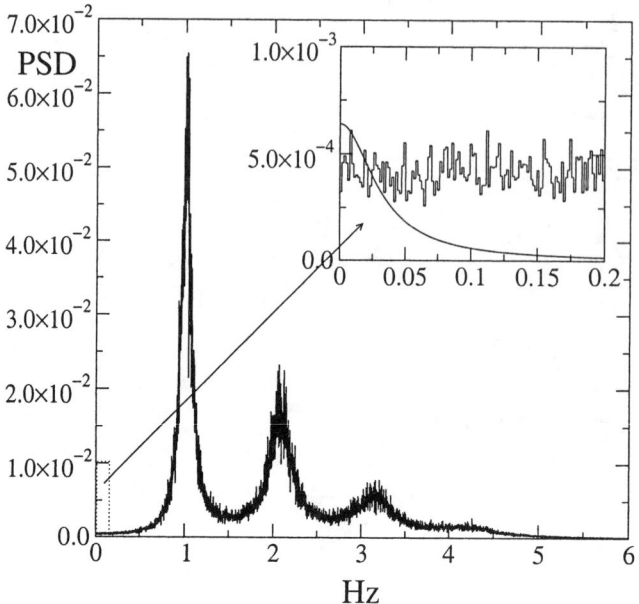

Fig. 4. PSD of a neuron for $D = 10^{-6}$. Inset: an expanded view of the low frequency region: the skyline curve is the neuron PSD, the solid line is the PSD of the input signal.

It should be stressed that NIL can occur in two different senses [7]. First, in the sense that the system dynamics can be described by a linear response function such as $\chi(\tau)$ in (1) and secondly, in the stronger sense, that dispersion effects are also removed and hence the response is simply given by,

$$\langle r(t) \rangle = Gs(t) + C \tag{8}$$

where G and C are constants. This is NIL in the "HiFi" sense. For (7) to hold we require NIL in this stronger sense. From (1) it is easy to see that for (8) to be valid, the response function has to be closely approximated by a delta function response with weighting G. This, via the fluctuation-dissipation theorem [4], implies the power spectral density (PSD) in the absence of the signal has to be flat over the range of frequencies occupied by the signal.

To test for this, we obtained the PSD in a single FHN element (see Fig. 4), in the absence of input signal. It is observed that, indeed, the PSD is approximately constant over the signal frequency range, thus explaining the validity of (8).

However, the fact that the PSD does have some frequency dependence (albeit weak) is important. If there were no dispersive effects, then, in principal, taking the limit $N \to \infty$ would yield the result $C_1 \to 1$ and hence AMI $\to \infty$. The small amount of distortion that does occur due to dispersion places an upper bound on the limiting value value of C_1 and hence the AMI. This is true regardless of how many elements are included in the array. Perfect signal transmission can only occur in the limit that the signal bandwidth goes to zero and $N \to \infty$.

¿From the above discussion it would be easy to conclude that SSR in the FHN network is simply due to a linearisation of the system dynamics with increasing noise. However, one has to be careful about making such statements. As discussed in [7], NIL is a general phenomenon and is to be anticipated in most systems with strongly nonlinear dynamics. Consequently, NIL will always occur if SSR occurs but the reverse is not true i.e. the occurrence of NIL (characterised by a noise induced maximum in C_1) is not a sufficient condition to predict the onset of SSR. This statement is also valid for classical SR.

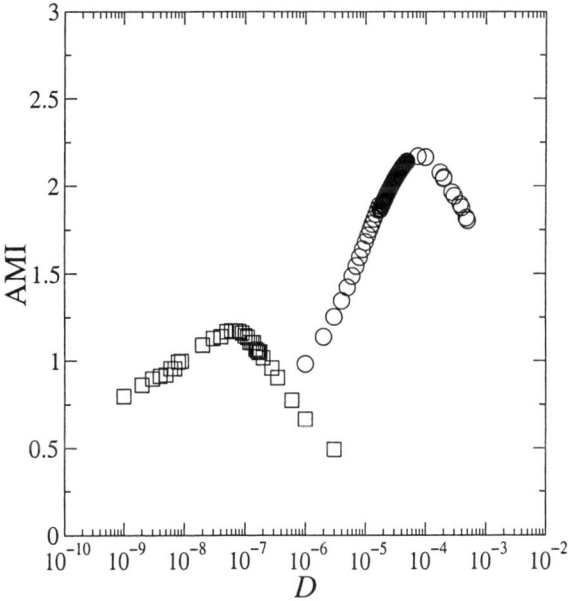

Fig. 5. AMI vs noise intensity in a network with $N = 64$ elements: circles, $b_i = 0$ and $\sigma_s = .00387$, squares, $b_i = -0.11$ and $\sigma = .0387$

Finally, as pointed out in [5], a single suprathreshold biased neuron (with respect to the bifurcation point) can outperform a whole array of subthreshold neurons. This obviously raises doubts as to whether neurons really do use classical SR (which requires the neurons to be subthreshold) as a means of encoding sensory stimuli. For this reason we studied the case similar to [5] with all the neurons biased into their constant firing mode. The results of simulations done for $b = -0.11$ (a value well above the Hopf bifurcation point) and $N = 64$ are shown in Fig. 5. We plot here the AMI as function of the noise intensity for $N = 64$ and for comparison the results for $b_i = 0$.

It is clear from the figure that the AMI increases dramatically, well above what is expected for a single element. This tends to suggest, that unlike classical SR, significant benefit can be gained by a network utilising SSR *even when the thresholds are fully adaptable*.

5 Conclusions

We have demonstrated that, similar to threshold networks, a network comprising of FHN model neurons also displays the SSR effect. Furthermore, enhancements of the transmitted information can be achieved well in excess of that attainable by a single individual neuron. In stark contrast to classical "subthreshold" SR, this conclusion remains valid even when the thresholds are fully adjustable. Consequently, the argument as to whether neurons could use SR or SSR (or both) to enhance information flow is really matter of determining to what extent neurons can adjust their thresholds. If sensory neurons can freely adapt their thresholds then it seems difficult to understand why they would rely on classical SR to enhance information flow. On the other hand SSR does not suffer from these objections.

The mechanism proposed (SSR) is clearly fairly robust, requires noise to be present, leads to information flows which are relatively large. Work in progress shows, furthermore, that the output SNR of a single neuron in an SSR network seems to be in agreement with that measured in many physiological studies (order of unity). It is possible, therefore, that SSR could be a mechanism by which sensory neurons encode information.

This work was supported by the INFM Parallel Computing Initiative.

References

1. Gammaitoni L., Hänggi P., Jung P., Marchesoni F. (1998) Rev. Mod. Phys. 70:223; Bulsara A., Gammaitoni L. (1996) Physics Today 49:39; *Proceedings of the International Workshop on Fluctuations in Physics and Biology: Stochastic Resonance, Signal Processing and Related Phenomena*, Bulsara A., Chillemi S., Kiss L., McClintock P. V. E., Mannella R., Marchesoni F., Nicolis K., Wiesenfeld K. (1995) Nuovo Cimento D 17:653
2. Collins J. J., Chow C. C., Imhoff T. T. (1995) Nature (London) 376:236
3. Neiman A., Schimansky-Geier L., Moss F. (1997) Phys. Rev. E 56:R9
4. Dykman M. I., Mannella R., McClintock P. V. E., Stocks N. G. (1990) Phys. Rev. Lett. 65:2606; Dykman M. I., Mannella R., McClintock P. V. E., Stocks N. G. (1990) Pis'ma Zh. Ezsp. Teor. Fiz. 52:780 [JETP Letters 52:141 (1990)]
5. Chialvo D. R., Longtin A., Müller-Gerking J. (1997) Phys. Rev. E 55:1798
6. Bialek W., de Weese M., Rieke F., Warland D. (1994) Physica A 200:581; DeWeese M. (1996) Network - Computation in Neural Systems 7:325; DeWeese M., Bialek W. (1995) Nuovo Cimento D 17:733
7. Dykman M. I., Luchinsky D. G., Mannella R., McClintock P. V. E., Short H. E., Stein N. D., Stocks N. G. (1994) Phys. Lett. A 193:61; Stocks N. G., Stein N. D., Short H. E., McClintock P. V. E., Mannella R., Luchinsky D. G., Dykman M. I. (1996). In: M Millonas (Ed.) Fluctuations and Order: the New Synthesis. Springer, Berlin, 53-68
8. N.G. Stocks, "Suprathreshold stochastic resonance in multilevel threshold networks" submitted to Phys. Rev. Lett. and N.G. Stocks, "Suprathreshold stochastic resonance" to appear in the AIP proceedings of the "Stochaos" conference, Ambleside, 16-20 August 1999.
9. Feinstein A. (1958) Foundations of Information Theory McGraw-Hill, New York
10. Bulsara A. R., Zador A. (1996) Phys. Rev. E 54:R2185
11. Apostolico F., Gammaitoni L., Marchesoni F., Santucci S. (1997) Phys. Rev. E 55:36
12. Longtin A. (1993) J. Stat. Phys. 70:309

13. Mannella R. (1997) Numerical Integration of Stochastic Differential equations. In: L Vázquez, F Tirando and I Martín (Eds.) Supercomputation in nonlinear and disordered systems: algorithms, applications and architectures. World Scientific, 100-130
14. Reza M. F. (1994) An introduction to information theory. Dover, New York

Part IV

Bioinformatics

Chapter 13. Information Science and Bioinformatics

Chris Brown*, Mark Schreiber, Bernice Chapman, Grant Jacobs

Department of Biochemistry, University of Otago,
P.O. Box 56 Dunedin, New Zealand.
*Corresponding author, email: chris.brown@stonebow.otago.ac.nz

Abstract. The new field of bioinformatics has resulted from a marriage between Biology, Statistics and Information Science. Biology and informatics are likely to be the two most important sciences for the next century, and this interdisciplinary field between them will have a crucial role. Computational analyses of vast amounts of DNA sequence data are currently being used to find genes, and to predict the roles of the encoded proteins in cells. Currently successful data-mining approaches include the use of Hidden Markov Models and Neural Networks. We are utilising these methods to accurately predict genes and also to understand how proteins are made from these coded instructions.

Keywords. Bioinformatics, molecular biology, genomics, protein structure prediction

1 What is Molecular Bioinformatics?

We define bioinformatics as 'the development and application of information technology to the analysis of large amounts of biological information [1]. An explosion in the amount and availability of many types of biological data, particularly DNA sequence data (the genetic blueprint), has brought about this revolutionary new field [2-5]. Here we will concentrate on the analysis of DNA, RNA and protein sequence and structure data, or 'molecular informatics'. Molecular informatics is considered critical in the future of biology and the focal point of much commercial and scientific investment. Other aspects of bioinformatics have recently been reviewed elsewhere [6-14].

The amount of DNA sequence data publicly available is increasing exponentially, currently doubling every 18 months [13]. What now faces the study of life at a molecular level (molecular biology) is a veritable "Embarrassment of Riches", although fortunately improvements in computer hardware and software technology have paralleled this growth [11]. Several advances have coincided to produce this wealth of DNA information. Recent breakthroughs in DNA sequencing technology and automation have greatly increased the output rate of the global sequencing consortiums [15, 16]. Further driving this race is competition between private and public sectors to complete the genomes of several important species [17]. Perhaps the most important of these is the human genome (the entire DNA-encoded set of instructions that make us human). Concurrent advances in information distribution via the World Wide Web have aided the dissemination of this data to researchers world-wide. The development of platform independent computer languages such as Perl and Java have alleviated many of the network establishment and maintenance problems faced by genome sequencing consortiums.

Molecular biology is founded on a mechanistic view of life. It is expected that a reductionist approach to studying the component parts of an organism, in isolation, will allow an understanding of the organism as a whole. One major goal of bioinformatics is to predict the nature of an entire organism from its genomic DNA sequence in silico (in a computer). It will be the greatest experiment in molecular biology, and will be augmented by biochemical experimental techniques in vitro (in glass) and in vivo (in life) which permit high though-put experimental data accumulation. This data will in turn require new informatics to handle it [13].

Major Questions Currently Being Addressed by Molecular Bioinformatics

The reasons for addressing these questions vary from the purely scientific to commercial, so can be phrased in different ways.

1. We have determined a DNA sequence; does it encode a protein? Shall we keep sequencing or have we discovered a potential goldmine?

2. In what cells will this protein be made? Is this protein important for memory in the brain, wood strength in tree trunks, or nutritional quality of soya beans?

3. What does the protein do? Is the protein likely to be involved in cancer? If trees make more of it will it make wood stronger?

4. What is the three dimensional shape of the protein molecule? I'd like to have a drug to change how the protein works, what drugs shall I try first?

This paper will address several of these questions. The collection and compilation DNA sequences has become almost routine, the difficulty now lies in interpreting the data to address questions like those above. It is here that much of the current effort is focussed. Our own research concerns this aspect of bioinformatics [18-21]. Although the biological questions can only be answered definitively by experiment, such techniques are too slow and expensive to examine the hundreds or thousands of genes being discovered every day.

First the types of data available will be considered, then current methods for predicting if a DNA sequence encodes a protein (Question 1 above). Next, current methods for predicting what a protein does and how it will be made (2 and 3) are examined. Finally, methods applicable to protein structures including the important problem of determining if a representative protein structure is available for a user's protein sequence will be addressed (4). Particular emphasis will be placed on the use of fuzzy systems for this interpretation.

2 Biological Data and Flow of Information

Although in cells the information is stored as DNA sequence, the characteristics of the organism are due mainly to the proteins that have been made using these instructions. For example, a developing human red blood cell mainly makes (or expresses) just two haemoglobin genes of the estimated 80,000-120,00 human genes, producing haemoglobin mRNA, and haemoglobin protein. The flow of information from DNA to RNA to protein is called the central dogma of molecular biology (Fig.1). This flow of information is accompanied by increasing fuzziness. We presently have a wealth of data at the DNA level, but less clear information at the RNA or protein level. One of the main goals of bioinformatics is to mimic or model this controlled flow of information by computer.

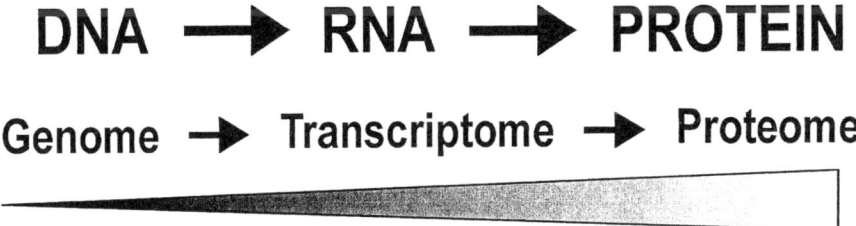

Fig. 1. The Central Dogma of Molecular Biology or the process of gene expression. Every cell in an organism contains essentially the same DNA (Genome), but each section of the DNA may be copied or transcribed into mRNA in several different ways, and in turn each of these mRNAs could produce several functional proteins. Thus there is increasing fuzziness in the data (indicated by increasing gradient).

The Structure of DNA, RNA and Proteins

DNA, RNA and proteins "encode" their biological structure and function and hence contain biological information. These molecules are made of chains of repeated units. These units are composed of two parts, an invariant repeated *backbone* and a protruding *side-chain* which varies in structure. For simplicity, these units are referred to by single-letter codes representing the chemical names of the varying side-chains. DNA is composed of chains having four sidechains or bases: Thymidine (T), Guanidine (G), Cytidine (C), and Adenine (A) linked by a deoxyribose sugar phosphate nucleic acid backbone. Two anti-parallel complementary chains make up the characteristic double helical secondary structure of DNA (Fig. 2). The information on a strand of DNA must be read in a one direction (the 5' end to the 3' end), but as there are two strands, two readings are possible, from one double stranded DNA. RNA has a similar structure, but with Uridine (U) taking the place of thymidine, it also uses a different sugar in the backbone. From an information point of view RNA may be considered to be a working copy of the DNA sequence, but as it is single stranded it can only be interpreted in one direction.

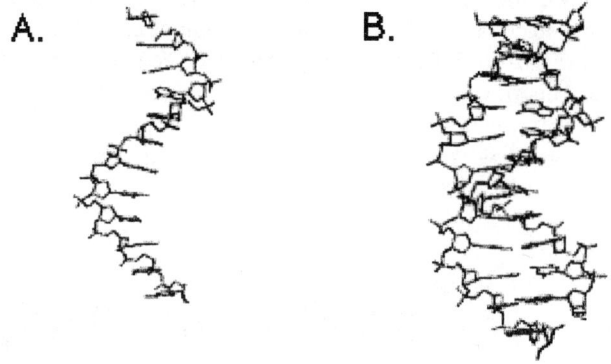

Fig. 2. DNA structure. (A) Primary structure of DNA, a deoxyribose backbone with bases protruding. The four bases may occur in any order. (B) Two anti-parallel complementary chains make up the double helical secondary structure of DNA.

Proteins are quite different. They are composed of chains of the twenty amino acids (A, C–H, I, K–N, P–T, V, W, Y, Fig. 3). Proteins are encoded by the informational molecules DNA and RNA but perform functions in the cell, thus determining a cell's characteristics.

In DNA, RNA and proteins the primary units can be chemically modified to give more variation, which we will only consider briefly below. The repeated units (bases or amino acids) make up the *primary structure* or *sequence* of the molecule, which is usually represented as a string using the single-letter codes. These units are arranged in space to form characteristic *secondary structure* shapes by forming chemical interactions between neighbouring units. This higher structure is particularly important in proteins, they have several characteristic repeated units of structure – for example: beta strands and alpha helices (Fig. 4).

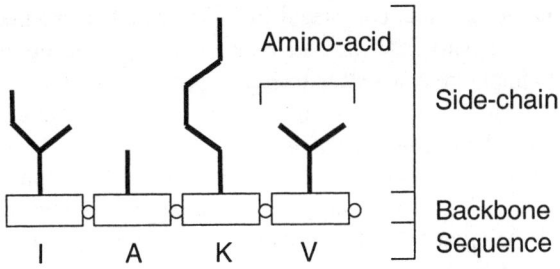

Fig 3. Schematic primary structure of a part of a protein with sequence IAKV.

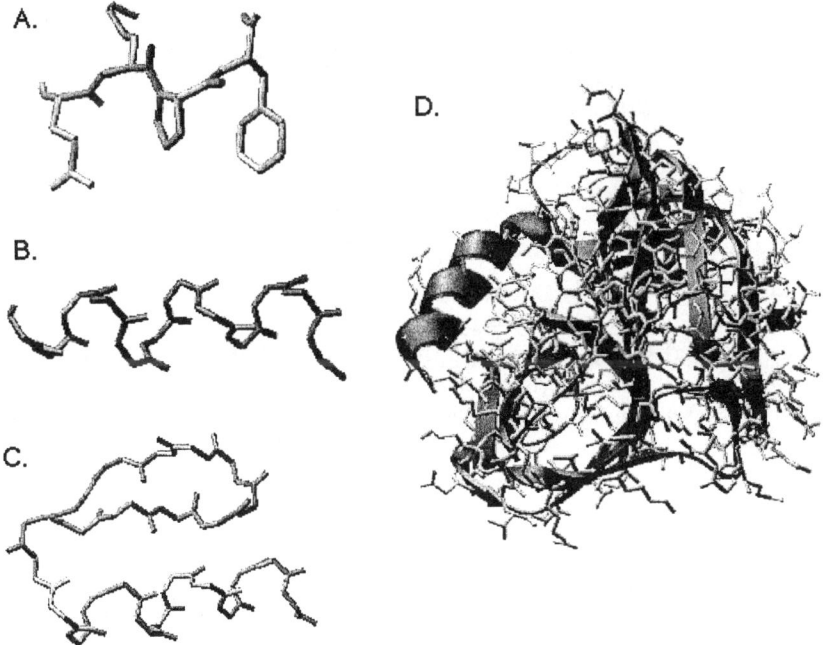

Fig. 4. Protein structure. (A) A β-strand secondary structure (B) An α-helix secondary structure (C) The structures in A and B combined to form a larger structure (D) An example of the complexity of a small protein (a serine protease required for human digestion). A "ribbon" is superimposed on the positions of the atoms to indicate the positions of secondary structures; helical ribbons are α-helices and flat ribbons β-strands.

Genome (DNA)

The complete set of DNA encoded instructions in an organism is called the genome. The first complete genome of an organism, that of the bacterial pathogen

Haemophilis influenzae, was completed in 1995 [22]. It contained a little fewer than 2 million base pairs (2Mbp) and took only a year to complete what had hitherto been an impossible task. It is interesting to compare this size to the size of current computer applications, which often require much more than 2MB to perform tasks much less complex than life.

In 1999 a private company, Celera Genomics, completed the 1.8 billion base sequence of the fruitfly (*Drosophila melanogaster*) in just three months. At time of writing this is being distilled by the largest civilian computer in the world into the final 400 million base genomic sequence. This 'model' organism is of commercial as well as scientific interest as interpretation of this data should aid in development of new drugs. With these advances in technology it is expected that a working draft of the human genome will be completed by late 2000. This draft sequence may have error rates of up to one per one thousand bases, due mainly to cost cutting. The final US$ 3 billion human genome project aims to have less error (1/10,000 bp). The presence of these errors needs to be borne in mind in developing methods to analyse the data.

Transcriptome (RNA)

DNA information is not directly translated into functional proteins. Rather it is first transcribed to a mRNA (messenger RNA) transcript. The subset of mRNAs produced in an organism or cell is known as the transcriptome. Each cell type has a unique transcriptome, differing both in identities and amounts of each mRNA. Although the transcriptome is a subset of the genome, it is difficult to predict which genes will be activated in a cell at a particular time or under what conditions. New technologies to analyse the transcriptome as a whole have recently been developed and these permit the simultaneous study of thousands of genes [23-26].

To further complicate computer analysis of the genes of higher organisms, regions that encode proteins (exons) are usually interrupted by non-coding regions (introns). In the cell a site-specific process called splicing removes the introns to make a functional mRNA (Fig. 5). Correct prediction of intron/exon structure by computer has been a very active area of research [27-29]. It is essential for the major higher organism sequencing projects that will be completed by the end of 2000 (Nematode Worm, Fruitfly, Thale cress plant, Human). In addition alternative splicing of exons can lead to variants of the original message, thus allowing a single gene to encode several variant mRNAs. These mRNAs may then encode different forms of a protein with different functions.

By extracting all the mRNAs from a cell and "reverse-transcribing" them to DNA, libraries containing the sequences of the mRNAs after splicing can be obtained. These libraries are called Expressed Sequence Tag (EST) libraries. Many genes may not be expressed in the cell or tissue used, and will therefore be absent from these libraries. Frequently only partial sequence is obtained, giving an incomplete EST with a large number of errors in the sequence. EST libraries have been made and mapped onto a genome to identify all genes involved in coding proteins in that tissue at that time [30, 31]. This type of approach has been used by Human Genome Sciences to focus on the 3% of the human genome that actually encodes proteins [32]. Human Genome Sciences claim to have identified 80% of all human genes. With thousands of patents in the pipeline, this data is not publicly available.

Proteome (Protein)

Proteins provide the majority of the structural and functional components of a cell. The total protein complement of a cell is called the Proteome. Because a mRNA can make several copies of a single protein with varying efficiency there is a poor correlation between the transcriptome and the proteome of a particular cell. The Proteome is often studied using protein gels that allow visualisation of protein levels [33]. Proteins gels also allow visualisation of major protein variants [34] Sequencing of short sections of proteins from a protein gel can be used to confirm the location of genes encoding those proteins in the genome [35,36].

Databases of Biological Data

Database technology is a critical aspect of Bioinformatics. Databases of each of the types of data above are currently freely available [37]. The primary databases are available to the international research community via the WWW, as are several cross-referenced databases. Major publicly funded efforts in the USA [38] and Europe [39] and Japan [13] have produced integrated databases. Other commercial equivalents are being developed. Relational database technology and developments in WWW technology are important in making these data sources accessible to the research community and to presenting tools to perform specialised analyses of the data (often of the kind not able to be practically encoded in a relational database). CORBA [40] and XML [41, 42] (particularly the latter) look set to have important roles in allowing data from the diverse databases to be used together. Both of the most popular CGI modules for Perl (CGI.pm and cgi-lib.pl [43]) – driving much of the "dynamic" WWW – were written by bioinformatics scientists. More recent developments include the BioPerl [44] and BioJava [45] projects aimed at presenting software modules for working with biological data. Secondary databases are built from the data within the primary databases, with the data presented in a manner useful for specific types for studies along with appropriate computational methods [37].

An excellent example of a secondary database is Transterm [46, 47]. This database contains information concerning the way translation of mRNAs is controlled. It is curated by experts in this field and provided to the world via the WWW. It contains a 'value added' subset of the sequences available in the major primary databases. This subset can be downloaded for further analysis or examined using online tools with user friendly interfaces. Its current implementation is a relational database (PostgresQL) with access provided by a CGI written using Perl [21]. It has been used successfully to predict biological functions [18, 19] which were subsequently demonstrated experimentally [48].

3 Hidden Markov Models, Neural Networks and Machine Learning to Find Genes

Finding Genes in Bacterial DNA Sequences

Accurately finding genes in newly generated DNA sequence is crucial to elucidating biology. The genome sequencing projects are generating more data that can be easily applied to experimental approaches to identify genes within DNA sequences. The challenge is to accurately identify and annotate genes *in silico* and thus "annotate" the DNA sequence with the (putative) location of biological features such as genes. Fortunately some biological rules can be applied to this process.

Messenger RNA (mRNA) sequence is translated into protein sequence by cellular machines called ribosomes. The ribosome reads mRNAs and produces a string of amino acids (protein) based on the information contained in the mRNA. Three adjacent ribonucleic acid (RNA) bases, called a codon, code for each amino acid. For example, the codon AUG codes for the amino acid Methionine (M). There are 64 possible codons (4^3) but only twenty amino acids, so the genetic code is redundant. AUG is also an initiation codon as it signals the ribosome to initiate translation. Two other codons can also be used to initiate translation (UUG and GUG) in bacteria. Three codons signal the end of translation, UAA, UGA, and UAG ("stop" or "termination" codons). Therefore, a region that might code for a protein must begin with an initiation codon and end with an in frame termination codon. These regions are called Open Reading Frames (ORFs). For double stranded DNA there are always six possible reading frames (three in each direction) many of these will be ORFs.

ORFs which encode a genuine protein usually contain sequence bias. Both the redundancy of the code and the three base periodicity of the codons give considerable sequence bias. This bias is often used in the prediction of coding regions and is referred to as coding potential [49] We will first examine gene prediction in bacteria were gene structure is simpler than in "higher" organisms (eukaryotes) (Fig. 5).

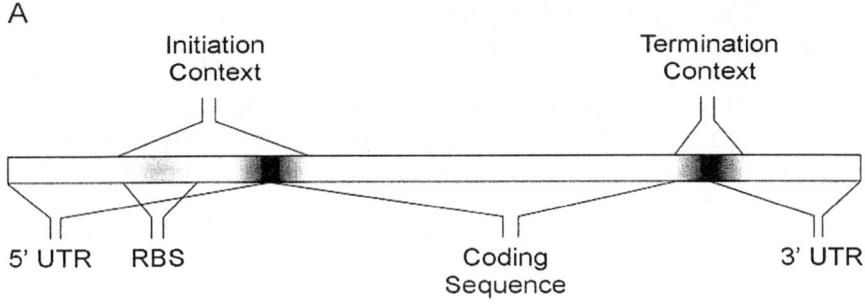

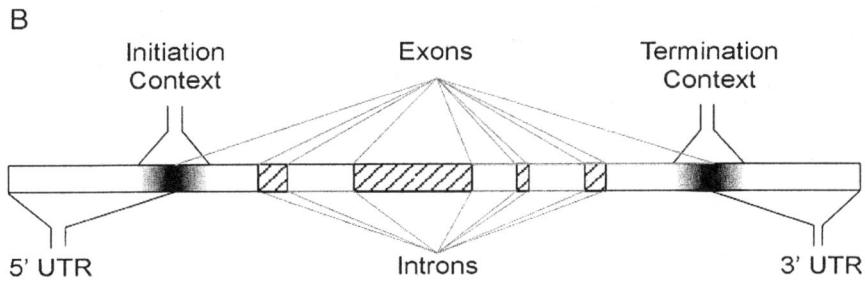

Fig. 5. Gene structure. Each long rectangle represents a typical mRNA. (A) Bacterial gene structure. The 5' UTR (untranslated region) is defined as all untranslated mRNA prior to the initiation codon. The RBS (ribosome binding site; light grey shading) is the motif recognised by the ribosome during selection of the correct initiation codon. Initiation context, the region of mRNA involved in initiation including the RBS, the initiation codon and surrounding bases. Coding region, the Open Reading Frame (ORF) from the initiation codon to the termination codon. Termination context, the termination codon and surrounding bases that regulate the efficiency of termination. The 3' UTR is defined as all untranslated mRNA after the termination codon. Regions shown in grey represent less well conserved motifs. Dark regions represent high conservation (bases that vary infrequently); light regions represent low conservation. (B) Higher organism (eukaryotic) gene structure. Eukaryote mRNAs do not have an RBS as the ribosome binds at the 5' terminus of the mRNA and initiates at the first initiation codon. Hatched boxes represent the introns that interrupt coding regions in higher organisms.

The Use of Hidden Markov Models to Find Genes

Gene structure can be thought of as following a Chomsky grammar [50-52]. It is therefore appropriate to model this grammar using Hidden Markov Models (HMMs). A HMM asks, what is the probability that the test sequence matches the model. In Bayesian terms:

$$P(M \mid S) = \frac{P(S \mid M)P(M)}{P(S)} \quad (1)$$

Where $P(S)$ is the probability of the sequence calculated from *a priori* observations of nucleotide (or amino acid) frequencies, $P(S \mid M)$ is the probability of the model emitting the observed sequence, and $P(M)$ is the probability of the model which is a constant (for this type of analysis). The models are usually applied over windows 90 - 120 bases in length. To find genes all ORFs are predicted and compared to the HMM to determine the probability that each ORF is a genuine gene.

HMMs for gene finding can be trained using sets of genes identified by experimental molecular biology. However, we now have over 20 complete genomes, often from organisms where little is known about their biology [14, 53, 54]. In these cases training sets are unavailable. This problem is usually solved by finding "long ORFs" to train the HMM. Long ORFs are very unlikely to occur randomly and can be assumed to be present for a biological reason. In practice three models are produced corresponding to: coding regions on one DNA strand, coding regions on the opposite strand, and non-coding regions.

The two most commonly used programs employing this approach are GeneMark [55, 56] and GLIMMER [5] GeneMark uses a non-homogeneous fifth order HMM which can be trained with a known set of genes or with long ORFs. GLIMMER applies an Interpolated Markov Model (IMM) which uses 1st through to 8th order Markov chains to score open reading frames. Both of these programs claim over 90% accuracy at finding genes.[5, 56] Note, however, that these claims may never be experimentally verified. On a more practical note, these methods have identified many potential drug targets, pharmaceuticals and new biochemical pathways which are of immense use to the biotechnology industry [1].

Audic and Claverie [57]have used an *ab initio* iterative Markov modelling procedure to overcome the lack of a training set. This method tolerates high sequencing error rates of 1-2% and will also work on unprocessed sequence data. Their approach also avoids the inherent bias in using long ORFs as a training set. The authors claim their unbiased approach shows no significant loss in performance when compared with GeneMark. The central principle of their algorithm is to cluster similar sequences. Unlike most clustering approaches in bioinformatics, pairwise comparisons are not used. Instead the sequences are clustered into N groups by comparison to N Markov models. The problem is to determine which model was most likely to have emitted the input sequence. This can be expressed as:

$$P(M_j \mid W) = \frac{P(W \mid M_j)}{\sum_{r=1,2,\dots,N} P(W \mid M_r) P(M_r)} \quad (2)$$

where *P(Mj)* is the probability of *Mj* corresponding to any sequence before input. In this case *P(Mj)* can be taken as equiprobable. *(Mj) = 1/N*. Again, three models need to be generated for coding, coding on the opposite strand, and non-coding subsets. Training sets are avoided by cutting the genomic sequence into non-overlapping fragments ≅ 100 bases in length and distributing these randomly amongst the three subsets. A model is then generated from each subset and applied to the genome using a sliding window. The most likely of the three emitters for each window is determined using the above equation (2). If two successive windows fail to be associated with a given matrix *(Mj)* a classification decision is made. If *n* or more successive windows were associated with *Mj*, the largest, *j*-consistent segment is collected into the *j* data set. Otherwise the region will remain unclassified. The three models are then rebuilt and reapplied iteratively until convergence is reached.

Prediction of the Beginning of Bacterial Genes

The accurate prediction of initiation sites in bacteria is particularly troublesome. A bacterial mRNA may be polycistronic (containing more than one protein coding region). This means that bacterial ribosomes must be capable of internally "docking" with the mRNA at the start of each ORF. The question is how does the ribosome differentiate between several potential initiation codons on a given mRNA? For example, 4000 true initiation codons must be distinguished from 47,000 false initiation codons by *E. coli* ribosomes [58] Identifying a transition from coding to non-coding regions can be achieved by applying HMMs. Often the coding region immediately after an initiation codon has unusual nucleotide composition and codon usage. This can cause HMMs to underestimate the length of the coding region [59].

Approaches which involve combining expert knowledge of several aspects of translation initiation are more successful. Two such systems have recently been described [59,60]. Two theoretical problems arise when applying such a system. How do you identify a training set of true starts in organisms that are not well studied? Which elements of the start site are actually "sensed" by the ribosome? A solution to the first problem is to identify sequences that are similar to known genes in other organisms. This is not always successful as similarity may only be preserved in functional regions and extensions to either end of the protein may not affect function. An alternative solution is to use a subset of predicted ORFs that, by chance, have only one possible initiation codon [61]. The second problem calls for a critical biological assessment of any models produced by machine learning approaches. Any prediction of putative control elements must be interpreted in light of biological processes.

Locating Genes in DNA Sequences from Higher Organisms

Finding genes in higher organisms (eukaryotes) is more challenging than for bacteria (prokaryotes). Whereas 90-98% of a bacterial genome codes for protein, only 1-3% of the human genome does. As mentioned previously, most eukaryotic genes are interrupted with introns (often several). Two additional approaches are used when defining the location of coding sequences in higher organisms. These are the use of experimental transcriptome data in the form of Expressed Sequence Tags (ESTs), and prediction of intron/exon boundaries by computer.

Intron Exon Boundary Prediction from Genomic DNA Sequence

A number of programs are available to predict intron/exon boundaries. These programs employ a diverse variety of methods to score potential boundaries. Their accuracy is usually tested by comparison to experimentally verified training sets, often specific to each organism [62].

The authors of this chapter are investigating the process of protein synthesis in plants, with the overall goal of increasing protein production in genetically modified plants and blocking viral protein production in infected plants. The upf1 gene product is an important protein involved in protein synthesis [63, 64]. It had not been found in any plant, but had been studied in yeast and man and these sequences were available. A major international consortium is currently over half way through sequencing the first plant genome (120Mbp, Arabidopsis thaliana, Thale Cress), with more sequence being made publicly available each day [65, 66]. A computer program was set up to automatically search for similar sequences each week, in the hope of finding a upf1 from plants. Eventually this yielded a match to part of a new 175, 000 base sequence of DNA deposited in the public databases by a Japanese research group. The data had been placed in the database simply as a string of bases without any annotation as to where genes and other features might be located. Using a number of computer packages, the gene structure was predicted within a day. Experimental verification of the structure by contrast took 2 months. This example illustrates well the role and importance of computational gene finding in modern molecular biology. The structure of the upf1 gene is unusually complicated due to the large number of small introns and exons (Fig. 6).

Both GRAIL [29,67] and NetGene2 v2.4 [68] use artificial neural networks to predict intron-exon boundaries. NetGene2 failed to predict many of the splice sites bordering the smaller exons probably due to a failure to recognise coding potential in the smaller exons. There are three versions of the GRAIL neural network. GRAIL 1 relies solely on the recognition of coding potential over a sliding window of 100 nucleotides. The lack of flexibility in window size is the probable reason why many of the small exons were missed. GRAIL 1a failed to improve performance. This is perhaps because each candidate exon is evaluated using information from the two 60 nucleotide regions adjacent to the candidate, which is inappropriate for a gene containing many small introns. Again a fixed window is used to score coding potential. GRAIL 2 failed to predict many of the exons but those it did predict were more accurate than for previous versions of this package. GRAIL 2 uses variable windows to score coding information as well as incorporating more genomic context information. GRAIL 2 can also incorporate EST information, which was unavailable in this case.

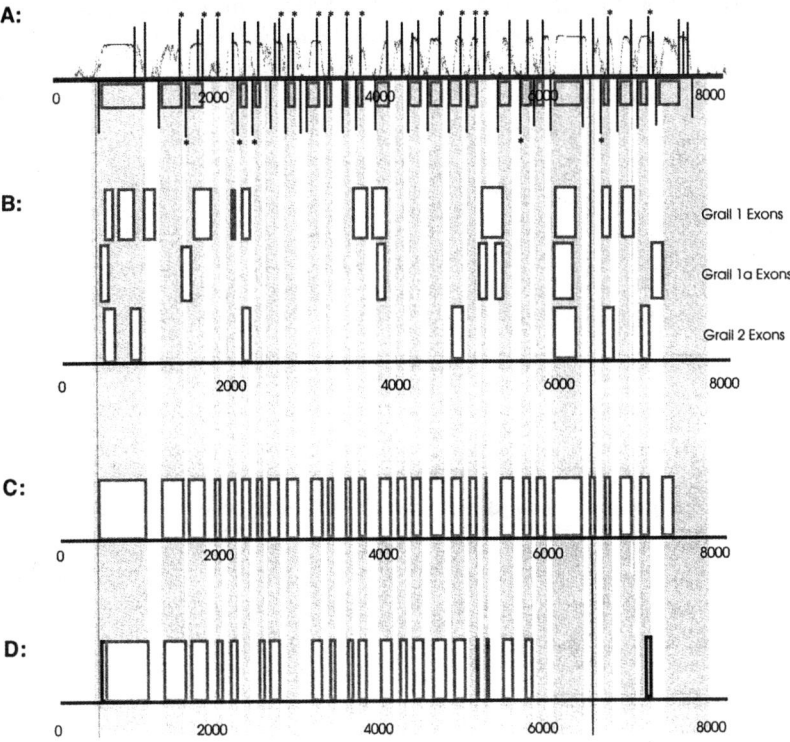

Fig. 6. The 0-8000 scale represents the section of *A. thaliana* genomic DNA encoding *upf*1. Shaded grey boxes represent experimentally determined exon structure. Open grey boxes represent exon predictions. The two vertical hairlines represent the initiation (left line) and termination (right line) codons A) NetGene2 v2.4 predictions. Upward lines correspond to predicted donor splice sites, downward lines correspond to predicted acceptor splice sites. The height of the lines is proportional to confidence, a * indicates a confidence of 1.0. Coding potential is represented in graph form. B) Grail predictions. Number represents the program version. C) GeneMark.hmm prediction. D) Procrustes prediction. Open grey boxes represent exons predicted by comparison to the yeast UPF1 protein, open black boxes represent exons uniquely identified when compared to human UPF1.

GeneMark.hmm [69, 70] was the most accurate in this case with only one boundary of a coding region being more than 10 nucleotides out. This program uses HMMs to detect coding potential. The model used was trained on *Arabidopsis thaliana* coding regions.

Procrustes [71] predicts exons based on a comparison with a supplied protein sequence. This relies on the protein being evolutionarily related to the gene with the DNA sequence. It is a powerful technique for gene-family and cross-species exon prediction but can be misleading or fail in regions of low similarity between the protein supplied and the DNA sequence being examined. Procrustes performed well when the *upf*1 sequence was compared with the protein products of the yeast and human *upf*1 genes. The failure to predict exons near the terminus of the gene was due to a lack of similarity between plants and other organisms in this region.

All of the programs failed to identify the beginning of the first exon. This is probably due to the low level of sequence bias near initiation codons that is also observed in bacteria.

Complications for Gene Finding

A number of confounding factors affect the accuracy of gene prediction. The remnants of disused genes are often found in genomes. These "pseudo-genes" look similar to functional genes but due to small changes (mutations) are now non-functional. Pseudo-genes confuse locating genes by searching for similarities. Furthermore, the biological rules on which the searches are based are not absolute. Some signals do not match the consensus but still function, some termination codons don't always function as termination codons, alternative initiation codons are permissible, and the "reading frame" (which of the three possible "frames" the three-base codons can be read in) is not always conserved. Although these rule bending events are rare they are often of biological significance. A major challenge for bioinformatics is the recognition of novelty. These factors in addition to errors in gene prediction complicate the interpretation of an organism's genome.

4 The Use of Alignments, Profiles and Motifs to Predict Function

Finding Homology

Once genes have been identified, predicting the function of the gene is the next step. In molecular biology proteins with very high similarity (i.e. those that are said to be homologous) are assumed to have a common ancestor and hence a related function. Note that a related function does not necessarily mean an identical function as two proteins may have followed slightly different evolutionary paths. Homologues that have identical functionality are called orthologues. Homologues with similar, but different, functions are known as paralogues.

Alignments and Motifs

To determine the possible function of an unknown gene, the gene can be compared to a database of gene sequences such as GenBank. The level of identity (or similarity) is determined by using local [72] and global [73, 74] alignment techniques. These algorithms make use of dynamic programming to identify potential homologues. When several homologues are identified it is possible to identify motifs corresponding to regions that are highly conserved between the homologues. Regions of high conservation are often functionally important. Motifs can be useful for finding paralogues and distantly related homologues. Motifs may be represented in many ways including regular expressions, fuzzy regular expressions, and weighted matrices [13]

Fuzzy Motifs

When searching a sequence database with an amino acid motif it is more sensitive to construct a fuzzy regular expression. The 20 possible amino acids can be classified into a number of overlapping subsets on the basis of the biochemical properties of that amino acid [75]. The use of fuzzy regular expressions allows the detection of sequences where the biochemical profile is conserved while the exact amino acid composition is not [76].

Correspondence Analysis

Correspondence analysis is a profiling technique that has its foundations in geometry. It has been applied in molecular biology to identify the principle trend (or trends) in codon usage bias. Of the 64 codons, three are termination codons and another two code for only one amino acid. This leaves 59 codons to code for the remaining 18 amino acids. The profile of codon usage for each gene can be used to plot that gene in a 59 dimensional space. To aid in visualisation this space is projected onto a two dimensional plot. Correspondence analysis (CA) can be used to cluster those genes in an organism that have a similar preference for a particular subset of codons. Ordination on the principle axis of the resulting CA plot can often be correlated to some biological property, such as expression level (the rate the gene is translated to protein), which is the cause of the biased codon selection [77-80].

Uses of Information Theory

DNA is a chemical store of information [81]. It is therefore not surprising that Shannon information theory can be used to analyse the information content of any motif, normally expressed as a log score. As DNA has an alphabet of four bases it is appropriate to use a log with a base of two and to express the score in bits. Any position in a motif can then contain up to two bits of information if it is completely conserved. If a priori frequencies are accounted for the bit score may be greater than two. The information content of a sequence motif can be visualised with either a conventional graph or a sequence logo [82-84]

Given that the information content of each position in a motif is additive, the total information content of that motif can be calculated. It is then possible to determine if a motif contains enough information for recognition by another macromolecule by using the equation:

$$R_{freq} = -\log_2 \frac{\gamma}{G} \quad (3)$$

Where γ is the number of true sites to be recognised, and G is the number of potential sites in the genome (generally the number of nucleotides in the genome). Schneider *et a* [85, 86] have used this approach to show that a conserved motif present in the genome of T7 contains a two-fold excess of information. T7 is a bacterial virus that encodes its own RNA polymerase. RNA polymerase recognises a motif known as a promoter. The T7 promoter contains 35 bits of information, although calculations show that it only requires 17 bits in order to be recognised. Experimental studies revealed that polymerase only require a sequence with 18 bits of information to achieve optimal function. This implies that the remaining information is used by an unknown molecule to recognise this region. Thus, information theory can be used to obtain information about the biological function of motifs, and predict new motifs with unknown biological roles.

In whole genome studies it has been shown that *E. coli* has enough information content in its ribosome binding sites for ribosome recognition [87] Recently we have determined that a novel initiation context present in the mRNAs of the blue-green algae bacterium *Synechocystis* sp. PCC6803 also contains enough information for ribosomal recognition (Schreiber and Brown, Unpublished data).

5 Data Structures and Algorithms for Examining the Three Dimensional Structure of Proteins

By studying the three dimensional structure of a protein biologists can determine the function of the protein in a cell. Here we examine some of the current computational problems of studying the detailed atomic structure of proteins and attempting to identify the structure of a protein sequence. There is clearly a role for advanced computational methodology in this area. While computing techniques are essential, much of the "power" of most bioinformatics methods lies with the biological principles implemented. This makes an understanding of both the biological principles and computational techniques important.

We first briefly recall the molecular structure of proteins (see also Biological Data and Flow of Information earlier in this chapter), then introduce some useful data structures to represent them. We present some example applications in brief, and describe one example, protein sequence-structure matching or threading, in more detail.

Proteins have multiple orders of structure from primary to quaternary. The primary structure of a protein is simply the order of amino acids that make up the protein. Secondary structure arise from atomic interactions between neighbouring amino acids to produce simple structures such as alpha helices and beta sheets. Secondary structures can often be predicted with useful reliability [88, 89]. Tertiary structure is the higher level "folding" of molecules and the packing of secondary structures against one-another. Folding of a protein is an exercise in energy minimisation and is influenced by the environment the protein folds in, such as water, or lipid for membrane bound proteins. Proteins are sometimes helped to fold or prevented from folding by the presence of chaperone molecules. Tertiary structure is normally solved experimentally by X-ray crystallography or NMR spectroscopy [90] As more protein structures are solved experimentally it is expected that our ability to assign a structure for a protein sequence will improve. Quaternary structure involves the interaction of two or more protein molecules. Proteins are often chemically modified post-translation. These modifications allow a mRNA to produce a diverse range of subtlety different products [34]. As one can see, by 'structure' we mean the arrangement of the units of a molecule in three-dimensional space. An excellent introduction to protein structure is given by Brändén and Tooze [91].

The atomic structures of molecules are averages over time and space: while each atom is given a precise position in the databases, these are in fact average positions.[92]. Additional variables that estimate the degree of motion of the atoms from this average position are available. This implies a certain "fuzziness" to the data, which needs to be considered.

The properties of different amino acids affect where they tend to occur in proteins and how they interact with other amino acids. Example properties are charge, acidity, willingness to be exposed to water (hydrophilicity). Individual amino acids associate via atomic interactions. Opposite charges will favourably interact while those of like charge will repel one-other when in close proximity. So-called "non-polar" side chains do not like being exposed to water (and thus tend to be found in the interior of the protein) whereas charged amino acids (of either charge) like being exposed to water.

Evolution often builds proteins by adding, removing, modifying and reordering modules called *domains*, which are (more or less) capable of forming stable structures on their own. Domains are typically 30-50 amino acids long. Complete proteins typically range from one hundred to several thousand amino acids in length.

Data Structures to Represent the Structure of Proteins

In a few cases, molecular structures can be approximated by strings in space. Individual characters (amino acids) of the strings (molecule chains) interact with one-another. This approach is used in so-called lattice models of folded proteins. In lattice models, the "amino acids" are placed on a regular grid of points in three-dimensional space [93] Simplification by approximation to a "3-D string" allows exhaustive theoretical studies that would be blocked by the complexity of a complete representation of the protein structure. Care must be taken not to remove all of the characteristics of the protein. As might be expected, some biologists are sceptical of the validity of work based on grossly simplified representations such as these lattice models.

More realistically, molecular structures are composed of atoms in three-dimensional space. Atoms are related to other atoms by physical (atomic) interactions. In this way all, or some of the atoms making up each amino acid of a protein can be represented, along with their inter-atomic distances, bond angles, and so on. One can imagine a variety of useful representations using graphs where the nodes are the atoms and the arcs represent the bonds and interactions between the atoms. More typically protein structures have been represented as (sparse) matrices of (x, y, z) co-ordinates and processed via matrix algebra, Fourier or Bessel functions and the like.

One can describe protein structures in terms of a network of interactions, without considering the points in space occupied by the atoms. Voronoi polygons can represent the volume in space occupied by the atoms of a molecule (Voronoi polygons are irregular polygons formed by "caging" an atom with intersecting planes formed at appropriate distances between all the neighbouring atoms of a given atom [94]). A different polygon caging is used to represent three-dimensional contours of electron density levels when determining the atomic structures of molecules by x-ray crystallography. Other structures are used to represent the surface of a molecule (like those used in 3-D graphics surface modelling; see Fig. 7) or the motion of the atoms in a simulation. Most of the forces between atoms act over short distances, so data structures designed to quickly locate spatially close atoms are useful.

Visualising and Studying Protein Structures: Typical Applications in Brief

Databases of Experimentally Determined Protein Structures

PDB is the main archive of molecular structures including proteins and available via the WWW [96]. Several secondary databases are available such as CATH,

SCOP, or The Protein Motions Database. Other specialised databases are available for specific types of molecular structures [37].

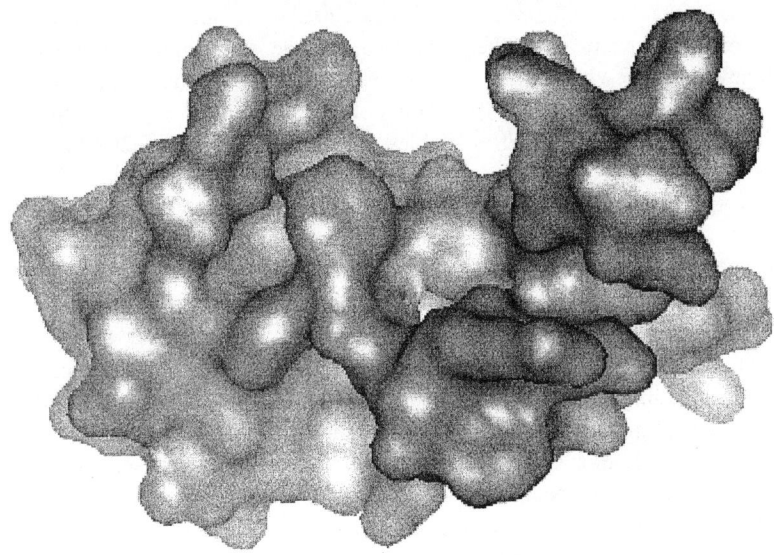

Fig. 7. Molecular surface of a small protein, cobra venom toxin.[95]

Visualising Proteins

MolScript is a macro language built on top of PostScript™, designed to render molecular structures into PostScript™. There are many "molecular graphics" packages designed specifically for three-dimensional display and manipulation of molecular structures running on many systems from Macintosh computers to top-end Silicon Graphics workstations. These include all the problems of rendering complex 3-D data [97]. These programs bring sophisticated molecular graphics within reach of all biological researchers. Recently several markup languages for representing 3-D graphics data for XML browsers have been developed [98].

While several attempts to map the characteristics of a the 3-D surface protein onto a 2-D map have been made, these are generally impractical as the 2-D maps are difficult to relate to the original molecule. Drawing maps of the interaction of a protein with small organic ligands is useful. LIGPLOT [99] draws maps of protein-ligand interactions, by "flattening" the organic ligand by rotating the atom-to-atom bonds of the ligand that are able to rotate, transforming the associated interactions of the protein with that portion of the ligand along with the bond rotation(s). This usefully presents the interactions to the ligand in a more simplified manner than the complex 3-D interactions of the original structure whilst retaining faithfulness to the reality.

Molecular Simulations

Organic chemists are able to use quantum mechanics to simulate small molecules "behaving" over time. Quantum mechanics calculations are too time consuming for proteins which are much larger. Instead, crude approximations are used in which several somewhat *ad-hoc* "terms", each representing one aspect of atomic interactions, are summed.[100-103] This collection of terms is called the energy function. Typically around 20 variables result, leading to a ~20 dimensional search space.

Simulations typically use a Monte Carlo strategy. The amount of energy in a molecule is related to the temperature. The simulation is set to a given temperature. Initially random vectors describing the direction and size of the force acting on the atom are assigned to each atom, proportional to the temperature of the simulated system. A trial "move" is made to the atom. If this improves (lowers) the energy of the structure, it is accepted, otherwise the change is accepted according to a probability scheme (which is related to the change in energy in accepting that change and the "temperature" of the simulation; $P = \exp(-\Delta E/T)$). Vectors are thus assigned for each atom in the structure, and all atoms are moved one (small) step according to the vectors. This process is repeated many (eg. 10^6) times to mimic continuous motion of the structure over time in small discrete steps. Every (say) 1000 steps, the current structure is recorded yielding a progression of structures. (The first few structures from an "equilibration period" are usually ignored.) These simulations require a detailed knowledge of the atomic structure of proteins and must be set up with care to yield sensible results.

"Refining" a Structure

Using the same approach, one can attempt to "refine" a protein structure to locate a low-energy variant with fewer "bad" features, as judged by the energy function, by varying the temperature as the simulation proceeds. As the temperature is lowered, less energy is available to allow the atoms to move "through" energetically unfavourable regions of the search space, forcing the movements of atoms to travel down the energy function (search space, energy landscape) towards local minima. Once a local minima is found, one can record the structure found and then raise the temperature to allow the simulation to explore other regions of the search space which might have other minima. Related approaches are frequently used to optimise an experimentally determined structure, by adding the experimental data as addition terms to the energy function (or "constraints" on the atoms) [104, 105].

Distance Geometry Structure Determination

Another quite different structural problem is to determine the "best" structure (or set of structures) for a set of estimated distances (or distance ranges). These "distance restraints" are the output of experimental data from Nuclear Magnetic Resonance (NMR) analysis of proteins and are one method used to determine the structure of small proteins.[106-108] Rather than literally solve this problem, most methods add "distance restraint energy terms" to molecular dynamics simulations to restrain the distance between atoms to within the prescribed range. Molecular simulation software conveniently also takes into consideration the restrictions imposed by the chemical nature of the molecule in question. Several attempts have been made to tackle this problem using Genetic Algorithms rather than Monte Carlo simulations [109].

Comparing Protein Structures

Comparison of protein structures traditionally uses a "rigid body root-mean-square fit" process. The atoms making up two molecules are treated as a single rigid collection of 3-D points which are rotated about their superimposed centres of masses to yield the best average fit of the atoms scored with a root-mean-square weighting. In practice, this only works well for proteins composed of a single domain. In proteins with several domains, even if the individual domains are closely related, adjacent domains can be in different positions with respect to one-another so the protein structures cannot be compared as single, rigid units. One alternative strategy is to compare the networks of atomic interactions [110-111] in the structures being compared. Biologically this makes more sense, but it is harder to develop rigorously-defined methods with scoring schemes that can validly be compared from one protein structure comparison to another. One can easily imagine graph theory applications for this approach.

Prediction of Interactions with Other Molecules

All biology reduces to interactions of molecules. Molecules have characteristic surface shapes (Fig. 7). The contact surfaces of interacting molecules form complex, "well-packed", interfaces. Many molecules have some shape and property complementary to the molecule(s) they interact with in their absence. Several clever geometric algorithms have been used to explore the surface shapes and properties of two molecules, attempting to locate complementary surfaces. Typically several assumptions are made. The two proteins are assumed not to deform in the process of interacting (which to at least some extent most, if not all, proteins do). Other small molecules (eg. water molecules, ions and, in some cases, larger organic molecules) are assumed not play a major part in forming the interacting surface. For many proteins these assumptions do not hold, so these docking attempts are somewhat futile. More recent approaches are tackling these issues.[112-115] Docking can, when used with appropriate caution and biological knowledge, provide initial ideas about how two molecules might interact.

Fitting a Sequence Onto a Structure Using an Environment Table

A protein structure can be transformed into a '1D-3D profile' [116], describing the environment of each position in the template protein. The environment of each position is characterised by a chosen set of features such as the secondary structure the position is in, the extent to which the position is buried (is the position hidden from the solvent surrounding the protein by other amino acids?), and so on. The user's sequence is aligned against this using a modified form of the classical sequence alignment algorithms [73, 74]. Because there is some residual "memory" of the sequence of the structure used as the template, this approach is best used to identify other members of the protein family in question.

This is not a "threading" method (see below) as it does not fit the sequence onto the structure and assess how this fitted sequence interacts with itself. Here an amino acid from the user's sequence is tested to see if it is compatible with the neighbouring amino acids of the *template* protein, rather than the amino acids of the user's sequence being fitted onto those positions in the template, as in threading (see below). This limits this method to relatively closely related proteins for several biological reasons.

Profile alignments are much faster at sequence-structure matching than threading methods. The match of a amino acid with a position in the template structure only requires knowledge about those two items, so the time bounds are the same as for optimal sequence alignment (at worst $O(n^2)$).

Protein Folding: Predicting the Structure of a Protein from Only Its Sequence

The essence of the "folding problem" is that the protein sequence itself specifies the tertiary structure it forms, so in principle if we knew the rules of how the atoms interact to create a protein structure we could "fold" them on a computer. Predicting protein structure is often regarded as the "holy grail" of computational biology, since the functional activity of a protein is "encoded" in its tertiary structure. Experimental solution of protein structure is very time-consuming. Despite decades of research, no method is able to predict protein structures reliably [117-119]. The essential problem is not computational, but a lack of true understanding of the physics of protein folding. (That said, there are interesting computational applications in this area although their use is at present only theoretical.) Using the molecular simulations described above by starting with an unfolded protein is (at present) too slow and when applied is unsuccessful.

Protein folding is reductionism at its best, aiming to describe the folding a protein purely by the laws of physics in the absence of any experimental data (and hence bias). One important new approach allows biologists to obtain an approximate structure for a protein sequence in many cases by testing the "fit" of protein sequences onto existing experimentally-determined protein structures. This approach is called "threading".

It has been estimated that there are only between 1000 and 100,000 protein folds used in all of nature. If an experimentally determined representative of every fold was known, biologists might then be able to assign a fold to most protein sequences using threading. Were this possible, threading would then displace the need to solve the protein folding problem.[120-122] Research groups have recently started to deliberately determine all the known folds [123]. Researchers are already assigning folds to as many proteins as possible for yeast [124] and *Mycoplasma genitalium* [125] and databases of sequence-structure alignments have been constructed [123,125].

6 Threading: Matching a Protein Sequence to a Three Dimensional Structure of a Protein

Because evolution builds on things previously made, biological molecules tend to fall into families with related structures. Evolution tends to make many small changes to the individual amino acids making up the sequence, most of which have small effects on the fold. As a result, the folds of molecules change more slowly than their sequences. Hence, biologists can see that two molecules have an evolutionary relationship more easily by comparing their structures (if they are available) than by comparing their sequences. This makes comparisons of structures with other structures (see above) and comparison of sequences with structures very valuable.

One can get an approximate idea of the structure of a molecule whose structure is unknown if one can identify a related molecule of known structure and "fit" the sequence of the unknown structure onto it. This process is called "threading": imagine taking a string of amino acids and threading it through a previously known structure, so that the amino acids of the new sequence are placed on the three-dimensional scaffold of the template and then assessing how good that "threading" looks (Fig. 8). First, a few definitions:

- Core – the densely packed "heart" of the structure/fold, excluding the loops. Loops are often poorly characterised by experimental methods to determine molecular structure, and are frequently altered during evolution. Hence threading methods generally ignore loops.

- Fold – the general shape of the protein; the layout of the secondary structures. Secondary structures can shift a little with respect to one another and retain the same fold, but the order of occurrence of the secondary structures and how they pack against one-another should not alter.

- Structure – the detailed "all-atom" atomic structure. A structure is required in order to understand the full chemistry of a protein; folds do not have enough detail to analyse at this level.

In this section we review one this task in more detail as a case study of (some of) the computational and biological issues involved. Threading is a complex problem, which we have highly simplified to let us focus on the computational aspects at the expense of molecular structural considerations [126-130].

The Aim of Threading

Threading methods aim to align (thread) a protein sequence on a core fold. These methods are used mainly to either: (a) identify from a library of template folds, which of these folds might correspond to that of the user's protein sequence (fold selection), or (b) generate an initial structural model of the user's protein sequence based on a particular fold/structure template (sequence-structure alignment). For brevity, "an alignment" in the text below refers to a sequence-structure alignment, not a sequence-sequence alignment, nor a structure-structure alignment.

The Importance of Threading

The function of a protein is encoded in its specific arrangement of amino acids in space, and while there are many proteins available from the genome projects experimental determination of protein structure, while precise, is slow and expensive. Therefore, the main value of threading is the "short-cut to structure and function" and this is the reason the biological and pharmaceutical industries are interested in threading-type methods. For example:

1. The shape of the region binding another molecule (a ligand) relates to the type of ligand bound [131]. Knowing the shape of so-called "binding pockets" can suggest the ligands identity. This strategy has been used on proteins encoded by the *E. coli* genome [132]. Such ligands might be potential drugs or inhibitors.

2. Protein structures are scaffolds to bring together in 3-D space, the (few) atoms involved in a chemical reaction in appropriate surroundings. The reaction provided by these atoms occurs much faster than by the random collisions inside a chemistry test tube. The layout of these "active sites" can sometimes suggest the chemistry of the protein.

3. "Self threading" of a sequence onto a model of its own structure (perhaps experimentally determined) can be used to judge the correctness of the model or to locate any grossly incorrect regions in the model.

While threading is potentially a shortcut to *structure*, in practice it only generates a model of the *fold* of the *core* region of a protein. Several difficult, and not always successful, steps are required to create a detailed structural model from a model of a fold of the core region of a protein. The lack of detail in models resulting from threading limits how much can be learnt from them.

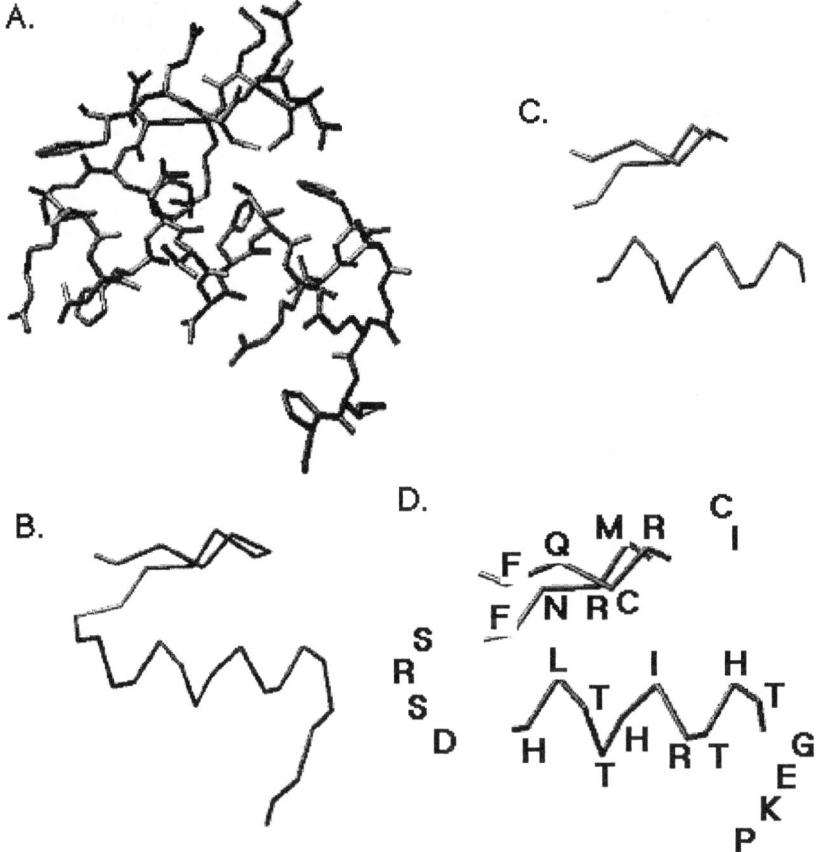

Fig. 8. The steps of protein sequence-structure threading. (A) The all atom structure (here of a small portion of a protein – a single domain) has (B) all but one of the atoms of each amino acid removed, removing the "sequence" of the structure. All amino acids in "loop" regions are removed also, leaving just the "core", shown in (C). Next, (D) the user's sequence (FQCRICMRNFSRSDHLTTHIRTHTGEKP in the example) is "threaded" onto the 3-D structure template. This sequence-structure alignment is tested for how well it fits this structure in that alignment, by assessing how complementary neighbouring amino acids are. Note some amino acids (IC, SRSD, GEKP) do not match amino acid positions in the template structure.

Components of Threading Methods

There are several components used in threading (Fig. 8). First, a template structure or template structure library must be made, then a scoring scheme to evaluate any one alignment of a sequence to a template must be applied. Finally some means of searching the possible alignments to locate the candidate matches must be used.

Making a Template Library

The exact nature of the structure template depends on what features of the structure are to be used in assessing the fit of the sequence to it. Threading methods do not contain the amino acid sequence of the protein used as the template. Likewise atomic details are simplified or erased. Side-chains are removed (but often the first (β) carbon of the side chain is retained as this indicates the region in space that the side-chain occupies). Loop regions are also removed, so that what remains is the core of the structure. When constructing a library of template structures, closely related ("duplicate") structures need to be removed (this is an issue in constructing any library or database of biological sequences or structures). There are many ways of representing structure templates, ranging from simple tables, to complex graph-theoretical structures.

Generating and Assessing the Potential Alignments of a Sequence to Template

The number of possible alignments of one sequence to any one structure are vast, even for small proteins. Three factors make selecting good alignments from the many alternatives difficult:

1. Variable length gaps are usually allowed
2. The need to measure the interaction of a protein sequence with itself when threaded onto the template structure
3. How to score an alignment, given that the score depends on the amino acid and position in question, and all the other amino acids and positions. This forces threading methods in principle to consider all possible alignments of the sequence to the structure

During evolution amino acids can be added or removed from a protein sequence. We need to allow unmatched amino acids if this improves the overall match of the sequence to the structure. In these cases, the shorter of the user's sequence or the template protein will contain gaps. As in any pattern matching method, dealing with variable-length matching (or, in this case, mismatching) regions gaps complicate things considerably.

For 100 amino acids, there are approximately 2^{100} or 10^{30} possible sequence-structure alignments (by comparison the estimated age of the universe is $\sim 10^{17}$ seconds). As a real example, there are 2.9 x 10^{16} possible threadings of the 30-amino acid core of the small protein BPTI. In general, for N secondary structural elements, with loops of average length and no gaps in the secondary structures allowed, there are $\sim 10^N$ possible threadings. Thus threading methods call for sophisticated search strategies.

"Branch and Bound" Procedures

Any one point in the search process has several alternatives (e.g. in "growing" the alignment of the sequence on the structure you can either create a gap in the sequence, a gap in the structure or no gap). At each alternative the alignment must "branch" and explore each of these possibilities until the current alternative is unlikely to generate a reasonable alignment (the "bound"). There are many possible branching strategies [133].

Monte Carlo Procedures (Metropolis-Type Scheme)

After a random alignment is generated a trial "move" or change is made. If this makes better the sequence-structure fit the change is accepted, otherwise the change is accepted or rejected according to a probability scheme. The probability of acceptance is related to the change in energy resulting from accepting that change and the "temperature" of the simulation; $P = \exp(-\bullet E/T)$. This process is repeated many (eg. 106) times. Periodically the current alignment is recorded. Thus an ensemble of sequence-structure alignments is generated. The first few from the "equilibration" period are generally discarded. This ensemble of structures is then assessed with the energy function of choice to locate the optimal alignment and to try detect "reliably aligned" regions [134].

Genetic Algorithms (GAs)

Individual alignments are generated at random. A collection of individuals is allowed to evolve in the face of some fitness function. In the case of threading, the scheme used to score the alignments is used to assess fitness. Mutation and recombination events are used to vary the character of the individuals to allow them to "evolve". The GA approaches to threading have been reviewed [135].

These random search-based approaches cannot be *guaranteed* to locate the best alignment for the scoring function used, which raises questions of how to assess what they do find (discussed below).

One can reduce the number of alignments to be considered by simplifying the methods. In general, these simplifications weaken the ability to create potentially matching alignments and often contradict what is known about protein structure and evolution. As an example, one could reduce the size of the search space by disallowing gaps in the regions with secondary structure, but this restriction effectively requires that the structure template know ahead of time all the places that a gap or loop might occur.

Usually, the alignment method used is a variant of 'double dynamic programming' [136, 137] named because it uses the classical dynamic programming method twice. Fixing an amino acid at some position on the alignment, the best alignment is found for that position. An alignment is calculated for every amino acid-position pair. The score of each alignment is stored in a "higher-level" matrix (of amino acid by position). Then the "best path" through this matrix is found to locate the optimal sequence-structure alignment.

Leaving the method as described is impractical [138]. There are too many alignments to consider and the time taken grows exponentially with the length of the alignments being made (eg. for 100 amino acids, it might take around 1000 CPU seconds to calculate a single double dynamic programming score). Jones resolved this by considering only overlapping windows of some length (5-31) [139]. Alternatively, other simpler methods of alignment can be employed [140], provide they can be demonstrated to generate biologically relevant alignments.

Scoring a Sequence-Structure Alignment

Developing a scoring function requires a balance between how much detail to include and how long it would take to consider this amount of detail. Features a scoring function can consider include:

- The properties of the amino acids proposed to be "interacting"
- Local or long-distance interactions between amino acids, usually using the β carbon. (Here, the "orientation" of the side chain using the C_α-C_β bond can also be considered.)
- The environment of the amino acids proposed to be interacting (eg., the secondary structure they lie in, whether they are buried or not)

Typically, experimentally determined, high resolution protein structures are used to calculate the frequencies of side chain interactions. These frequencies are converted into a Boltzmann-type (or Gibbs-type) "energy" functions. (To be more accurate, an inverse Boltzmann is usually used, ie: $E = -kT \ln(P)$. The kT term is often omitted.) Sometimes direct interactions (typically if the distance between the β carbons is less than, say, 8.0–10Å) are distinguished from longer interactions. Because these statistics are divided into many different subsets, the actual frequencies recorded are rather small, so many of the statistics will only be marginally reliable. Despite this, the frequencies obtained do, to some extent, reflect the nature of the sidechains involved. For example, long, charged side chains have more favourable energies at larger separations than do shorter ones.

Do Threading Methods Work?

Sub-optimal alignments frequently score better than the "true" alignment, even if the scoring function is optimised for the known structure of the test sequence. Even with this "ideal" scoring function there is sufficient noise in the current scoring methods to allow incorrect alignments to score better than the "real" alignment. Assessing the results of threading is an area in need of more research [141].

Some researchers [134] believe proteins which in practice have good global structural similarity, align well irrespective of the exact method chosen. Therefore either the scoring method must be very accurate or the unknown structure of the user's sequence must be very similar to the template to compensate for the noise in the scoring methods.

Threading appears to be better at fold recognition than sequence-structure alignment. The former requires fewer alignments, which may differ from one-another more than the many slightly different alignments based on one template fold. As a result, fold recognition can tolerate errors in alignment and scoring functions.

Removing the details of sidechain interactions, means there will always be a certain amount of noise in the assessment of the alignment. This is something of a catch-22 as one needs to remove the side chains of the template structure so that the alignment is to structure and not sequence. The algorithms implemented need to have a tolerance for error or "fuzziness" to cater for this.

It is difficult to judge the correctness of a sequence-structure alignment. The structure of the user's sequence the user is seeking a fold for will differ from the template unless the sequence of the template is essentially identical to that of the user's sequence, which is unlikely. Again, this implies a certain "fuzziness" is needed in sequence-to-structure matching.

7 Prospects

Much of current bioinformatics efforts are focussed on deciphering the DNA sequences of genomes. By the end of 2000 we will have working drafts or complete DNA sequences of representatives of most major forms of life, including humans and plants. There has been a scramble to mine these genomes for the 'hottest' genes, with in some cases too little time or resources for careful analysis. There will be opportunities for more careful re-analysis using steadily improving methods.

There is currently a great need for methods in 'post-genomics' or 'functional genomics'. These methods would be aimed at understanding the transcriptome and proteome, and the differences between cells and individuals in a population. Perhaps the greatest hurdle in this development will be melding biological data from high-throughput experiments with computational methods. The ultimate goal will be to model the flow of information in a cell *in silico*.

References

1. Lyall, A. (1996): Bioinformatics in the pharmaceutical industry. Trends In Biotechnology 14, 308-312.
2. Baldi, P., Brunak, S. 1999: Bioinformatics: The Machine Learning Approach. MIT Press.
3. Gusfield, D. 1997: Algorithms on Strings, Trees, and Seqeunces: Computer Science and Computational Biology. Cambridge University Press.
4. Waterman, M. S. 1995: Introdcution to Computational Biology. Maps, Sequences, and Genomes. Chapman and Hall.
5. Salzberg, S. L., Searls, D. B., Kasif, S. (1998): Computaional Methods in Molecular Biology. In G., B. (Ed.): *New Comprehensive Biochemistry*, Elsevier.
6. Rodbell, M. (1994): Bioinformatics: an emerging means of assessing environmental health [editorial]. Environmental Health Perspectives 102, 136.
7. Niederberger, C. (1996): Computational tools for the modern andrologist. Journal of Andrology 17, 462-466.
8. Boyer, C., Baujard, O., Baujard, V., Aurel, S., Selby, M., Appel, R. D. (1997): Health On the Net automated database of health and medical information. International Journal of Medical Informatics 47, 27-29.
9. Bezdek, J. C., Hall, L. O., Clark, M. C., Goldgof, D. B., Clarke, L. P. (1997): Medical image analysis with fuzzy models. Statistical Methods in Medical Research 6, 191-214.
10. Jacob, H. J. (1999): Physiological genetics: application to hypertension research. Clinical & Experimental Pharmacology & Physiology 26, 530-535.
11. Boguski, M. S. (1994): Bioinformatics. Current Opinion in Genetics & Development 4, 383-388.
12. Lane, M. (1999): Biological Informatics - Weaving a Web of Wealth, Australian Academy of Science, Canderra.
13. Boguski, M. (1998): Bioinformatics - a new era. In: Trends Guide to Bionformatics (Brenner, S. & Lewitter, F. (Eds). Elsevier Science
14. Brenner, S., Lewitter, F. (1998): Trends Guide to Bioinformatics, Elsevier Science.
15. Venter, J. C., Adams, M. D., Sutton, G. G., Kerlavage, A. R., Smith, H. O., Hunkapiller, M. (1998): Shotgun sequencing of the human genome. Science 280, 1540-1542.
16. Venter, J. C., Smith, H. O., Hood, L. (1996): A new strategy for genome sequencing. Nature 381, 364-366.
17. Collins, F. S., Patrinos, A., Jordan, E., Chakravarti, A., Gesteland, R., Walters, L. (1998): New goals for the U.S. Human Genome Project: 1998-2003. Science 282, 682-689.
18. Brown, C. M., Stockwell, P. A., Trotman, C. N., Tate, W. P. (1990): Sequence analysis suggests that tetra-nucleotides signal the termination of protein synthesis in eukaryotes. Nucleic Acids Research 18, 6339-6345.

19. Brown, C. M., Stockwell, P. A., Trotman, C. N., Tate, W. P. (1990): The signal for the termination of protein synthesis in procaryotes. Nucleic Acids Research 18, 2079-2086.
20. Jacobs, G. H. (1992): Determination of the base recognition positions of zinc fingers from sequence analysis. EMBO Journal 11, 4507-4517.
21. Dalphin, M. E., Stockwell, P. A., Tate, W. P., Brown, C. M. (1999): TransTerm, the translational signal database, extended to include full coding sequences and untranslated regions. Nucleic Acids Research 27, 293-294.
22. Fleischmann, R. D., Adams, M. D., White, O., Clayton, R. A., Kirkness, E. F., Kerlavage, A. R., Bult, C. J., Tomb, J. F., Dougherty, B. A., Merrick, J. M. (1995): Whole-genome random sequencing and assembly of Haemophilus influenzae Rd. Science 269, 496-512.
23. Velculescu, V. E., Zhang, L., Zhou, W., Vogelstein, J., Basrai, M. A., Bassett, D. E., Jr., Hieter, P., Vogelstein, B., Kinzler, K. W. (1997): Characterization of the yeast transcriptome. Cell 88, 243-251.
24. Watson, S. J., Akil, H. (1999): Gene chips and arrays revealed: a primer on their power and their uses. Biological Psychiatry 45, 533-543.
25. Sikora, K. (1999): Developing a global strategy for cancer. European Journal of Cancer 35, 24-31.
26. Johnston, M. (1998): Gene chips: array of hope for understanding gene regulation. Current Biology 8, R171-174.
27. Brunak, S., Engelbrecht, J., Knudsen, S. (1991): Prediction of human mRNA donor and acceptor sites from the DNA sequence. Journal of Molecular Biology 220, 49-65.
28. Hebsgaard, S. M., Korning, P. G., Tolstrup, N., Engelbrecht, J., Rouze, P., Brunak, S. (1996): Splice site prediction in Arabidopsis thaliana pre-mRNA by combining local and global sequence information. Nucleic Acids Research 24, 3439-3452.
29. Xu, Y., Mural, R., Shah, M., Uberbacher, E. (1994): Recognizing exons in genomic sequence using GRAIL II. Genetic Engineering (New York) 16, 241-253.
30. Rounsley, S. D., Glodek, A., Sutton, G., Adams, M. D., Somerville, C. R., Venter, J. C., Kerlavage, A. R. (1996): The construction of Arabidopsis expressed sequence tag assemblies. A new resource to facilitate gene identification. Plant Physiology 112, 1177-1183.
31. Eckman, B. A., Aaronson, J. S., Borkowski, J. A., Bailey, W. J., Elliston, K. O., Williamson, A. R., Blevins, R. A. (1998): The Merck Gene Index browser: an extensible data integration system for gene finding, gene characterization and EST data mining. Bioinformatics 14, 2-13.
32. Wickelgren, I. (1999): The man who would spin genes into gold [news]. Science 285, 999.
33. James, P. (1997): Of genomes and proteomes. Biochemical & Biophysical Research Communications 231, 1-6.
34. Packer, N. H., Pawlak, A., Kett, W. C., Gooley, A. A., Redmond, J. W., Williams, K. L. (1997): Proteome analysis of glycoforms: a review of strategies for the microcharacterisation of glycoproteins separated by two-dimensional polyacrylamide gel electrophoresis. Electrophoresis 18, 452-460.
35. Gooley, A. A., Ou, K., Russell, J., Wilkins, M. R., Sanchez, J. C., Hochstrasser, D. F., Williams, K. L. (1997): A role for Edman degradation in proteome studies. Electrophoresis 18, 1068-1072.

36. Sazuka, T., Ohara, O. (1997): Towards a proteome project of cyanobacterium Synechocystis sp. strain PCC6803: linking 130 protein spots with their respective genes. Electrophoresis 18, 1252-1258.
37. Roberts, R. J., Gait, M. J. 1999: Nucleic Acids Research Database Issue.
38. NCBI. Entrez. Available: http://www.ncbi.nlm.nih.gov/Entrez/.
39. EBI. (1999): Sequence Retrieval Service. http://srs.ebi.ac.uk/.
40. CORBA. (1999): OMG Home Page. Located: http://www.omg.org/. .
41. W3C XML. (1999): W3C XML Home Page. Located: http://www.w3.org/XML/. .
42. Garshol, L. M. (1999): XML Tutorial. Located: http://www.stud.ifi.uio.no/~lmariusg/download/xml/xml_eng.html. .
43. Comprehensive Perl Archive Network. (1999): CPAN Modules Home Page. Located: http;//www.perl.com/CPAN/modules. .
44. BioPerl. (1999): BioPerl Home Page. Located: http://bio.perl.org/. .
45. BioJava. (1999): BioJava Home Page. Located: http://biojava.org/. .
46. Dalphin, M. E., Stockwell, P. A., Tate, W. P., Brown, C. M. (1999): Transterm. Available: http://biochem.otago.ac.nz/Transterm/home_page.html.
47. Jacobs, G. H., Stockwell, P. A., Tate, W. P., Brown, C. M. (2000): (In Preperation). Nucleic Acids Research 28.
48. Poole, E. S., Brown, C. M., Tate, W. P. (1995): The identity of the base following the stop codon determines the efficiency of in vivo translational termination in Escherichia coli. EMBO Journal 14, 151-158.
49. Borodovsky, M., Peresetsky, A. (1994): Deriving non-homogeneous DNA Markov chain models by cluster analysis algorithm minimizing multiple alignment entropy. Computers & Chemistry 18, 259-267.
50. Searls, D. B. (1997): Linguistic approaches to biological sequences. Computer Applications in the Biosciences 13, 333-344.
51. Searls, D. B. (1992): The Linguistics of DNA. American Scientist 80, 579-591.
52. Dong, S., Searls, D. B. (1994): Gene structure prediction by linguistic methods. Genomics 23, 540-551.
53. Klenk, H. P., Clayton, R. A., Tomb, J. F., White, O., Nelson, K. E., Ketchum, K. A., Dodson, R. J., Gwinn, M., Hickey, E. K., Peterson, J. D., Richardson, D. L., Kerlavage, A. R., Graham, D. E., Kyrpides, N. C., Fleischmann, R. D., Quackenbush, J., Lee, N. H., Sutton, G. G., Gill, S., Kirkness, E. F., Dougherty, B. A., McKenney, K., Adams, M. D., Loftus, B., Venter, J. C. (1997): The complete genome sequence of the hyperthermophilic, sulphate-reducing archaeon Archaeoglobus fulgidus [published erratum appears in Nature 1998 Jul 2;394(6688):101]. Nature 390, 364-370.
54. Bult, C. J., White, O., Olsen, G. J., Zhou, L., Fleischmann, R. D., Sutton, G. G., Blake, J. A., FitzGerald, L. M., Clayton, R. A., Gocayne, J. D., Kerlavage, A. R., Dougherty, B. A., Tomb, J. F., Adams, M. D., Reich, C. I., Overbeek, R., Kirkness, E. F., Weinstock, K. G., Merrick, J. M., Glodek, A., Scott, J. L., Geoghagen, N. S. M., Venter, J. C. (1996): Complete genome sequence of the methanogenic archaeon, Methanococcus jannaschii. Science 273, 1058-1073.
55. Borodovsky, M., McIninch, J. (1993): Recognition of genes in DNA sequence with ambiguities. Biosystems 30, 161-171.
56. Borodovsky, M., McIninch, J. D., Koonin, E. V., Rudd, K. E., Medigue, C., Danchin, A. (1995): Detection of new genes in a bacterial genome using Markov models for three gene classes. Nucleic Acids Research 23, 3554-3562.

57. Audic, S., Claverie, J. M. (1998): Self-identification of protein-coding regions in microbial genomes. Proceedings of the National Academy of Sciences of the United States of America 95, 10026-10031.
58. Futschik, M., Schreiber, M., Brown, C., Kasabov, N. (1999): Comparative Studies of Neural Network Models for mRNA Analysis (Unpublished). .
59. Hannenhalli, S. S., Hayes, W. S., Hatzigeorgiou, A. G., Fickett, J. W. (1999): Batcerial start site prediction. Nucleic Acids Research 27, 3577-3582.
60. Frishman, D., Mironov, A., Mewes, H. W., Gelfand, M. (1998): Combining diverse evidence for gene recognition in completely sequenced bacterial genomes [published erratum appears in Nucleic Acids Res 1998 Aug 15;26(16):following 3870]. Nucleic Acids Research 26, 2941-2947.
61. Hayes, W. S., Borodovsky, M. (1998): Deriving ribosomal binding site (RBS) statistical models from unannotated DNA sequences and the use of the RBS model for N-terminal prediction. Pacific Symposium on Biocomputing, 279-290.
62. Farber, R., Lapedes, A., Sirotkin, K. (1992): Determination of eukaryotic protein coding regions using neural networks and information theory. Journal of Molecular Biology 226, 471-479.
63. Leeds, P., Peltz, S. W., Jacobson, A., Culbertson, M. R. (1991): The product of the yeast UPF1 gene is required for rapid turnover of mRNAs containing a premature translational termination codon. Genes & Development 5, 2303-2314.
64. Czaplinski, K., Weng, Y., Hagan, K. W., Peltz, S. W. (1995): Purification and characterization of the Upf1 protein: a factor involved in translation and mRNA degradation. Rna 1, 610-623.
65. Goodman, H. M., Ecker, J. R., Dean, C. (1995): The genome of Arabidopsis thaliana. Proceedings of the National Academy of Sciences of the United States of America 92, 10831-10835.
66. Terryn, N., Rouze, P., Van Montagu, M. (1999): Plant genomics. FEBS Letters 452, 3-6.
67. ORNL. (1999): Oak Ridge National Laboratory. GRAIL. Available: http://compbio.ornl.gov/Grail-bin/EmptyGrailForm [15 September 1999].
68. Hebsgaard, S. M., Korning, P. G., Tolstrup, N., Engelbrecht, J., Rouze, P., Brunak, S. (1999): NetGene2. Available: v2.4 http://www.cbs.dtu.dk/services/NetGene2/ [15 September 1999].
69. Lukashin, A. V., Borodovsky, M. (1998): GeneMark.hmm: new solutions for gene finding. Nucleic Acids Research 26, 1107-1115.
70. Borodovsky, M., Lukashin, A. V. (1999): Eukaryotic GeneMark.hmm. Located: http://dixie.biology.gatech.edu/GeneMark/eukhmm.cgi [15th September 1999].
71. Gelfand, M., Mironov, A., Pevzner, P., Roytberg, M., Sze, S.-H. (1997): Procrustes. Available: http://www-hto.usc.edu/software/procrustes/index.html [15 September 1999].
72. Smith, T. F., Waterman, M. S. (1981): Identification of common molecular subsequences. Journal of Molecular Biology 147, 195-197.
73. Needleman, S. B., Wunsch, C. D. (1970): A general method applicable to the search for similarities in the amino acid sequence of two proteins. Journal of Molecular Biology 48, 443-453.
74. Gotoh, O. (1990): Optimal sequence alignment allowing for long gaps. Bulletin of Mathematical Biology 52, 359-373.

75. Taylor, W. R. (1986): The classification of amino acid conservation. Journal of Theoretical Biology 119, 205-218.
76. Nevill-Manning, C. G., Wu, T. D., Brutlag, D. L. (1998): Highly specific protein sequence motifs for genome analysis. Proceedings of the National Academy of Sciences of the United States of America 95, 5865-5871.
77. Shields, D. C., Sharp, P. M. (1987): Synonymous codon usage in Bacillus subtilis reflects both translational selection and mutational biases. Nucleic Acids Research 15, 8023-8040.
78. Lloyd, A. T., Sharp, P. M. (1993): Synonymous codon usage in Kluyveromyces lactis. Yeast 9, 1219-1228.
79. Fennoy, S. L., Bailey-Serres, J. (1993): Synonymous codon usage in Zea mays L. nuclear genes is varied by levels of C and G-ending codons. Nucleic Acids Research 21, 5294-5300.
80. Chiapello, H., Lisacek, F., Caboche, M., Henaut, A. (1998): Codon usage and gene function are related in sequences of Arabidopsis thaliana. Gene 209, GC1-GC38.
81. Davies, P. (1999): Life Force. New Scientist 18 September, 27-30.
82. Schneider, T. D., Stephens, R. M. (1990): Sequence logos: a new way to display consensus sequences. Nucleic Acids Research 18, 6097-6100.
83. Schneider, T. D. (1996): Reading of DNA sequence logos: prediction of major groove binding by information theory. Methods in Enzymology 274, 445-455.
84. Schneider, T. D. (1997): Sequence walkers: a graphical method to display how binding proteins interact with DNA or RNA sequences [published erratum appears in Nucleic Acids Res 1998 Feb 15;26(4):following 1134]. Nucleic Acids Research 25, 4408-4415.
85. Schneider, T. D., Stormo, G. D., Gold, L., Ehrenfeucht, A. (1986): Information content of binding sites on nucleotide sequences. Journal of Molecular Biology 188, 415-431.
86. Schneider, T. D., Stormo, G. D. (1989): Excess information at bacteriophage T7 genomic promoters detected by a random cloning technique. Nucleic Acids Research 17, 659-674.
87. Barrick, D., Villanueba, K., Childs, J., Kalil, R., Schneider, T. D., Lawrence, C. E., Gold, L., Stormo, G. D. (1994): Quantitative analysis of ribosome binding sites in E.coli. Nucleic Acids Research 22, 1287-1295.
88. Bohm, G. (1996): New approaches in molecular structure prediction. Biophysical Chemistry 59, 1-32.
89. Eisenhaber, F., Frommel, C., Argos, P. (1996): Prediction of secondary structural content of proteins from their amino acid composition alone. II. The paradox with secondary structural class. Proteins 25, 169-179.
90. Kraemer, E. T., Ferrin, T. E. (1998): Molecules to maps: tools for visualization and interaction in support of computational biology. Bioinformatics 14, 764-771.
91. Branden, C.-I. 1998: Introduction to Protein Structure. Garland Publishers.
92. Janin, J. (1990): Errors in three dimensions. J. Biochemie 72, 705-709.
93. Skolnick, J., Kolinski, A. (1991): Dynamic Monte Carlo simulation of a new lattice model of globular protein folding, structure and dynamics. J. Mol. Biol. 221, 499-531.
94. Gerstein, M., Tsai, J., Levitt, M. (1995): The volume of atoms on the protein surface: calculated from simulation, using Voronoi polyhedra. Journal of Molecular Biology 249, 955-966.

95. Rees, B., Samama, J. P., Thierry, J. C., Gilibert, M., Fischer, J., Schweitz, H., Lazdunski, M., Moras, D. (1987): Crystal structure of a snake venom cardiotoxin. Proceedings of the National Academy of Sciences of the United States of America 84, 3132-3136.
96. Research Collaboratory for Structural Bioinformatics. (1999): PDB. Available: http://rscd.org/pdb [27/9/99]. .
97. Savary, F. (1995): The Representation of Molecular Models Rendering Techniques. Available: http://scsg9.unige.ch/fln/eng/toc.html [27/9/99]. .
98. Zara, S. (1999): What is CML? Available: http://xml-cml.org/ [26/9/99]. .
99. Wallace, A. C., Laskowski, R. A., Thornton, J. M. (1995): LIGPLOT: a program to generate schematic diagrams of protein-ligand interactions. Protein Engineering 8, 127-134.
100. McCammon, J. A., Harvey, S. C. 1987: Dynamics of proteins and nucleic acids. Cambridge University Press.
101. Gerstein, M., Levitt, M. (1998): Simulating water and the molecules of life. Scientific American 279, 100-105.
102. Doniach, S., Eastman, P. (1999): Protein dynamics simulations from nanoseconds to microseconds. Current Opinion in Structural Biology 9, 157-163.
103. Karplus, M., Petsko, G. A. (1990): Molecular dynamics simulations in biology. Nature 347, 631-639.
104. Westhof, E., Dumas, P. (1996): Refinement of protein and nucleic acid structures. Methods in Molecular Biology 56, 227-244.
105. Brunger, A. T. (1996): Recent developments for crystallographic refinement of macromolecules. Methods in Molecular Biology 56, 245-266.
106. Cooke, R. M. (1997): Protein NMR extends into new fields of structural biology. Current Opinion in Chemical Biology 1, 359-364.
107. Cooke, R. M., Campbell, I. D. (1988): Protein structure determination by nuclear magnetic resonance. Bioessays 8, 52-56.
108. Braun, W. (1987): Distance geometry and related methods for protein structure determination from NMR data. Quarterly Reviews of Biophysics 19, 115-157.
109. Gippert, G. P., Yip, P. F., Wright, P. E., Case, D. A. (1990): Computational methods for determining protein structures from NMR data. Biochemical Pharmacology 40, 15-22.
110. Sali, A., Blundell, T. L. (1990): Definition of general topological equivalence in protein structures. A procedure involving comparison of properties and relationships through simulated annealing and dynamic programming. Journal of Molecular Biology 212, 403-428.
111. Zhu, Z. Y., Sali, A., Blundell, T. L. (1992): A variable gap penalty function and feature weights for protein 3-D structure comparisons. Protein Engineering 5, 43-51.
112. Sternberg, M. J., Gabb, H. A., Jackson, R. M. (1998): Predictive docking of protein-protein and protein-DNA complexes. Current Opinion in Structural Biology 8, 250-256.
113. Sternberg, M. J., Aloy, P., Gabb, H. A., Jackson, R. M., Moont, G., Querol, E., Aviles, F. X. (1998): A computational system for modelling flexible protein-protein and protein-DNA docking. ISMB 6, 183-192.
114. Jackson, R. M., Gabb, H. A., Sternberg, M. J. (1998): Rapid refinement of protein interfaces incorporating solvation: application to the docking problem. Journal of Molecular Biology 276, 265-285.

115. Wang, J., Kollman, P. A., Kuntz, I. D. (1999): Flexible ligand docking: a multistep strategy approach. Proteins 36, 1-19.
116. Bowie, J. U., Luthy, R., Eisenberg, D. (1991): A method to identify protein sequences that fold into a known three-dimensional structure. Science 253, 164-170.
117. Sternberg, M. J., Bates, P. A., Kelley, L. A., MacCallum, R. M. (1999): Progress in protein structure prediction: assessment of CASP3. Current Opinion in Structural Biology 9, 368-373.
118. Shortle, D. (1999): Structure prediction: The state of the art. Current Biology 9, R205-209.
119. Creighton, T. E. (1988): The protein folding problem. Science 240, 267.
120. Teichmann, S. A., Chothia, C., Gerstein, M. (1999): Advances in structural genomics. Current Opinion in Structural Biology 9, 390-399.
121. Fischer, D., Eisenberg, D. (1999): Predicting structures for genome proteins. Current Opinion in Structural Biology 9, 208-211.
122. Frishman, D., Mewes, H. W. (1999): Genome-based structural biology. Progress in Biophysics & Molecular Biology 72, 1-17.
123. Terwilliger, T. C., Waldo, G., Peat, T. S., Newman, J. M., Chu, K., Berendzen, J. (1998): Class-directed structure determination: foundation for a protein structure initiative. Protein Science 7, 1851-1856.
124. Sanchez, R., Sali, A. (1998): Large-scale protein structure modeling of the Saccharomyces cerevisiae genome. Proceedings of the National Academy of Sciences of the United States of America 95, 13597-13602.
125. Huynen, M., Doerks, T., Eisenhaber, F., Orengo, C., Sunyaev, S., Yuan, Y., Bork, P. (1998): Homology-based fold predictions for Mycoplasma genitalium proteins. Journal of Molecular Biology 280, 323-326.
126. Jones, D. T. (1999): GenTHREADER: an efficient and reliable protein fold recognition method for genomic sequences. Journal of Molecular Biology 287, 797-815.
127. Jones, D. T., Taylor, W. R., Thornton, J. M. (1992): A new approach to protein fold recognition. Nature 358, 86-89.
128. Jones, D. T., Thornton, J. M. (1996): Potential energy functions for threading. Current Opinion in Structural Biology 6, 210-216.
129. Miller, R. T., Jones, D. T., Thornton, J. M. (1996): Protein fold recognition by sequence threading: tools and assessment techniques. FASEB Journal 10, 171-178.
130. Torda, A. E. (1997): Perspectives in protein-fold recognition. Current Opinion in Structural Biology 7, 200-205.
131. Martin, A. C., Orengo, C. A., Hutchinson, E. G., Jones, S., Karmirantzou, M., Laskowski, R. A., Mitchell, J. B., Taroni, C., Thornton, J. M. (1998): Protein folds and functions. Structure 6, 875-884.
132. Fetrow, J. S., Godzik, A., Skolnick, J. (1998): Functional analysis of the Escherichia coli genome using the sequence-to-structure-to-function paradigm: identification of proteins exhibiting the glutaredoxin/thioredoxin disulfide oxidoreductase activity. Journal of Molecular Biology 282, 703-711.
133. Lathrop, R. H., Smith, T. F. (1996): Global optimum protein threading with gapped alignment and empirical pair score functions. Journal of Molecular Biology 255, 641-665.
134. Mirny, L. A., Shakhnovich, E. I. (1998): Protein structure prediction by threading. Why it works and why it does not. Journal of Molecular Biology 283, 507-526.

135. Yadgari, J., Amir, A., Unger, R. (1998): Genetic algorithms for protein threading. ISMB 6, 193-202.
136. Orengo, C. A., Taylor, W. R. (1990): A rapid method of protein structure alignment. Journal of Theoretical Biology 147, 517-551.
137. Taylor, W. R., Orengo, C. A. (1989): Protein structure alignment. Journal of Molecular Biology 208, 1-22.
138. Lathrop, R. H. (1994): The protein threading problem with sequence amino acid interaction preferences is NP-complete. Protein Engineering 7, 1059-1068.
139. Jones, D. (1998): THREADER: protein sequence threading by double dynamic programming. In: Computaional Methods in Molecular Biology, Vol. 32 (Salzberg, S. L., Searls, D. B. & Kasif, S. (Eds). Elsevier
140. Gerstein, M., Levitt, M. (1996): Using iterative dynamic programming to obtain accurate pairwise and multiple alignments of protein structures. ISMB 4, 59-67.
141. de la Cruz, X., Thornton, J. M. (1999): Factors limiting the performance of prediction-based fold recognition methods. Protein Science 8, 750-759.

Chapter 14. Neural Network System for Promoter Recognition

Vladimir B. Bajić[1] and Ivan V. Bajić[2]

[1] Centre for Engineering Research, Technikon Natal, Durban, South Africa
[2] Rensselaer Polytechnic Institute, Troy, NY, USA

Abstract. *The computational prediction of regulatory components in genomic DNA is an attractive and complex research field. The main interest is in finding protein coding genes in long stretches of non-mapped DNA. A particularly important segment of gene finding is the location of promoters - a specific group of regulatory components that are just at the beginning of the gene and which initiate the DNA transcription process. The computational methods for promoter recognition are not sufficiently developed yet. Current methods are prone to produce a large number of false predictions. We present a new method based on clustering the PCA transformed DNA data with further signal processing of the clustered data. The basic technical system consists of eleven neural networks (one SOM ANN and ten GRNNs). On an independent test set the system shows an increased accuracy of recognition with a reduced level of false positive reporting. A special method of data separation into the training set and test set is used. The results achieved with the extended system appear to be currently the best in the class of those that use neural networks for promoter recognition.*

1 Introduction

One of the most fascinating research fields today is the computational investigation, discovery and prediction of biological functions of different parts of DNA/RNA and protein sequences. This is the essence of bioinformatics, a relatively new scientific discipline that spans over, and utilizes results of, several other research fields including, for example, molecular biology, biochemistry, genetics, biotechnology, parallel computing, internet technology, artificial intelligence, and information systems in general. Bioinformatics is already making an enormous impact on the pharmaceutical and agricultural industries and it strongly supports genetic engineering. Bioinformatics and developments of computer technology have opened new avenues for large-scale research in computational genetics. The practical goal of bioinformatics is to reduce the need for laboratory experiments, as these are expensive and time consuming.

Most of the problems investigated in bioinformatics are related to the efficient utilization of databases of biological sequences, to the analysis of such data and to the development of theoretical and practical methods so as to conduct such analyses. One of the general problems that has to be solved is the precise mapping of the genome of specific species, as well as discovering and locating regions of the DNA/RNA sequences responsible for specific biological functions. This problem relates strongly to identification and location of genes in long DNA strings, for which one of the most crucial aspects is the recognition and location of promoters that indicate starting ends of genes.

The predominant interest is in finding genes that code for proteins, and recognition of related promoters is a part of the problem.

1.1 Promoter Recognition

There are several reasons why we are interested in searching for promoters in genomic DNA (see [45], [15]). For example:

1. Promoters have a regulatory role for a gene and thus the recognizing and locating of region(s) of promoter(s) in the genomic DNA is an important part of DNA mapping.
2. Finding the promoter determines more precisely where the transcription start site (TSS) is located.
3. We may have an interest in looking for specific types of genes and then locating specific promoters which will help in gene identification.

Numerous techniques exist for promoter recognition and location. They are all characterized by different success rates. For promoter recognition in prokaryotic organisms see [1], [11], [19], [21], [22], [28], [35], [36], [41], [48], [51], [54].

The techniques for the recognition of eukaryotic promoters are much less efficient. The eukaryotic promoters are far more complex and with very individual microstructure that is specialized for different conditions of gene expression. It is thus much more difficult to devise a general promoter recognition and location method for eukaryotes. As in the case of prokaryotic promoter recognition, different techniques have been used to deal with the recognition of eukaryotic promoters of different classes (see [2], [4], [6], [8], [9], [15], [16], [20], [23], [26], [29], [30], [31], [33], [40], [44], [45], [46], [47], [50], [49], [53], [55], [61]). A recent evaluation study [15] of publicly available computer programs has indicated that the tools for eukaryotic promoter recognitions are still in the embryonic stage.

1.2 ANN Based Promoter Recognition

Some of the techniques mentioned are based on the utilization of artificial neural networks (ANNs). Results on the recognition of promoters by ANNs in prokaryotes can be found in [11], [21], [22], [28], [48]), and for those for eukaryotic organisms in [4], [20], [25], [29], [30], [31], [33], [48], [50], [49].

Our interest will be in promoter recognition in eukaryotes. Results of four programs based on ANNs were available for comparison: NNPP program [48], [50], [49], Promoter2.0 [25] which is an improved version of the program built in GeneID package, SPANN [4] and SPANN2 (this paper). The programs mentioned operate on different principles and use different input information. For example, some of ANN based programs are designed to recognize specific characteristic subregions of eukaryotic promoters, such as is the case with the NNPP program [49], [50], Promoter2.0 [25], promoter finding part of GRAIL program [31] and program of [20]. On the other hand program SPANN [4] uses integral information about the promoter region. The scores achieved by some of these programs are shown later and indicate that the ANN based methods for eukaryotic promoter recognition rate favorably with regard to the other, non-ANN based, tested programs.

1.3 A New ANN System for Promoter Recognition

In this paper we propose a possible approach to promoter recognition. It combines several techniques, including signal processing and ANNs. It results in a basic system that has eleven ANNs, one of which performs the clustering of data, and the other ten are utilized in promoter recognition (one ANN per cluster). In the expanded version, the system utilizes one more ANN for clustering. The final promoter recognition and location is made by combining the score related to a possible presence of hypothetical TATA-like components (see later in the text) in the neighborhood of the transcription start site (TSS). This system show some interesting properties and compares favorably with other ANN based promoter recognition systems. The solution proposed here is based on [3]. This solution shows good rating and improvements in comparison with the other ANN based methods.

2 A Bit of Biology

We present here some very brief material on the biological background necessary for promoter recognition.

The DNA molecule has two strands. Each strand consists of a sequence of four types of monomers called nucleotides. Each nucleotide in DNA consists of a phosphate group, a pentose sugar (2'-deoxyribose) and one of four bases: adenine (A), thymine (T), cytosine (C) or guanine (G). Nucleotides appear in different ordering in the DNA strand. The ordering of nucleotides is assumed to be the most responsible for biological functions of specific parts of the DNA sequences. Two DNA strands are combined in a 3-D structure of a twisted helical type. Two such strands are connected together by weak bonds between the bases on each strand. Specific pairing of nucleotides in the strands is arranged in such a way that only A-T and C-G pairing is possible. Such paired nucleotides form the so-called base pairs (bp). In addition to the DNA there is another nucleic acid called ribonucleic acid (RNA). This plays a crucial role in the biochemistry of cells. RNA molecules are also composed of four nucleotides, each of which contains a phosphate group, pentose sugar (ribose) and one of four nitrogenous bases A, C, G or uracil (U). In RNA uracil replaces thymine (T).

The DNA molecule contains numerous organizational units called genes (see Fig. 1). From a functional point of view, a gene can be defined as a specific sequence of nucleotides in genomic DNA that is necessary for the production of an RNA molecule [7].

Generally, in eukaryotes, the sections in genes responsible for protein-coding are not in continuous segments, but comprise protein-coding parts called exons and parts that do not code for proteins called introns. Roughly, about 2% − 5% of the genome provides coding for proteins [37], [7]. The current opinion is that the remaining parts of the genome contain control regions and intergenic regions. It should be pointed out that the present understanding of the genome structure, and the functions of specific parts of it, is not yet complete.

The process of synthesis of a protein, based on information coded in DNA, can be presented in a simplified manner by the following global steps. This process is different for prokaryotic and for eukaryotic organisms and goes through the following phases (note that prokaryotes do not have the second phase):

DNA strand

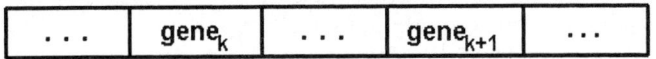

Fig. 1. Consecutive genes in DNA

1. Initially we have a phase of transcription, which is the process of synthesizing RNA under the control of the DNA template. The complementary section of the DNA sense strand is not transcribed. The resultant RNA is the direct complement of the DNA template, i.e. it is identical to the sense strand that corresponds to the transcribed section of the DNA template, with the exception that T is replaced by U. In prokaryotic organisms, the result of transcription is the so-called messenger RNA (mRNA), while in the eukaryotic organisms the result of transcription is the so-called pre-mRNA.
2. The second phase consists of the so-called RNA processing, which is characteristic only for eukaryotic organisms. Its main purpose is to eliminate sections of the pre-mRNA molecules that correspond to the intron regions in the DNA template, splicing the sections that correspond to the consecutive exon regions in the DNA template, and consequently producing mRNA.
3. The final phase is called translation. This is the process of the actual synthesis of a sequence of amino-acids based on information in mRNA. Each one of 20 possible amino-acids is determined by a group of three consecutive nucleotides called codons.

In this way the process is, in simplistic terms, described by a conversion:

- DNA template $\longrightarrow$ mRNA $\longrightarrow$ protein (Fig. 2).

From DNA template to PROTEIN

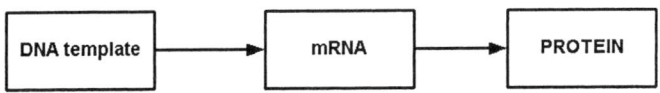

Fig. 2. Process of conversion of information in DNA to protein

2.1 Promoters and Their Characterizations

Promoters are the parts of genomic DNA that are intimately related to the initiation of the transcription process. In the process of the initiation of transcription, specific proteins, enzymes called RNA polymerases, attempt to bind to promoter regions. It is important to note that molecules of RNA polymerase cannot recognize and bind to the promoter region directly. Before they can do this, it is necessary that some other proteins, called transcription factors (TFs), bind to specific subregions of promoters, called the transcription factor binding sites. Only then will RNA polymerase molecules be able to recognize the complex between the transcription factors and DNA, and to bind to the promoter. Promoters generally indicate and contain the starting point of transcription, the TSS, and they regulate the rate of initiation of transcription.

A promoter in eukaryotes is defined somewhat loosely as a portion of the DNA sequence around the transcription initiation site (see [14]). Eukaryotic promoters may contain different subregions such as TATA-box, CAAT-box, Inr, GC-box, together with other different transcription factors binding sites. The problem with these subregions in eukaryotic promoters is that they vary a lot, they may appear in different combinations, their relative locations with respect to the TSS are different for different promoters, and not all of these specific subregions need to exist in a particular promoter. The high complexity of eukaryotic organisms led to specialization of the genes, so that promoters in eukaryotes are adjusted to their different conditions of expressions; for example in different cell types or tissues. Thus, the variability of internal eukaryotic promoter structure can be huge and the characteristics of the eukaryotic promoter are rather individual for the promoter than common for a larger promoter group (see [12], [24], [27], [32], [34], [38], [45], [52], [56], [57], [58]). For this reason it is not easy to precisely define a promoter in eukaryotic organisms. And this is also one of the reasons why at this moment there is no adequate computer tool to accurately detect different types of promoters in a large-scale search through DNA databases, whilst at the same time being very selective and not producing a large number of false reporting.

The simplistic version of the process of the initiation of transcription mentioned at the beginning of this section implies a possible model for eukaryotic promoters. It should have certain numbers of binding sites. However, having in mind that

- there is a large number (several hundreds) of TFs,
- TF binding sites (more than 2000 described, [18]) for different promoters may be at different relative distance from the TSS,
- for functional promoters the order of TF binding sites may be important,
- for different promoters, not all of the TF binding sites and other indicated subregions need to be present, and
- that the composition of TFs for a particular promoter is essentially specific and not shared by a majority of other promoters,

it is obvious that it is very difficult to make a general computer tool for promoter recognition that uses information based on TFs and their composition within the eukaryotic promoter (see discussion in [42], [4]). A possible model of an eukaryotic mRNA gene in relation to its promoter is given in Fig. 3.

There are three types of RNA polymerase molecules in eukaryotes that bind to promoter regions. Our specific interest will be for RNA Polymerase II and their corresponding promoters whose associated genes provide codes for proteins. Many eukaryotic Pol II promoters have some specific subregions that have reasonably high consensus. A typical example is the TATA-box ([5], [6], [10], [39], [43]). The TATA-box is a short region rich with thymine (T) and adenine (A) and located about -25 to -28 bp upstream of TSS (generally in the region of -11 bp to -40 bp upstream). But there are also others like the CCAAT-box ([5], [13]), Inr, etc. A typical arrangement of a Pol II promoter region may be as in Fig. 4.

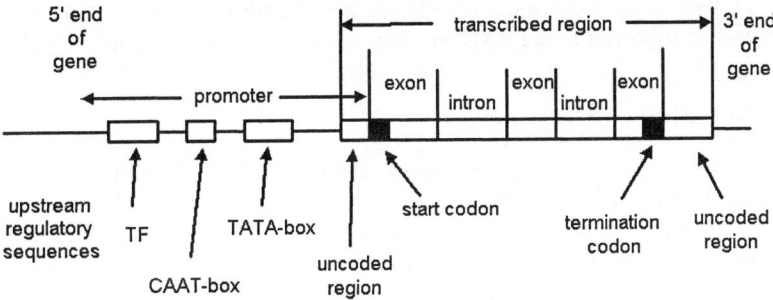

Fig. 3. A possible promoter-gene relation structure for mRNA eukaryotic gene

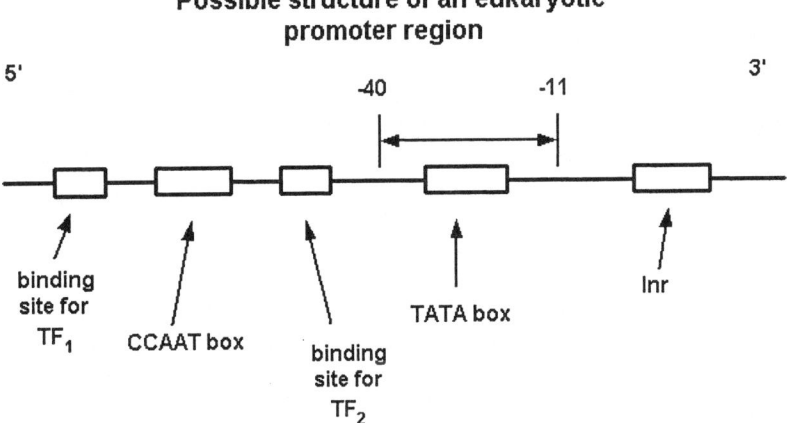

Fig. 4. A possible structure of a Pol II promoter

Although arriving at a general model for an eukaryotic promoter is difficult, for specific classes of promoters, however, suitable models can be derived. For example, the mammalian muscle-actin-specific promoters are modelled reasonably well for extremely accurate prediction, as reported in [17]. Such modelling of specific narrow groups of promoters makes a lot of sense in a search for specific genes. The point is, however, that computer tools for the general mapping of DNA are aimed at the large-scale scanning and searching of DNA databases so as to recognize and locate **as many as possible** different promoters, and not to make large numbers of false recognitions. Obviously, such tools cannot be based on highly specific structures of very narrow types of promoters.

3 Proposed Solution

The approach we use relies on the initial conversion of DNA sequences by means of assigning the so-called Electron-Ion-Interaction-Potential(EIIP) values [59], [60], to different nucleotides in the sequence. After that, for the purpose of further analysis, the

DNA sequence is treated as a discrete signal. In what follows we will outline the basic and extended structures of the ANN based system for promoter recognition, SPANN2.

3.1 Training and Test Sets

Promoter sequences are selected from the Eukaryotic Promoter Database (EPD). They are divided into two groups: training and test. The training group G_t^p has 565 non-related vertebrate promoters, while the test group G_{tst}^p had 302 promoters. All promoter sequences were with 51 nucleotides spanning the -40 to $+10$ region relative to the TSS. The non-promoter sequences were taken from CDS and intron regions, also having 51 nucleotides in length, with the training set G_t^n containing 6000 sequences and the test set G_{tst}^n containing 2000 sequences.

3.2 System Structure

The structure of the basic system for promoter recognition, as well as the extended one, are depicted in Fig.5 - Fig.8. From a long DNA sequence, windows of 51 nucleotides are extracted and presented to the system one by one (each window shifted by 1 nucleotide from the previous one toward the 3' end) and processed according to the procedure to be presented in the next subsection. In the block A the basic signal processing is made and the processed data are clustered by a SOM network into 10 groups. During the presentation of one sequence of 51 nucleotides, only one of the clusters will be activated. Blocks B_j, $j = 1, ..., 10$, represent processing within the clusters. In each of the blocks B_j there is a generalized regression ANN (GRNN) that makes decision if the presented sequence is candidate for promoter or not. That result is combined in the logic block with the score obtained from block C. The score in block C is obtained by assessing the potential of the input sequence to contain a hypothetical TATA-like element. The logic block makes the final decision.

In the extended system, block D makes the same type of processing as blocks A and B-s in the basic system, except that the input sequence is only a portion of the sequence that enters the basic system, and there are no neural networks in the clusters in block D. The sequences that enter block D contain the first 41 nucleotides of the sequences that are processed by the basic system. The logic block combines the results of the basic system and the filtering results of block D into the final result.

3.3 Training the System

First, all sequences are converted into discrete signals by means of the EIIP values forming the sets S_t^p, S_{tst}^p, S_t^n and S_{tst}^n, respectively. The sequences from S_{tst}^p are further modified by adding to them a small level of noise, making thus the set $S_{tst}^{p\text{-}noise}$. Then, sequences from $S_t^p \cup S_{tst}^{p\text{-}noise}$ are converted by the PCA transform matrix M_{PCA}, in symbolic notation $M_{PCA} \circ (S_t^p \cup S_{tst}^{p\text{-}noise})$. The SOM ANN (ANN$_{C1}$) is trained during the 200,000 epochs on the set $M_{PCA} \circ (S_t^p \cup S_{tst}^{p\text{-}noise})$ to cluster data into 10 groups. After the training of ANN$_{C1}$ the data from $M_{PCA} \circ (S_t^p \cup S_{tst}^{p\text{-}noise})$ and from $M_{PCA} \circ S_t^n$ are passed through ANN$_{C1}$. Let T_j denote the promoter and non-promoter data contained in the j-th cluster. Each sequence s_i in T_j is represented by a vector with 51

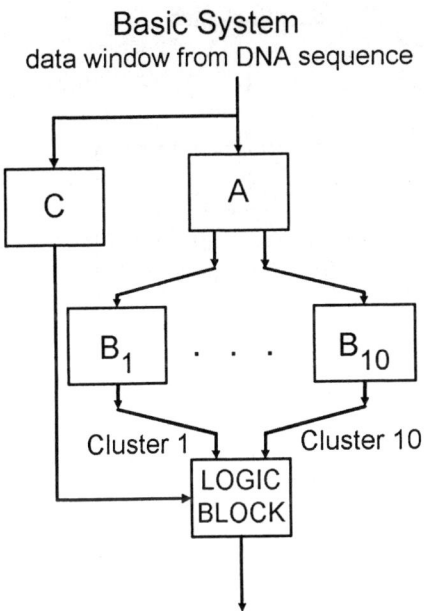

Fig. 5. Structure of the basic system for promoter recognition

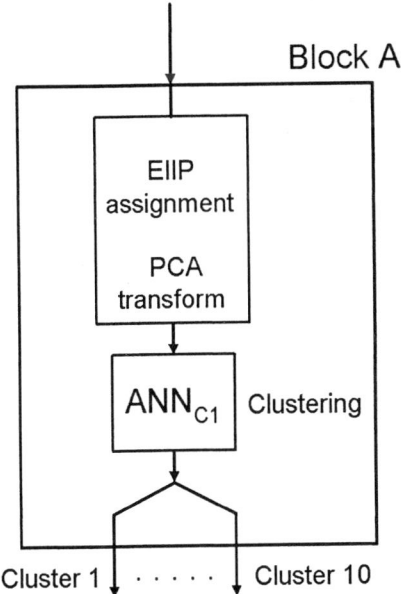

Fig. 6. Structure of the block A

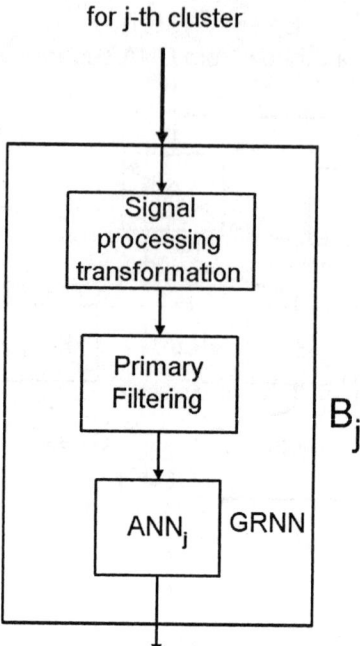

Fig. 7. Structure of the block that processes data in the j-th cluster

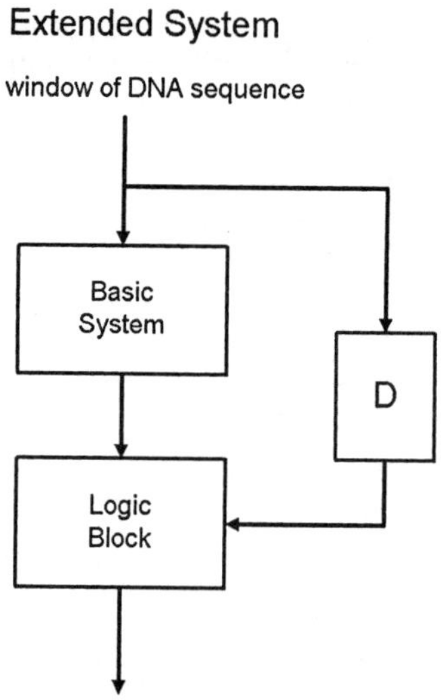

Fig. 8. Structure of the extended system for promoter recognition

components. We use the last two coordinates for discrimination purpose between the promoters and non-promoters. Thus, the data in T_j were subjected to an optimization procedure aimed at maximal compression of promoter data in the plane of coordinates number 50 and 51 and maximal dispersion of non-promoter data from the set $M_{PCA} \circ S_t^n$ in the same plane. A conversion matrix M_j is determined by this optimization. The procedure is repeated for each cluster.

At this point the converted data in the plane of coordinates number 50 and 51 had promoter sequences highly concentrated around the origin, while the non-promoter data were dispersed. A very simple geometric type of criterion was devised to separate non-promoter and promoter data. This criterion essentially isolated the domain where all promoter data is, and filtered out all sequences outside that domain. In this way, on average, about 70% of non-promoter sequences were eliminated, while the whole 100% of promoter sequences were retained. The remaining 30% of the non-promoter sequences, and the promoter sequences from the set $M_{PCA} \circ S_t^p$ served as a training set for a GRNN within each cluster. These networks were trained to distinguish between the promoter and non-promoter sequences. The outputs of GRNNs produced continuous signals, so the cut-off values were determined for the network in each cluster. The criterion for this determination was the maximal discrimination D between the correctly recognized promoters (the so-called TP - true positive recognitions) from $M_j \circ T_j$ and falsely recognized non-promoters as promoters from $M_j \circ T_j$ (the so-called FP - false positive recognitions). This discrimination $D = TP - FP$, where TP and FP are expressed as percentage.

For the extended system, the whole process is repeated for sequences with a length of 41 nucleotides, that span the region of -40 to 0 relative to the TSS. The only difference is that no ANN was used in any of the clusters in this extension and only the geometric criteria was employed for filtering in the clusters.

4 Results

The results with the basic and extended systems on the test sequences with lengths of 51 and 41 nucleotides, respectively, were similar to those achieved on the training sets. The real test of quality has been made on the independent test set (used in [15]) that contains 18 sequences of total length of 33120 bp which contained 24 promoters. The results for the basic system are depicted in Fig.9 to Fig.12.

The basic system has achieved 13 correct recognitions (54%) and 68 false recognitions. This is the same level of TPs as achieved by the NNPP program, but the level of FPs is better. Three measures are used to represent the quality of the system performance with regards to other tested programs in [15], or the other reported results [4], [25], [40]. Systems that used ANNs and whose scores are compared were Promoter2, NNPP, SPANN and SPANN2.

The extended system achieved 8 correct recognition (33%) on the test set of [15], and 16 false recognitions. The comparison results are shown in Fig.13 to Fig.16.

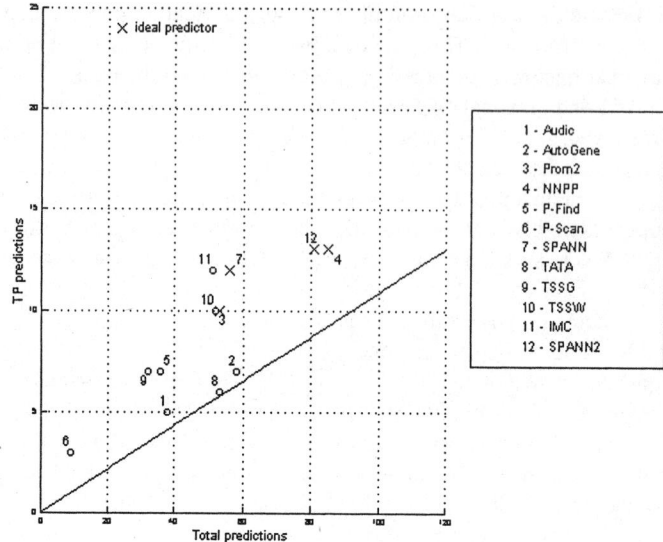

Fig. 9. Results achieved with the basic system on the Fickett & Hatzigeorgiou test set

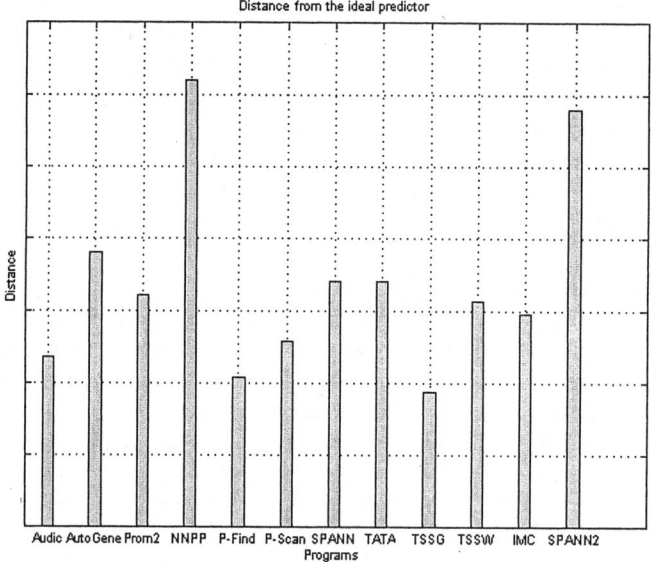

Fig. 10. Results achieved with the basic system on the Fickett & Hatzigeorgiou test set

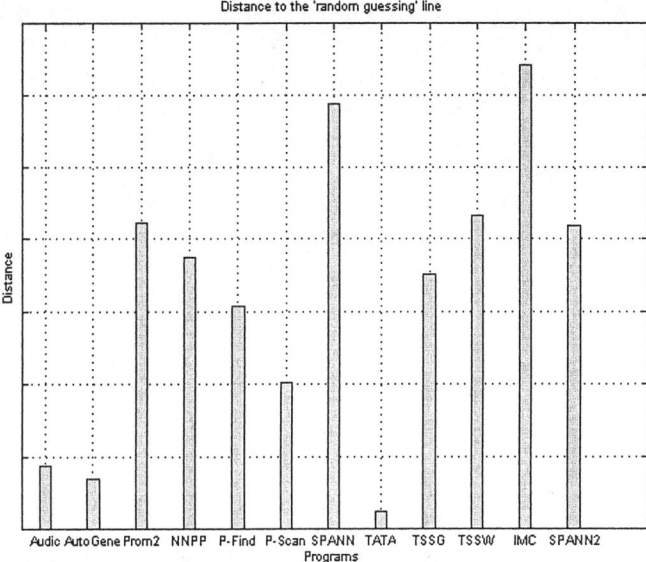

Fig. 11. Results achieved with the basic system on the Fickett & Hatzigeorgiou test set

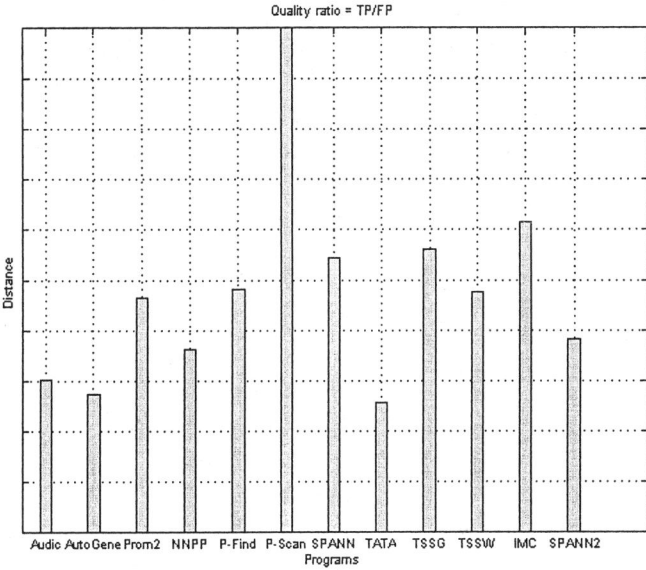

Fig. 12. Results achieved with the basic system on the Fickett & Hatzigeorgiou test set

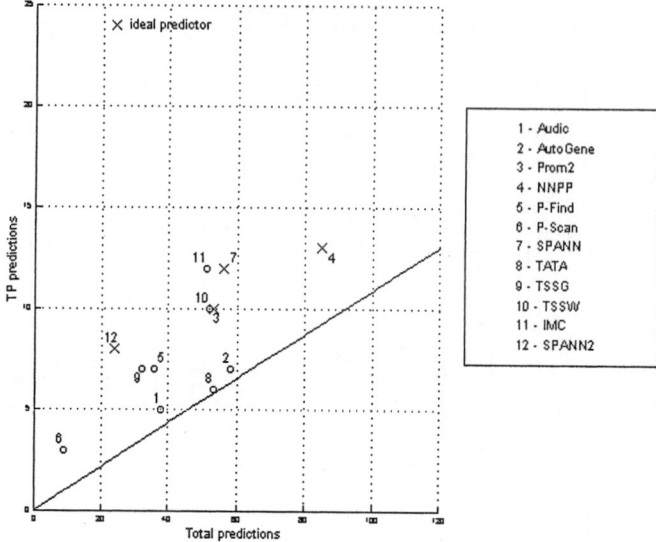

Fig. 13. Results achieved with the extended system on the Fickett & Hatzigeorgiou test set

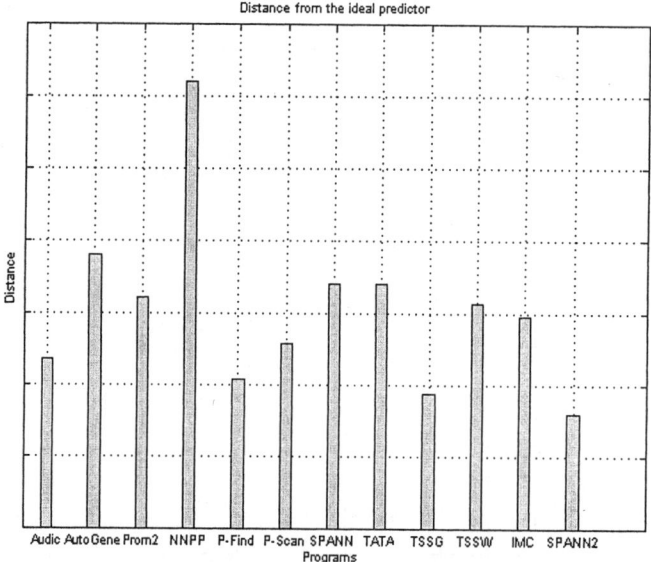

Fig. 14. Results achieved with the extended system on the Fickett & Hatzigeorgiou test set

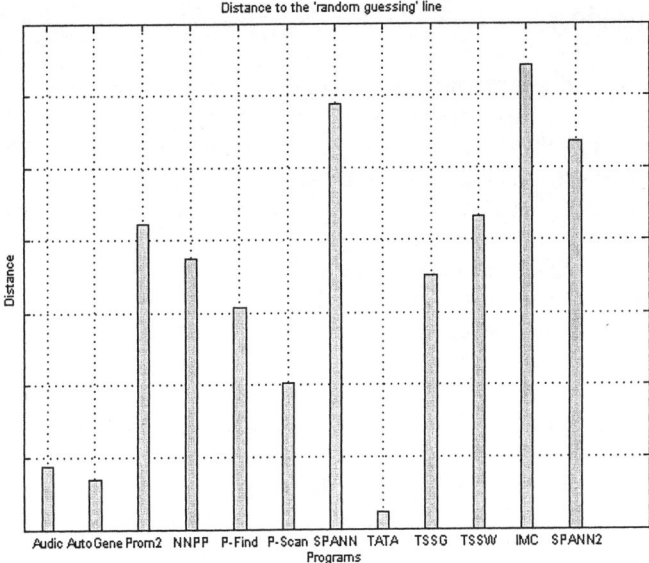

Fig. 15. Results achieved with the extended system on the Fickett & Hatzigeorgiou test set

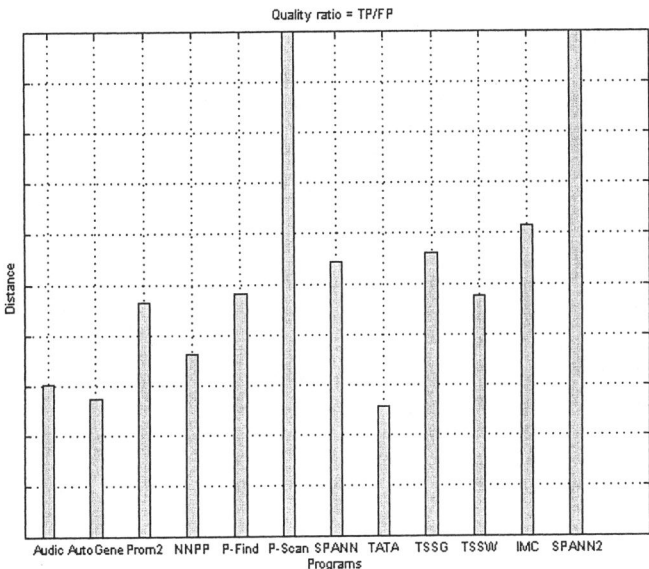

Fig. 16. Results achieved with the extended system on the Fickett & Hatzigeorgiou test set

5 Discussion

The scores achieved with the new system are shown for the basic system and for the extended system. The basic system, equally as the NNPP program, makes the largest number of TPs (13 of 24). SPANN2 also performs slightly better than the NNPP program as it has smaller number of FPs. However, its overall performance is average compared to all other tested systems. Moreover, score of SPANN2 in the basic version ranks consistently on the 3rd place among the scores achieved by other tested ANN based programs using the measures that we adopted.

The extended system shows different and much better overall score. As can be noticed the extended system score ranks very well among all 12 tested programs: it has the shortest distance from the ideal predictor score, it shares the first place with PromoterScan score [44], but achieves 8 TP recognitions as compared to 3 TP recognitions of PromoterScan, and this is a considerable improvement. Moreover, SPANN2 in the extended form achieves the third largest distance from the random guessing line from all 12 programs. These results indicate that clustering of data combined with the proper signal processing have positive impact on the overall result, and also that SPANN2 ranks the best of all tested ANN based programs, based on quality measures that we used.

6 Conclusions

We presented a modular ANN based system for recognition of eukaryotic promoters. The system uses 11 neural networks in the basic form, or 12 neural networks in the extended form. The results achieved by both versions of the system show improvements with regard to some other ANN based solutions. The score of the extended form of the system outperforms the other tested programs.

References

1. Alexandrov, N. N. and A. A. Mironov, Application of a new method of pattern recognition in DNA sequence analysis: a study of E.coli promoters. *Nucl. Acids Res.* 18, 1847-1852, 1990.
2. Audic, S. and J.-M. Claverie, Detection of eukaryotic promoters using Markov transition matrices, *Computer & Chemistry*, 21(4), 223-227, 1997.
3. Bajić, V. B. and I. V. Bajić, A challenging problem of bioinformatics: Artificial neural networks for promoter recognition, Fourth KZN Research Conference on Computer Science, Information Systems and Systems Engineering, Durban, South Africa, June 9, 1999.
4. Bajić, V. B. and I. V. Bajić, ANN in DNA regulatory region recognitions: The case of promoters, Tutorial, CD, International Joint Conference on Neural Networks, Washington, DC, USA, July 10-16, 1999.
5. Benoist, C., K. O'Hare, R. Breathnach and P. Chambon, The ovalbumin gene - sequence of putative control regions. *Nucl. Acids Res.* 8, 127-142, 1980.
6. Bucher, P., Weight matrix descriptions of four eukaryotic RNA polymerase II promoter derived from 502 unrelated promoter sequences, *J. Mol. Biol.*, 212, 563-578, 1990.
7. Campbell, N. A., *Biology*, 4th edition, The Benjamin/Cummings Publishing Company Ltd., Menlo Park, California, US, 1996.

8. Chen, Q., G. Z. Hertz and G. D. Stormo, PromFD 1.0: a computer program that predicts eukaryotic pol II promoters using strings and IMD matrices. *Computer Applic. Biosci.,* 13, 29-35, 1997.
9. Claverie, J. and I. Sauvaget, Assessing the biological significance of primary structure consensus patterns using sequence databanks. I. Heat-shock and glucocorticoid control elements in eukaryotic promoters. *Computer Applic. Biosci.,* 1, 95-104, 1985.
10. Corden, J., B. Wasylyk, A. Buchwalder, P. Sassone-Corsi, C. Kedinger and P. Chambon, Promoter sequence of eukaryotic protein-coding genes. *Science* 209, 1406-1414, 1980.
11. Demeler, B. and G. W. Zhou, Neural network optimization for E.coli promoter prediction. *Nucl. Acids Res.* 19, 1593-1599, 1991.
12. Dynan, W. S. and R. Tjian, Control of eukaryotic messenger RNA synthesis by sequence-specific DNA-binding proteins. *Nature,* 316, 774-778, 1985.
13. Efstratiadis, A., J. W. Posakony, T. Maniatis, R. M. Lawn, C. O'Connell, R. A. Spritz, J. K. DeRiel, B. G. Forget, S. M. Weissman, J. L. Slightom, A. E. Blechl, O. Smithies, F. E. Baralle, C. C. Shoulders and N. J. Proudfoot, The structure and evolution of the human beta-globin gene family, *Cell,* 21: 653-668, 1980.
14. http://www.epd.isb-sib.ch/
15. Fickett, J. W. and A. G. Hatzigeorgiou, Eukaryotic promoter recognition, *Genome Research,* 7(9), 861-878, 1997.
16. Frech, K. and T. Werner, Specific modelling of regulatory units in DNA sequences. *Proceedings of the 1997 Pacific Symposium on Biocomputing,* World Scientific Publishing Co. Pty.. Ltd., Singapore, 151-162, 1997.
17. Frech, K., K. Quandt and T. Werner, Muscle actin genes: A first step towards computational classification of tissue specific promoters, *In Silico Biol.,* 1, 29-38, 1998.
18. Ghosh, D., Status of the transcription factors database. *Nucl. Acids Res.,* 21, 2091-2093, 1993.
19. Grob, U. and K. Stuber, Recognition of ill-defined signals in nucleic acid sequences. *Computer Appl. Biosci.,* 4, 79-88, 1988.
20. Hatzigeorgiou, A., N. Mache and M. Reczko, Functional site prediction of the DNA sequence by artificial neural networks, *Proc. IEEE Int. Joint Symposia on Intelligence and Systems,* 12-17, 1996.
21. Hirst, J. D. and M. J. Sternberg, Prediction of structural and functional features of protein and nucleic acid sequences by artificial neural networks. *Biochemistry* 31, 7211-7218, 1992.
22. Horton, P. B. and M. Kanehisa, An assessment of neural network and statistical approaches for prediction of E.coli promoter sites. *Nucl. Acids Res.* 20, 4331-4338, 1992.
23. Hutchinson, G. B., The prediction of vertebrate promoter regions using differential hexamer frequency analysis. *Computer Applic. Biosci.,* 12, 391-398, 1996.
24. Jones, N. C., P. W. J. Rigby and E. B. Ziff, Trans-acting protein factors and the regulation of eukaryotic transcription: lessons from studies on DNA tumor viruses. *Genes Dev.* 2, 267-281, 1988.
25. Knudsen, S., Promoter2.0: for the recognition of Pol II promoter sequences, *Bioinformatics,* Vol. 15, No. 5, pp. 356-361, 1999.
26. Kondrakhin, Y. V., A. E. Kel, N. A. Kolchanov, A. G. Romashchenko and L. Milanesi, Eukaryotic promoter recognition by binding sites for transcription factors. *Computer Applic. Biosci.* 11, 477-488, 1995.
27. Latchman, D. S., *Eukaryotic transcription factors,* Academic Press, New York, 1991.
28. Lukashin, A. V., V. V. Anshelevich, B. R. Amirikyan, A. I. Gragerov and M. D. Frank-Kamenetskii, Neural network models for promoter recognition. *J. Biomol. Struct. Dyn.* 6, 1123-1133, 1989.
29. Mache, N. and P. Levi, Detection of eukaryotic POL II promoters with multi-state time-delay neural network, *Proc. of the German conference on Bioinformatics GCB'96,* IMISE

Report No. 1, Inst. fuer Medizinische Informatik, Statistik und Epidemilogie, Leipzig, ISB 3-000000872-1, 1996
30. Mache, N., M. Reczko and A. Hatzigeorgiou, Multistate time-delay neural networks for the recognition of POL II promoter sequences, http://www.informatik.uni-stuttgart.de/ipvr/bv/personen/mache.html
31. Matis, S., Y. Xu, M. Shah, X. Guan, J. R. Einstein, R. Mural and E. Uberbacher, Detection of RNA polymerase II promoters and polyadenylation sites in human DNA sequence. *Computers Chem.* 20, 135-140, 1996.
32. McKnight, S. and R. Tjian, Transcriptional selectivity of viral genes in mammalian cells. *Cell,* 46, 795-805, 1986.
33. Milanesi, L., M. Muselli and P. Arrigo, Hamming-Clustering method for signal prediction in 5' and 3' regions of eukaryotic genes, *Comput. Applic. Biosci.*, 12: 399-404, 1996.
34. Mitchell, P. J. and R. Tjian, Transcriptional regulation in mammalian cells by sequence-specific DNA binding proteins. *Science,* 245, 371-245, 1989.
35. Mulligan, M. E. and W. R. McClure, Analysis of the occurrence of promoter-sites in DNA. *Nucl. Acids Res.,* 14, 109-126, 1986.
36. Nakata, K., M. Kanehisa and J. V. Maizel, Discriminant analysis of promoter regions in Escherichia coli sequences. *Computer Applic. Biosci.,* 4, 367-371, 1988.
37. http://www.ncbi.nlm.nih.gov/disease/
38. Novina, C.D. and A. L. Roy, Core promoters and transcriptional control. *Trends Genet.,* 9, 351-355, 1996.
39. Nussinov, R., J. Owens and J. V. Maizel, Sequence signals in eukaryotic upstream regions. *Biochim. Biophys. Acta,* 866, 109-119, 1986.
40. Ohler, U., S. Harbeck, H. Niemann, E. Noth and M. G. Reese, Interpolated Markov chains for eukaryotic promoter recognition, *Bioinformatics*, Vol. 15, No. 5, pp. 362-369, 1999.
41. O'Neil, M. C., Consensus Methods for Finding and Ranking DNA Binding Sites. *J. Mol. Biol.* 213, 37-52, 1989.
42. Pedersen, A. G., P. Baldi, Y. Chauvin and S. Brunak, The biology of eukaryotic promoter prediction - a review, *Computers & Chemistry*, Vol. 23, pp. 191-207, 1999.
43. Penotii, F., Human DNA TATA boxes and transcription initiation sites. *J. Mol. Biol.* 213, 37-52, 1990.
44. Prestridge, D. S. Predicting Pol II promoter sequences using transcription factor binding sites, *J. Mol. Biol.*, 249:923-32, 1995.
45. Prestridge, D. S., Computer software for eukaryotic promoter analysis, (published over internet) 1999, http://biosci.umn.edu/class/bioc/8140/Promoter.html
46. Quandt, K., K. Grote and T. Werner, GenomeInspector: a new approach to detect correlation patterns of elements on genomic sequences. *Computer Applic. Biosci.* 12, 405-413, 1996.
47. Quandt, K., K. Grote and T. Werner, GenomeInspector: basic software tools for analysis of spatial correlations between genomic structures within megabase sequences. *Genomics,* 33, 301-304, 1996.
48. Reese, M. Erkennung von Promotoren in pro- und eukaryontischen DNA-Sequenzen durch Künstliche Neuronale Netze, Diploma work, University of Heidelberg, Germany, 1994.
49. Reese, M. G. and F. H. Eeckman, Time-delay neural networks for eukaryotic promoter prediction, submitted, 1999.
50. Reese, M., NNPP program internet address.
http://www-hgc.lbl.gov/projects/promoter.html
51. Rosenblueth, D. A., D. Thieffry, A. M. Huerta, H. Salgado and J. Collado-Vides, Syntactic recognition of regulatory regions in Escherichia coli, *Computer Applic. Biosci.,* 12(5): 415-422, 1996.
52. Smale, S. T., Generality of a functional initiator consensus sequence. *Gene,* 182, 13-22, 1997.

53. Solovyev, V. and A. Salamov, The Gene-Finder computer tools for analysis of human and model organisms genome sequences, in *Proc. of the Fifth Int. Conf. on Intelligent Systems for Molecular Biology* (T. Gaaserland, P. Karp, K. Karplus, C. Ouzounis, K. Sander and A. Valencia, Eds.), ISMB97, 294-302, AAAI Press, Menlo Park, CA, 1997.
54. Staden, R., Computer methods to locate signals in nucleic acid sequences. *Nucl. Acids Res.* 12, 505-519, 1984.
55. Staden, R., Methods to define and locate patterns of motifs in sequences. *Computer Applic. Biosci.*, 4, 53-60, 1988.
56. Stargell, L. A. and K. Struhl, Mechanisms of transcriptional activation in vivo: two steps forward. *Trends Genet.* 8, 311-315, 1996.
57. Wasylyk, B., Transcription elements and factors of RNA polymerase B promoters of higher eukaryotes. *Crit. Rev. Biochem.* 23, 77-120, 1988.
58. Wingender, E., Transcription regulating proteins and their recognition sequences. *CRC Crit. Rev. in Eukaryotic Gene Expression* 1, 11-48, 1990.
59. Veljković, V. and I. Slavić, Simple General-Model Pseudopotential, *Phys. Rev. Lett.*, Vol. 29, No. 5, pp. 105-107, 1972.
60. Veljković, V., I. Ćosić, B. Dimitrijević and D. Lalović, "Is It Possible to Analyze DNA and Protein Sequences by the Methods of Digital Signal Processing?", *IEEE Trans. Biomed. Eng.*, Vol. 32, No. 5, pp. 337-341, 1985.
61. Zhang, M. Q., Identification of Human Gene Core Promoters in Silico, *Genome Research*, 8: 319-326, 1998.

Part V

Knowledge Representation, Knowledge Processing, Knowledge Discovery and Some Applications

Chapter 15. Granular Computing : An Introduction

Witold Pedrycz

Department of Electrical & Computer Engineering,
University of Alberta, Edmonton, Canada, and

Systems Research Institute, Polish Academy of Sciences
01-447 Warsaw, Poland

pedrycz@ee.ualberta.ca

Abstract. The study is concerned with the fundamentals of granular computing. Granular computing, as the name itself stipulates, deals with representing information in the form of some aggregates (embracing a number of individual entitites) and their processing. We elaborate on the rationale behind granular computing. Next, a number of formal frameworks of information granulation are discussed including several alternatives such as fuzzy sets, interval analysis, rough sets, and probability. The notion of granularity itself is defined and quantified. A design agenda of granular computing is formulated and the key design problems are raised. A number of granular architectures are also discussed with an objective of dealinating the fundamental algorithmic and conceptual challenges. It is shown that the use of information granules of different size (granularity) lends itself to general pyramid architectures of information processing. The role of encoding and decoding mechanisms visible in this setting is also discussed in detail along with some particular solutions. The intent of this paper is to elaborate on the fundamentals and put the entire area in a certain perspective while not moving into specific algorithmic details.

Keywords. Information granules, information granulation, fuzzy sets, interval analysis, rough sets, shadowed sets, pyramid architectures, encoding and decoding

1 Introduction

Last years saw a rapid growth of interest in so-called granular computing and computing with words as one among its realizations [13]. In a nutshell, granular computing is geared toward representing and processing basic chunks of information – information granules [5][12][14]. Information granules, as the name itself stipulates, are collections of entities, usually originating at the numeric level, that are arranged together due to their similarity, functional adjacency, indistinguishability or alike. The process of forming information granules is referred to as information granulation. No matter how this granulation proceeds and what fundamental technology becomes involved therein, there are several essential factors that drive all pursuits of information granulation. These factors include

- A need to split the problem into a sequence of more manageable and smaller subtasks. Here granulation serves as an efficient vehicle to modularize the problem. The primary intent is to reduce an overall computing effort
- A need to comprehend the problem and provide with a better insight into its essence rather than get buried in all unnecessary details. In this sense, granulation serves as an abstraction mechanism that reduces an entire conceptual burden. As a matter of fact, by changing the "size" of the information granules, we can hide or reveal a certain amount of details one intends to deal with during a certain design phase.

The long-lasting tradition of computing using some specific information granules is a visible testimony that some specific versions of granular computing are omnipresent indeed. As a matter of fact, as we discuss in this study, digital – to- analog transformation leading to digital computing for analog world is just a highly representative (albeit quite specific) instance of granular computing. By tradition (and the associated technology dominant at that time), we have embarked on the digital world of computing. To interact with the continuous (analog) world, we use set-based granulation (more specifically, interval-valued granulation). This specific type of granulation comes under the name of analog-to-digital conversion.

Information granules may arise as a phenomenon of inherent nonuniqueness associated with the problem at hand. As a simple example, one can resort himself to any inverse problem; the type of characteristics involved (as the functions may be non-invertible) gives rise to relations and as a result, a collection of information granules rather than single numeric quantities. Dropping some input variable in a model may also lead to the same effect of granular information.

We may witness (maybe not always that clearly and profoundly) that the concept of granular computing tends to permeate a number of significant endeavors. The reason is quite straightforward. Granular computing as opposed to numeric computing is *knowledge-oriented*. Numeric computing is *data-oriented*. Undoubtedly, knowledge-inclined processing is a cornerstone of data mining, intelligent databases, hierarchical control, etc.

While the idea of granular computing has been advocated and spelled out in the realm of fuzzy sets (and seems to be a bit biased in this way), there are a number of fundamental formal frameworks that can be exploited as well. Several alternative paths to follow including interval analysis, rough sets and probabilistic environments, to name a few dominant and most visible options.

The diversity of the formal means used for information granulation and further processing of the resulting information granules has a common denominator. All of these environments share the same research agenda that attempts to address the fundamentals of granular computing.

A way of constructing information granules and describing them in an analytical fashion is a common problem no matter which path (probability or set-theoretic) we follow. The question as to the definition of the "size", "capacity" or "dimension" of the information granule is of primordial interest. How to measure granularity of the constructed information granules? How to relate this granularity with computational complexity? Those are open questions in the framework of granular computing that still await solid answers.

What are sound methodologies when operating on information granules? How to evaluate (validate and verify) granular constructs? What would be appropriate measures of relevance of granular models? These are fundamental issues posed in the case of numeric modeling and discussed in detail. The same suite of questions expressed in the case of granular architectures begs for further thorough investigations.

There is an intriguing question as to a way of navigating between constructs (models) developed at various levels of information granularity. Is the structure developed with the use of "large" information granules useful when more specific results are required? It is apparent that when forming information granules, the contributing elements loose their identity that is essentially a non-recoverable process. Now, how this could effect the results of computing involving bigger information granules? If we want to recover the details, how efficient could be our attempt? What are the limits of this reconstruction? These aspects boil down to the mechanisms of encoding and decoding granular information. When any datum enters a system operating at a certain level of information granularity, it becomes encoded. As a result it becomes "accepted" (tuned) to the level of information granularity present within this system. Once the system tends to communicate its results, these need to be decoded. In other words, encoding and decoding are interfaces between worlds (systems) operating at various levels of information granularity. We have already encountered this scheme in digital processing: encoding corresponds to the analog-to-digital (A/D) conversion whereas the decoding comes under the name of digital-to-analog (D/A) conversion.

Fuzzy modeling has emerged as an interesting, attractive, and powerful modeling environment applied to numerous system identification tasks. Granular computing forms a useful environment supporting all modeling pursuits and adding another dimension to the modeling itself. The key features being emphasized very often in this setting concern a way in which fuzzy sets enhance or supplement the existing identification schemes. It was Zadeh first [12] who has introduced the concept of fuzzy models and fuzzy modeling. The enhancements of system modeling conceived within this framework take place at the conceptual level as well as at the phase of detailed algorithms. In a nutshell, fuzzy models are concerned with the modeling pursuit that occurs at the level of linguistic granules (fuzzy sets or fuzzy relations) rather than the one that happens at a detailed and purely numeric level encountered in other modeling approaches. What fuzzy sets offer in system modeling is another more general and holistic view at the resulting model that gives rise to their augmented interpretation and better utilization. From a computational point of view, fuzzy sets are inherently nonlinear (viz. their membership functions are nonlinear mappings). As a consequence of such nonlinear character, one may anticipate that this feature augments the representation power of the fuzzy models. There have been a substantial number of various schemes of fuzzy modeling along with specific algorithmic variations that help eventually capture some specificity of the problem at hand and contribute to the efficiency of the overall identification schemes, cf. [4][6][8]. Quite often, in order to take advantage of numeric experimental data, the modeling algorithms resort themselves to a vast spectrum of neurofuzzy techniques, see e.g., [1][2][3][5][7][11].

Granulation is a necessary prerequisite that is required to take advantage of discrete models (where by "discrete" we mean granular) such as finite-state machines, Petri nets, and alike.

The objective of this study is to raise fundamental issues of granular computing as a new and unified paradigm of information processing, elaborate on a family of possible formal frameworks and formulate the key design problems associated with this form of computing.

2 Granular Computing: an Information Processing Pyramid

In granular computing we operate on information granules. Information granules exhibit different levels of granularity. Depending upon the problem at hand, we usually group granules of similar "size" (that is granularity) together in a single layer. If more detailed (and computationally intensive) processing is required, smaller information granules are sought. Then these granules are arranged in another layer. In total, the arrangement of this nature gives rise to the information pyramid. As portrayed schematically in Figure 1, in granular processing we encounter a number of conceptual and algorithmic layers indexed by the "size" of information granules. Information granularity implies the usage of various techniques that are relevant for the specific level of granularity. Alluding to system modeling, we can refine Figure 1 by associating the layers of the information processing pyramid with the pertinent most commonly used classes of processing and resulting models

- at the lowest level we are concerned with numeric processing. This is a domain completely overwhelmed by numeric models such as differential equations, regression models, neural networks, etc.

- at the intermediate level we encounter larger information granules (viz. those embracing more individual elements)

- the highest level can be solely devoted to symbol-based processing and as such invokes well-known concepts of finite state machines, bond graphs, Petri nets, qualitative simulation, etc. Note that some of these classes emerge at the intermediate level of information granularity and at that level their conceptual and symbolic fabric is usually augmented with some numeric component.

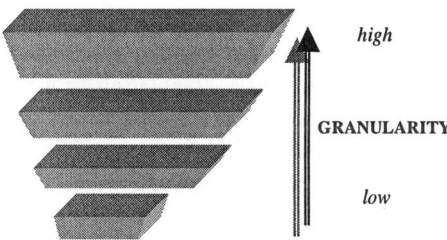

Figure 1. An information-processing pyramid (the respective layers are indexed by the corresponding level of information granularity)

The general characteristics of the principle of granular computing can be enumerated as shown in Table 1.

Allow for multiple abstraction levels (granularity levels)
Allow for several methods of traversing various levels of hierarchy (encoding – decoding mechanisms)
Allow for nonhomogeneous methods (differential or difference equations, Petri nets, finite state machines)

Table 1. The fundamental features of granular computing

3 Information Granulation

In this section, we look into the underlying rationale behind information granulation and discuss various means supporting the construction of information granules. The starting point is to look at any linguistic model as an association of information granules (linguistic terms) defined over some variables of the system. Quite descriptively, one may allude to such linguistic granules or linguistic landmarks as being focal point of all modeling activities. Linguistic granules are viewed as linked collections of objects (data points, in particular) drawn together by the criteria of indistinguishability, similarity or functionality. Such collections can be modeled in several formal environments including set theory, rough sets, random sets, shadowed sets or fuzzy sets.

3.1 Defining Information Granules

Informally speaking, information granules [9][12][14] are viewed as linked collections of objects (data points, in particular) being drawn together by the criteria of indistinguishability, similarity or functionality. Information granules and the ensuing process of information granulation is a vehicle of abstraction leading to the emergence of concepts.

Granulation of information is an inherent and omnipresent activity of human beings carried out with intent of better understanding of the problem. In particular, granulation of information is aimed at splitting the problem into several manageable chunks. In this way, we partition the problem into a series of well-defined subproblems (modules) of a far lower computational complexity than the original one.

Granulation occurs everywhere; the examples are numerous and they originate from various areas

- we granulate information over time by forming information granules over predefined time intervals. For instance, one computes a moving average with its confidence intervals
- in any computer model we granulate memory resources by subscribing to the notion of pages of memory as its basic operational chunks (then we may consider various swapping techniques to facilitate an efficient access to individual data items)
- We granulate information available in the form of digital images – the individual pixels are arranged into larger entities and processed as such. This leads us to various issues of scene description and analysis
- In describing any problem, we tend to shy away from numbers but rather start using aggregates and building rules (*if-then statements*) that dwell on them.
- We live in an inherently analog world. Computers, by tradition and technology, perform processing in a digital world. Digitization of this nature (that dwells on set theory - interval analysis) is an example of information granulation
- All mechanisms of data compression are examples of information granulation that is carried in a certain sense

Overall, there is a profound diversity of the situations that call for information granulation. There is also panoply of possible formal vehicles to be used to capture the notion of granularity and provide with a suitable algorithmic framework in which all granular computing can be efficiently completed. In the ensuing section, we elaborate on those commonly encountered in the literature. Examples of such formal environments include set theory, rough sets, random sets, shadowed sets or fuzzy sets.

3.2 Models of Granular Information

As mentioned, there are a number of possible formal avenues of information granulation. The most commonly used (and well known) include sets (more specifically, interval analysis, fuzzy sets, rough sets, random sets and probability theory. Far less known, even though they exhibit interesting and useful features are shadowed sets

- Intervals are examples of sets defined in the real lines. In essence, they are easy to comprehend and process. The formal underlying model encountered there is also straightforward: A set A is a two-valued mapping from a given universe of discourse **X** to $\{0,1\}$ where A(x) denotes a characteristic function of A

$$A: \mathbf{X} \to \{0,1\}$$

Interestingly, all elements included in A loose their identity in the sense they are completely indistinguishable once they enter the same set. In this sense, we can associate a single symbol (name, label) to the set at hand to provide with its meaningful naming. We usually use notation P(**X**) or P(**R**) to denote the family of sets defined over the corresponding universe of discourse

- Fuzzy sets are generalizations of sets (intervals) in the sense we admit a partial membership of some elements to the concept represented by a given fuzzy set. Formally, fuzzy sets are described by their membership functions

$$A: \mathbf{X} \to [0, 1]$$

The higher the membership value A(x), the stronger the binding of the given element (x) with the fuzzy set. Fuzzy sets exhibit a sort of duality: on one hand they can be associated with symbols (and in this venue they coincide with the symbolic character of any ensuing processing). On the other hand, numeric characteristics of the membership function provide us with a more refined, numeric look into the concept. Similarly, we use the standard notation F(**X**) to describe a family of fuzzy sets in **X**.

- Rough sets are geared toward a synthesis of approximation of concepts from the experimental data. Given the indiscernability relation R defined in **X** × **X**, a rough set A is expressed by its lower (A_*) and upper approximation A^*.

Denote by [x]R the equivalence class of x induced by the indiscernability relation, $[x]_R = \{y \in X | xRy \text{ holds}\}$ The lower approximation A_* of A reads as

$$A_* = \{x | [x]_R \subseteq A\}$$

The upper approximation A* is expressed as

$$A^* = \{x | [x]_R \cap A \neq \emptyset \}$$

Ultimately, the more distant the lower and upper bound, the more evident roughness of the discussed rough set. The notation being used is R(**X**).

- Shadowed sets A shadowed set A [10] defined in **X** is a three-valued mapping

$$A: X \rightarrow \{0, a, 1\}$$

The elements where A(x) = 1 produce the core of A. The elements of **X** whose membership values are equal to zero, A(x)=0, are excluded from A. The shadows capture elements with no specific single membership value assigned to them, A(x)=a. Here a could be defined as a certain interval in [0,1] or a collection of values included therein. In other words, for the shadow of A, sh(A) we envision a set-like (viz. multivalued) mapping. Note that for the remaining elements of **X** over which A is defined we are dealing with standard pointwise valued mapping (that is A(x) is a single numeric assignment).This type of membership allocation for the shadow of A reflects and help quantify an effect of "hesitation" that comes with the specific elements of the universe of discourse. Again, a family of shadowed sets in **X** will be denoted by S(**X**). As shadowed sets are quite new (and conceptually straightforward and computationally appealing), it is prudent to elaborate on more examples to illustrate the usage of the framework. Table 2 includes a number of representative applications coming from selected classes of problems.

- Probability density functions are examples of probabilistic information granules. They encapsulate elements whose occurrence is associated with the occurrence of some concept. For instance, in pattern recognition we may identify information granules for individual features where the conceptualization processes is guided by establishing a certain class (category) of patterns ω, say p(x|ω).

Class of problems	Detailed description	Shadowed set: a model
Quantization	Quantization error and its distribution with regard to the distribution of the elements of the codebook	
Decision making	Decide: - for class-1 - for class-2 - postpone decision – further investigations required	
Approximation	Modeling a distribution of approximation error; e.g. the one resulting from a piecewise linear approximation	
Relational nature of mapping	Some regions cannot be modeled through *any* function but need to be represented as relations (or fuzzy relations)	

Table 2. Examples of shadowed sets: general problems and ensuing models of shadowed sets

3.3 Quantifying Information Granularity

It is needless to say that the area of granular computing is still under intensive development. While some fundamental issues have been already identified and various existing methodologies unified to some degree, there is still a long way to go. For the purpose of this study, it is indispensable to elaborate on the notion of a size of information granules. Intuitively, we may equate a size of the given information granule with the number of elements encapsulated by this information granule. The larger the number of the elements, the lower the granularity of the construct. Evidently, numeric datum (information granule) exhibits the highest level of information granularity. Counting the number of elements can very much depend upon the formal setting in which such information granules are developed. As an example, let us consider set theory (more specifically, interval analysis). For the given universe of discourse $\mathbf{X} \subset \mathbf{R}$, the granularity of A can be quantified by counting the number of its elements, Card(A). To relate this to the entire universe of discourse, it is legitimate to consider its normalized version, that is Card(A)/Card($\mathbf{X}$). This is a normalized version of the previous cardinality measure.

In the case of fuzzy sets, as their elements contribute to the concept to a different extent, a notion of so-called σ-count (fuzzy cardinality) is discussed

$$\sigma\text{-count (A)} = \sum_{x \in X} A(x)$$

Following the same argumentation as before, the normalized version σ-count(A)/card($\mathbf{X}$) can be be used as well. In the case of information granulation completed in the framework of rough sets, an accuracy of approximation can be sought

$$\alpha(A) = \frac{\text{card}(A^*)}{\text{card}(A_*)}$$

Where A^* and A_* are the upper and lower approximation of A, respectively. As far as probabilistic granules are concerned, a standard deviation (variance) can be viewed as a useful vehicle to express a size of the probabilistic information granule. The higher the standard deviation, the lower the granularity of the corresponding information granule (the more elements become admitted to the information granule).

In the case of probabilistic granules we take their statistical characteristics. A variance (standard deviation, st-dev) of A described by a probability density function (p(x)) serves as a measure of its granularity,

$$\text{st-dev}(A) = \sqrt{\int_X [x-m]^2 p(x) dx}$$

where m stands for a mean value of A.

By working with the information granules, we do not commit ourselves to any detailed processing as it becomes visible in other approaches. This, in particular, permeates fundamental features of system design, Table 3. Note that the level of granularity of the information granules exhibit a profound effect on the processing style.

Level of granularity Design criterion	High	Low
Amount of data processed	LARGE	SMALL
Computing time	LONG	SHORT
Details	RICH & MANY	VERY FEW
Generality	BAD	GOOD

Table 3. Levels of information granularity vis-a-vis fundamental system design criteria

4 Quantifying Links Between Information Granules

Granular computing involves processing information granules. Information granules realize a principle of information hiding meaning that we do not see any details below this level. In particular, if there are intensive numeric layer providing experimental data, this experimental evidence is completely hidden. It is then of interest to quantify links occurring at these information granules. The quantification itself is carried out based on available experimental evidence. We introduce two measures: a relevance measure based on the cardinality of data invoked by information granules and a fuzzy correlation that is viewed as a generalized version of the well-known concept of correlation encountered in data analysis. In what follows, we consider experimental data (x_k, y_k), k=1, 2, ..., N where $x_k \in \mathbf{X}$ and $y_k \in \mathbf{Y}$. Moreover given are two fuzzy sets A and B defined in $\mathbf{X}$ and $\mathbf{Y}$, respectively.

4.1 Relevance Measure of Linkage Between Information Granules

The rationale behind the proposed relevance measure is the following. Each x_i's falling under the "umbrella" of A has its image (in terms of the corresponding element in the data set) in some other space **Y**. This image eventually becomes associated with some other fuzzy set B defined in **Y**. Denote by n_1 a cardinality of the data set falling under an α-cut of A,

$$n_1(\alpha) = \text{card}(x_k \mid A(x_k) \geq \alpha)$$

Similarly, we use the notation

$$n_2(\alpha) = \text{card}(y_k \mid B(y_k) \geq \alpha \text{ and } A(x_k) \geq \alpha)$$

to denote the cardinality of the elements "matching" B among those that have already been activated by A. We use notation $n_1(\alpha)$ and $n_2(\alpha)$ to underline the fact that these two cardinalities depend upon the threshold values (membership grades). As shown in Figure 2, it becomes apparent that n_1 exceeds n_2 (as there could be more data points which even present in A may not show up in any processing invoked by B).

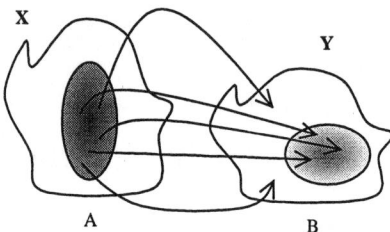

Figure 2. Expressing relevance of the directional link between A and B; fuzzy sets A and B play a role of information granules confining all computations to data compatible (activated) by them

Obviously, if cardinality of B gets higher, the values of n_2 become higher. The ratio

$$\text{rel}(A, B)(\alpha) = \frac{n_2(\alpha)}{n_1(\alpha)}$$

can be regarded as a measure of relevance (strength) of the link between A and B. It is worth emphasizing that this relevance is measured based on an experimental evidence provided by the available numeric data (x_k, y_k). The notion of relevance concerns a quantification of a directional link between A and B. Moreover the values of rel(A, B) treated as a function of α, are a good indicator of the changes of the strength of the link expressed in terms of the threshold values (α). By analogy, we define the relevance of the directional link between B and A in the form

$$\text{rel}(B, A)(\alpha) = \frac{n_2'(\alpha)}{n_1'(\alpha)}$$

where now we get

$$n_1'(\alpha) = \text{card}(y_k \mid B(y_k) \geq \alpha)$$

and

$$n_2'(\alpha) = \text{card}(x_k \mid A(x_k) \geq \alpha \text{ and } B(y_k) \geq \alpha)$$

This relevance index concerns the quantification of the strength of the link originating from **Y** to **X** where the link is determined for the same information granules as before. It could well be that rel(A, B) is very different in terms of its values from those achieved for rel(B, A). This is not surprising as the proposed index concentrates on a specific direction of the mapping. To form a direction-free index that may be regarded as a measure of association between **X** and **Y** confined to the information granules A and B, we need to aggregate rel(A, B) and rel(B, A). Any *and*-like combination is a rational choice, namely, rel(A, B) & rel(B, A). In a nutshell, such combination is another means to capture the meaning of the association and as such is related with the idea of fuzzy correlation.

4.2 Fuzzy Correlation

Correlation analysis has been a cornerstone of regression models and data analysis, in general. It reveals and quantifies an effect of linear dependency between two variables. A simple indicator of strength of association between the variables comes in the form of a so-called correlation coefficient. This coefficient, usually denoted by ρ, assumes values in [-1, 1]. The higher its absolute values, the stronger the dependency between the variables. The analysis of correlation arises as the first essential step of data analysis before proceeding with more refined designs in the form of full-fledged regression models.

To emphasize the origin of a *lack* of correlation between two variables, one may underline the *global* character of the correlation coefficient. It is a global measure of association between the variables expressed across the full range of their numeric values.

Our underlying approach is to abandon a *global* look at the overall data and concentrate on revealing meaningful relationships on a *local* level. When we introduce fuzzy sets A and B defined in **X** and **Y**, we endorse the local point of view. Fuzzy sets act as information granules that help confine ourselves to the section of the data set. We become interested in carrying out the correlation analysis on in the context of the linguistic labels. The values of the correlation coefficient are higher as reflecting a strong correlation effect for the more consistent (correlated) regions of the data. Because of the use of fuzzy sets in revealing the local dependencies, the resulting correlation coefficient will be referred to as a *fuzzy correlation*.

The notion of correlation is inherently associated with a descriptive statistical measure that expresses the degree of association (relationship) between two or more variables encountered in the problem at hand. The basic definition of the correlation coefficient (Pearson product-moment correlation coefficient) reads as

$$r = \frac{\sum_{k=1}^{N} x(k)y(k) - \frac{(\sum_{k=1}^{N} x(k))(\sum_{k=1}^{N} y(k))}{N}}{\sqrt{\left(\sum_{k=1}^{N} x^2(k) - \frac{\left(\sum_{k=1}^{N} x(k)\right)^2}{N}\right)\left(\sum_{k=1}^{N} y^2(k) - \frac{\left(\sum_{k=1}^{N} y(k)\right)^2}{N}\right)}}$$

(1)

where $\{x(k)\}$ and $\{y(k)\}$, k=1, 2, ..., N are the values of the corresponding pairs of data. The correlation coefficient assumes values in the [-1, 1] range. The higher the absolute value of r, the stronger is the relationship between the two variables under consideration. Low values of r stipulate no relationship between the variables. Once the correlation coefficient has been computed, it is common practice to determine whether its absolute value is large enough to make any conclusion as to the dependencies between the variables meaningful.

In the setting proposed in [9], the fuzzy correlation is a fuzzy set defined in [0, 1]. We determine its membership values by computing (numeric) correlation based on data embraced by successive α-cuts of the information granules A and B. This leads us to the idea of the correlation coefficient that becomes a fuzzy set defined in the unit interval and indexed by the values of α. In other words, the correlation coefficient is regarded as a function of α, $r(\alpha)$.

The algorithm of computing the fuzzy set of the correlation coefficient consists of several steps. They are repeated for a fixed value of the threshold. The correlation coefficient is computed for all data elements embraced by the corresponding α-cut.

Start with $\alpha = 0.0$ and increase it by a certain increment until it reaches the value equal to 1. Carry out the following steps:
➢ determine all elements of X for which x(k) belongs to A_α and y(k) belongs to B_α; denote the resulting subset of X by X_α.
➢ for X_α compute the correlation coefficient r; more specifically its value $r(\alpha)$.
➢ (optional) determine the number of the elements involved in the computations of the correlation coefficient (this is required to complete some hypothesis testing to figure out the statistical relevance of the correlation between the variables).

The fuzzy set of correlation provides us with a complete picture as to the overall form of association of the variables and delivers a better insight into the structure of the data. This is primarily revealed as moving across the range of the values of α. Moreover, for any fixed value of α, a hypothesis can be tested as to the strength of correlation captured within the range of the α-cuts of the linguistic granules. Several characteristic patterns can be envisioned as shown in Figure 3(a), (b) and (c):

(a) In general, the data in Figure 3(a) exhibit a very low level of correlation. By decreasing the size of the information granules (A and B), our analysis is confined to a subset of the data in which the linear relationships could become more profound. The curve portrayed in this figure underlines this effect: the less data invoked in the analysis, the higher the value of the correlation coefficient

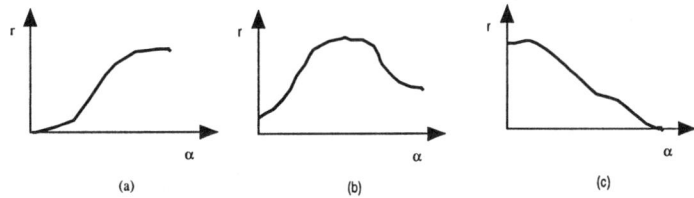

Figure 3. Examples of fuzzy sets of correlation; see detailed description in text.

(b) In Figure 3(b), more specific linguistic granules give rise to the higher values of the correlation coefficient meaning that the linear relationships become more profound in the subset of data "highlighted" by the fuzzy sets under consideration. There is an optimal value of α leading to the highest values of "r". With the increase of the threshold, a decrease in the strength of correlation is observed This effect can be explained as follows: as a small region of the data set is being concentrated on, some noise being inevitably associated there may be amplified. Noticeably, the maximum corresponds to a point where by selecting a proper level of granularity of linguistic terms, a sound balance is achieved between a global nonlinear pattern in data and the noise factor present there.

(c) In Figure 3(c), the correlation is high even for quite large information granules (low values of α) and any further increase of the values of α leads to the lower values of the correlation coefficient. This is due to the high level of noise associated with the data. Too specific information granules generate a "magnification" effect of noise associated with the data. In this sense, the second scenario, Figure 3(b), is an intermediate situation between the case of high nonlinearity and high noise.

One should stress that the membership functions of the fuzzy correlation depend on the location and size of the linguistic granules (fuzzy sets) encountered in data analysis.

5 Fundamental Issues of Traversing Information Pyramid: Encoding and Decoding

Granular computing supports modeling activities carried out at various levels of information granularity, refer again to Figure 4.

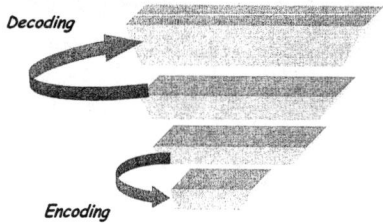

Figure 4. Decoding and encoding information granules as a vehicle of traversing the information pyramid

The ability to traverse through the layers characterized by different sizes of information granules is one of the dominant features of the modeling pursuits discussed in this framework. Each modeling layer indexed by the assumed level of granularity, comes with its own repertoire of modeling techniques. For instance, for the highest level of information granularity, viz. numeric data, we are dealing with differential equations and regression models as basic vehicles of system modeling. Commonly used neural networks fall under the same category. When moving towards nonnumeric layer where some information granules of lower granularity are formed, we encounter a diversity of models such as Petri nets, finite state machines, bond graphs, constraint-based, etc. Depending on the specific form of granulation, we subsequently allude to fuzzy Petri nets, probabilistic Petri nets, etc.

The layers communicate between themselves. They receive data from other layers, complete computing (processing) and return the results to some other layers. These communication mechanisms are referred to as encoding and decoding, respectively. The role of the encoder is to transform the input information entering the given layer. The objective of the decoder is to convert the information granules produced by the given layer into the format acceptable by the destination layer. Depending on the problem at hand and the formalism of information granulation being used, a specific naming comes into play. We will discuss this in a while.

The general formulation of the encoding – decoding problem can be delineated as follows

- develop encoding (Enc) and associated decoding (Dec) algorithms such that the following relationship is satisfied

$$\text{Dec}(\text{Enc}(X)) = X$$

for all information granules (X) defined in a certain formal framework of information granulation and for a broad range of sizes of the information granules involved. In a limit case numeric granules are also included. Note, however, that the decoding-encoding scheme could be very demanding and one may not be able to meet the equality. More practically, we request that the design of these transformation should minimize the associated transformation error meaning that we are interested in minimizing the expression involving the distance ||.|| between the original information granule and its transformation

$$\|\text{Dec}(\text{Enc}(X)) - X\| \to \text{Min}$$

over a given range of granularity of X's involved there and for a fixed granulation environment.

The A/D and D/A conversions form an interesting illustration to the formulation of the problem given above, see Figure 5. We have

A/D: $\text{Enc}(X) : X = \{x\} \in \mathbf{R} \to X \in P(\mathbf{R})$ (the resulting granules are intervals in $\mathbf{R}$; depending how the intervals are formed, one encounters either uniform quantization or a non-uniform one)

D/A: $\text{Dec}(X) : X \in P(\mathbf{R}) \to X = \{x'\} \in \mathbf{R}$ (usually a quantization error occurs so we never obtain the original numeric entity, $x \neq x'$).

The A/D and D/A conversions can be revisited and generalized in the framework of fuzzy sets, $F(\mathbf{X})$. This leads to the following formulation of the problem

A/D: $\text{Enc}(X) : X = \{x\} \in \mathbf{R} \to X \in F(\mathbf{R})$ (the resulting granules are fuzzy sets in $\mathbf{R}$; depending how they are formed, one encounters either uniform quantization or a non-uniform linguistic discretization of $\mathbf{X}$)

D/A: $\text{Dec}(X) : X \in P(\mathbf{R}) \to X = \{x'\} \in \mathbf{R}$ (usually a quantization error it can be avoided by selecting a proper family of fuzzy sets. The zero error occurs for the triangular fuzzy sets with ½ overlap between successive membership functions).

In fuzzy controllers, the process of converting numeric data into the format accepted by the inference engine is called *fuzzification*. This is the name used for the encoding mechanism. The decoding is referred to as a *defuzzification* scheme.

One may also envision also a mixed form of information granules, namely they may originate from different formal environments of information granulation.

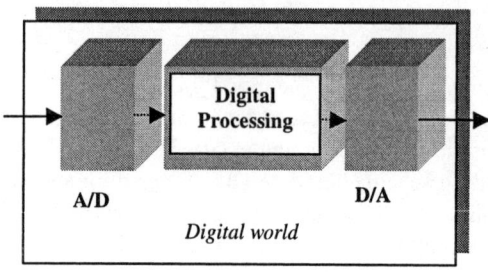

Figure 5. Digital processing as an example of commonly encountered granular computing; note a role of A/D and D/A converters utilized as the encoding and decoding modules

6 Conclusions

We have discussed the fundamentals of granular computing viewed as a new unified paradigm of processing information granules. Granular computing subsumes commonly encountered numeric processing as its special (limit) case.

The research agenda of granular computing includes a series of key and well-defined methodological and algorithmic issues

- Construction of information granules. This deals both with the selection of the formal framework of information granulation and detailed estimation procedure producing information granules. The latter dwells on the usage of the setting in which the granules are constructed
- Characterization of dimension (granularity) of information granules. This task is crucial as providing us with a better insight as to the essence of the granulation process and its implications both at the level of
- The development of the encoding and decoding mechanisms. These are essential to the functioning of any granular architecture. The encoding and decoding schemes are essential to the performance of granular computing. Interestingly enough, the essence of information compatibility expressed in terms of its granularity is inherently related with granular computing and nonexistent with

The study is an introduction to the rapidly growing research area. It concentrates on the methodology, attempts to identify the common features and help put the existing somewhat scattered approaches under the same conceptual umbrella.

Acknowledgment

The support from the Natural Sciences and Engineering Research Council of Canada (NSERC) is gratefully acknowledged.

References

1. Buckley, J., Hayashi, Y. (1994) Fuzzy neural networks: a survey, *Fuzzy Sets and Systems*, 66, 1-14.
2. Harris, C.J., Moore, C.G., Brown, M. (1993) *Intelligent Control - Aspects of Fuzzy Logic and Neural Nets,* World Scientific, Singapore.
3. Jang,J. S. R., Sun, C.T., Mizutani, E. (1997) *Neuro-Fuzzy and Soft Computing*, Prentice Hall, Upper Saddle River, NJ.
4. Kandel, A. (1986) *Fuzzy Mathematical Techniques with Applications*, Addison-Wesley, Reading, MA.
5. Kasabov, N. (1996) *Foundations of Neural Networks, Fuzzy Systems, and Knowledge Engineering*, MIT Press, Cambridge, MA.
6. Kruse, R., Gebhardt, J., Klawonn, F. (1994) *Foundations of Fuzzy Systems*, J. Wiley, Chichester.
7. Pedrycz, W (1997). *Computational Intelligence: An Introduction*, CRC Press, Boca Raton, FL.
8. Pedrycz, W., Gomide, F. (1998) *An Introduction to Fuzzy Sets*, Cambridge, MIT Press, Cambridge, MA.
9. Pedrycz, W., Smith, M. H. 1999. Granular correlation analysis in data mining, *Proc. 18th Int Conf of the North American Fuzzy Information Processing Society (NAFIPS),* New York, June 1-12, pp. 715-719.
10. Pedrycz, W. Vukovich, G. (1999) Quantification of fuzzy mappings: a relevance of rule-based architectures, *Proc. 18th Int Conf of the North American Fuzzy Information Processing Society (NAFIPS),* New York, June 1-12, pp. 105-109.
11. Tsoukalas, L.H., Uhrig, R.E. (1997) *Fuzzy and Neural Approaches in Engineering*, J. Wiley, New York.
12. Zadeh, L. A. (1979) Fuzzy sets and information granularity, In: M.M. Gupta, R.K. Ragade, R.R. Yager, eds., *Advances in Fuzzy Set Theory and Applications*, North Holland, Amsterdam, 3-18.
13. Zadeh, L. A. (1996) Fuzzy logic = Computing with words, *IEEE Trans. on Fuzzy Systems*, vol. 4, 2, 1996, 103-111.

14. Zadeh, L. A. (1997) Toward a theory of fuzzy information granulation and its centrality in human reasoning and fuzzy logic, *Fuzzy Sets and Systems*, 90, 1997, 111-117.

Chapter 16. A New Paradigm Shift from Computation on Numbers to Computation on Words on an Example of Linguistic Database Summarization

Janusz Kacprzyk

Systems Research Institute, Polish Academy of Sciences
Ul. Newelska 6, 01-447 Warsaw, Poland
kacprzyk@ibspan.waw.pl

Abstract. We present a fuzzy logic based approach to the derivation of linguistic summaries of sets of data (databases), and show that it may be viewed as an example of a new paradigm shift from computing on numbers to computing on words that has been recently strongly advocated by Zadeh. We present an implementation of linguistic database summaries for sales data of a computer retailer that clearly shows that the new approach is viable and yields a new quality by providing human consistent results.

Keywords. Data mining, knowledge discovery, linguistic summaries, fuzzy logic, database, querying, fuzzy querying, fuzzy linguistic quantifier, natural language, interface

1 Introduction

Over the last couple of years, Zadeh has been actively advocated what might be called a new paradigm shift that may be characterized by replacing the calculations on numbers by the calculations on perceptions [cf. Zadeh (1996, 1999) for recent accounts, or Zadeh and Kacprzyk (1999a, b) for a comprehensive coverage].

This new direction is centered on the concept of a perception (of time, distance, force, direction, shape, intent, etc.) which plays a pivotal role in human cognition. Humans have a formidable ability to manipulate perceptions, and to reason with perceptions, and this makes possible the performing of a variety of tasks that cannot be solved, or even modeled, using conventional techniques based on numbers and their handling using conventional mathematical tools. Such tasks may be exemplified by parking a car, driving in a dense traffic, summarizing a story, etc.

Perceptions are inherently imprecise which is implied by the finiteness of a human ability to differentiate details and store information. In other words, perceprions are both fuzzy and granular which is to be meant in the sense that, on the one hand, boundaries of the perceived classes are not sharply defined, and, on the other hand, elements of classes are grouped into granules consisting of elements drawn together by indistinguishibility, similarity, etc.

Traditionally, scientists have been following the traditional paradigm that might be described by the transition "from perceptions to numbers (measurements)". And, indeed, it has resulted in many successes though mostly in what might be described as "hard" sciences and technology in which a human element is not playing a significant role. However, practically no success has been encountered in those "human centered" tasks (driving a car, summarizing a story, etc.) that are performed by humans without any measurements (e.g. numeric data).

This situation, and a shortcoming of conventional tools, do indicate that some paradigm shift might be needed to model and solve a variety of human centered problems and tasks. Such a paradigm shift might be described by the transition "from numbers (measurements) to perceptions".

Such a paradigm shift needs some computational theory of perceptions. Unfortunately, in spite of a huge research effort in all perception-related matters (e.g. in cognitive sciences), there have been no computational approaches in that respect. It seems that fuzzy logic can play here a key role by providing relatively simple and constructive tools.

In general, perceptions are expressed by humans in a verbal form. First, values of some quantities in question are given as linguistic terms as, e.g., high, low, etc. It is lear that these linguistic descriptions are intrinsically imprecise, and can be equated with fuzzy sets. Such descriptions may be subjected to linguistic modifiers as, e.g., very, more or less, and fuzzy logic provides here tools too. Relations between perceptions are expressed by some propositions in a (qiasi)natural language that might be generally written as "X isr R" where R is a constraining (fuzzy) relation that is called a generalized value of X, and "isr" defines the way in which R constrains X as, e.g., by equality, usuality, (fuzzy) probability, etc. Finally, fuzzy logic makes it possible to devise human-consistent and realistic reasoning schemes that makes the reasoning with perception possible.

From a technical point of view, the above mentioned paradigm shift "from measurements (numbers) to perceptions" may be viewed as being equivalent to the application of "computing with words" instead of the conventional "computing with numbers". For a comprehensive coverage of the fundamentals and major applications of computing with words see Zadeh and Kacprzyk (1999a, b).

In this paper we do not deal in more detail with computing with words, or – more generally speaking– with the newly advocated paradigm shift from computing with measurements (numbers) to computing with perceptions (words) referring the interested readers to the source papers by Zadeh (1996/9), or better yet to Zadeh and Kacprzyk's (1999a, b) volumes.

Instead we will show how the above paradigm shift may be implemented in linguistic summarization of databases, and what new qualities it will provide.

2 Data Mining via Linguistic Database Summaries

Information Technology (IT) has become one those reas of science and technology that have been developing at rapid pace. One of implications is an increased availability of a huge amount of data (from diverse, often remote databases).

Unfortunately, the availability of (raw) data is not by itself of any use, and we usually are not in a position to productively use those data. In general, more interesting and useful are relevant, nontrivial dependencies that are encoded in those data. Unfortunately, they are usually hidden, and their discovery is not trivial, and requires some intelligence.

Data mining is a multifaceted field of science and technology, and we deal here with linguistic summaries od tata stored in databases.

We propose an approach to the derivation of linguistic summaries of large sets of data in the sense of Yager (1982, 1991), i.e. derived as linguistically quantified propositions as, e.g., by "most of the employees are young and well paid", with a degree of validity (truth).

We would also be interested in more conceptually sophisticated linguistic summaries as, e.g., "most orders are *difficult*". We advocate the view that the linguistic summaries of the type mentioned above may only be practically formulated interactively, i.e. via an interaction with the user. This interaction will proceed via Kacprzyk and Zadrożny's (1994a - 1995c) FQUERY for Access, a fuzzy querying add-on to Access. New perspectives related to querying via WWW will be mentioned too.

Notice that our approach is not equivalent to using a natural-language-based user interface, whose essence is shown in Figure 1, i.e. that we have a nonfuzzy database (i.e. "numbers"), the user formulates a query using (quasi) natural language, i.e. "words", and the result is a set of records, i.e. "numbers" (our database contains "numbers").

The application of the new paradigm could be, on the other hand, shown as in Figure 2, i.e. that we have a nonfuzzy database (i.e. "numbers"), the user formulates a request for linguistic summarization in a (quasi) natural language (i.e. in "words"), and the result is a linguistic summary of the contents of the database (i.e. "words").

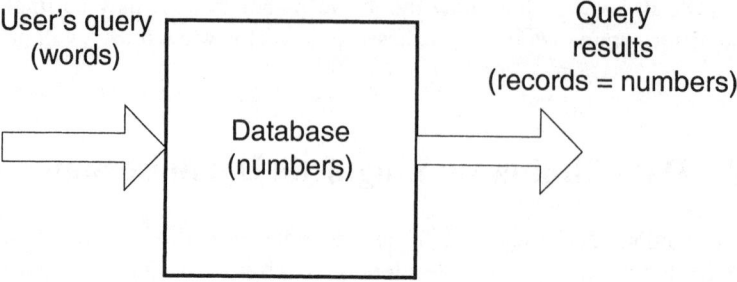

Fig. 1. The essence of fuzzy querying from our perspective

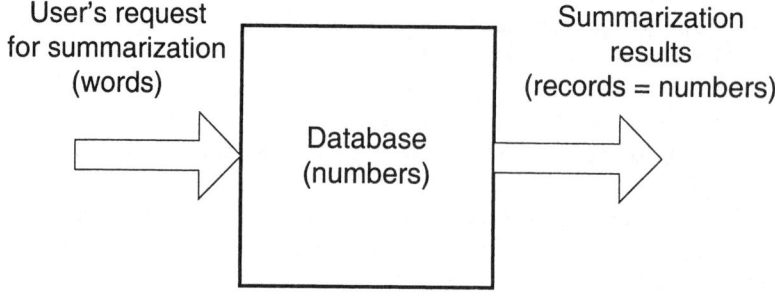

Fig. 2. The essence of linguistic summarization from our perspective

As an example we will show an implementation of the data summarization scheme mentioned above for the derivation of linguistic data summaries in a sales database of a computer retailer.

3 Linguistic Summaries Using Fuzzy Logic with Linguistic Quantifiers

We adopt the concept of a linguistic summary in the sense of Yager (1982, 1991). Basically, we suppose that we have:

- V - a quality (attribute) of interest, with numeric and non-numeric (e.g. linguistic) values - e.g. salary in a database of workers,

- $Y = \{y_1, \ldots, y_n\}$ - a set of objects (records) that manifest quality V, e.g. the set of workers; $V(y_i)$ - values of quality V for object y_i,

- $D = \{V(y_1), \ldots, V(yn_1)\}$ - a set of data (database)

A *summary* of data set consists of:

- a summarizer S (e.g. young),
- a quantity in agreement Q (e.g. most),
- truth (validity) T - e.g. 0.7,

and may exemplified by "T(*most* of employees are *young*)=0.7".

More specifically, if we have a summary, say "*most* (Q) of the employees (y_i's) are *young* (S)", where "*most*" is a fuzzy linguistic quantifier $\mu_Q(x)$, $x \in [0,1]$, "*young*" is a fuzzy quality S, and $\mu_S(y_i), y_i \in Y$, then using the classic Zadeh's (1983) calculus of linguistically quantified propositions, we obtain

$$T = \mu_Q[\frac{1}{n}\sum_{i=1}^{n}\mu_S(y_i)] \tag{1}$$

For more sophisticated summaries as, e.g., "*most* (Q) employees (y_i's) are *young* (S) and *well paid* (F)", the reasoning is similar, and we obtain

$$T = \mu_Q(\sum_{i=1}^{m}(\mu_P(x_i) \wedge \mu_B(x_i)) / \sum_{i=1}^{m}\mu_B(x_i)) \tag{2}$$

where, $\sum_{i=1}^{N}\mu_B(x_i) \neq 0$, and $\wedge$ is a *t*-norm.

The above calculus may be replaced by, e.g., the OWA operators [cf. Yager and Kacprzyk (1997)].

4 Fuzzy Logic Based Linguistic Data Summarization

The simple approach, in its source version, it is meant for one-attribute simplified summarizers (concepts) - e.g., young. It can be extended to cover more sophisticated summaries as, e.g, "*young* and *well paid*", but this should be done "manually", and leads to some combinatorial problems as a huge number of summaries should be generated and validated to find the most proper one. For instance [cf.. George and Srikanth (1996)] a genetic algorithm to find the most appropriate summary, with quite a sophisticated fitness function.

The validity criterion is not trivial either, and various measures of specificity, informativeness, etc. may be used. This relevant issue will not be discussed here, and we will refer the reader to, say, Yager (1982, 1991) or Yager and Kacprzyk (1999).

The basic idea of fuzzy logic based linguistic data summarization adopted in this paper consists in using a linguistically quantified proposition, as originated by Yager (1982, 1991), and here we extend it to using a fuzzy querying package.

First we reinterpret (1) and (2) for data summarization. Thus, (1) is meant as formally expressing a statement

"Most records match query S" (3)

We allow for fuzzy terms in a query which implies a degree of matching from [0,1] rather than a yes/no matching. Effectively, a query S defines a fuzzy subset (fuzzy property) on the set of the records, where the membership of them is determined by their matching degree with the query.

Similarly, (2) may be interpreted as expressing a statement

"Most records meeting conditions F match query S" (4)

Therefore, in database terminology, F corresponds to a filter and (4) claims that most records passing through F match query S. Moreover, since F may be fuzzy, a record may pass through it to a degree from [0,1]. As this is more general than (3), we will assume (4) as a basis.

We seek, for a given database, propositions of the type (1), interpreted as (4), which are highly valid (true).

Basically, a proposition sought consists of three elements:

- a fuzzy filter F (optional),
- a query S, and
- a linguistic quantifier Q.

There are two limit cases, where we:

- do not assume anything about the form of any of these elements,
- assume fixed forms of the fuzzy filter and query, and only seek a linguistic quantifier Q.

In the first case data summarization will be extremely time-consuming, though may produce interesting results. In the second case the user has to guess a good candidate formula for summarization, but the evaluation is fairly simple [more or less as answering a (fuzzy) query]. The second case refers to the summarization known as *ad hoc queries*, extended with an automatic determination of a linguistic quantifier. Between these two extremes there are different types of summaries, with various assumptions on what is given and what is sought.

In our implementation of linguistic summaries, there are supported two types of summaries. First, the so-called Type 1 that is a simple extension of fuzzy querying as in FQUERY for Access. The user has to conceive a query which may be true for some population of records in the database. As a result of this type of summarization he or she receives some estimate of the cardinality of this population as a linguistic quantifier. The primary target of this type of summarization is to propose such a query that a large proportion as, e.g., *most*, of the records satisfy it Second, Type 2 summaries that are concerned with the detrmination of typical (exceptional) values of a field; then, a query S consists of only one simple condition referring to the field under consideration, and the summarizer tries to find a value, possibly fuzzy, such that the query is true for Q records.

5 FQUERY for Access: A Fuzzy Querying Add-on

FQUERY for Access is an add-on (add-in) to Microsoft Access that provides fuzzy querying capabilities [cf. Kacprzyk and Zadrożny (1994 - 1997c), Zadrożny and Kacprzyk (1995)].

FQUERY for Access makes it possible to use fuzzy terms in regular queries, then submitted to the Microsoft Access's querying engine. The result is a set of records matching the query, but obviously to a *degree* from [0,1].

Briefly speaking, the following types of fuzzy terms are available:

- fuzzy values, exemplified by *low* in "profitability is *low*",
- fuzzy relations, exemplified by *much greater than* in "income is *much greater than* spending", and
- fuzzy linguistic quantifiers, exemplified by *most* in "*most* conditions have to be met".

If a field is to be used in a query in connection with a fuzzy value, it has to be defined as an *attribute*. The definition of an attribute consists of two numbers: the attribute's lower (LL) and upper (UL) limit. They set the interval which the field's values are assumed to belong to, according to the user. This interval depends on the meaning of the given field. For example, for *age* (of a person), the reasonable interval would be, e.g., [18,75], in a particular context, i.e. for a specific group. Such a concept of an attribute makes it possible to universally define fuzzy values.

Fuzzy values are defined as fuzzy sets on [-10, +10]. Then, *the matching degree* $md(\cdot,\cdot)$ of a simple condition referring to attribute AT and fuzzy value FV in a record R is calculated by

$$md(\text{AT} = \text{FV}, \text{R}) = \mu_{\text{FV}}(\tau(\text{R}(\text{AT}))) \qquad (5)$$

where: R(AT) is the value of attribute AT in record R, μ_{FV} is the membership function of fuzzy value FV, τ: $[\text{LL}_{\text{AT}}, \text{UL}_{\text{AT}}] \to [-10, 10]$ is the mapping from the interval defining AT onto [-10,10] so that we may use the same fuzzy values for different fields. A meaningful interpretation is secured by τ which makes it possible to treat all field domains as ranging over the unified interval [-10,10].

For simplicity, it is assumed that the membership functions of fuzzy values are trapezoidal τ is assumed linear.

Linguistic quantifiers provide for a flexible aggregation of simple conditions. In FQUERY for Access the fuzzy linguistic quantifiers are defined in Zadeh's (1983) sense, i.e. as fuzzy sets on the [0, 10] interval instead of the original [0, 1]. They may be interpreted either using the original Zadeh's approach or via the OWA operators [cf. Yager and Kacprzyk (1997)]; Zadeh's interpretation will be used here. The membership functions of fuzzy linguistic quantifiers are assumed piecewise linear, hence two numbers from [0,10] are needed. Again, a mapping from [0,N], where N is the number of conditions aggregated, to [0,10] is employed to calculate the matching degree of a query. More precisely, the matching degree, $md(\cdot,\cdot)$, for the *query* "Q of N conditions are satisfied" for record R is equal to

$$md(Q \text{ condition}_i, \text{R}) = \mu_Q(\frac{1}{\tau(N)} \tau(\sum_i md(\text{condition}_i, \text{R}))) \qquad (6)$$

We can also assign different importance degrees for particular conditions. Then, the aggregation formula is equivalent to (1). The importance is identified with a fuzzy set on [0,1], and then treated as property F in (1).

In FQUERY for Access, queries containing fuzzy terms are still syntactically correct Access's queries through the use of parameters. Access represents the queries using SQL. Parameters, expressed as strings limited with brackets, make it possible to embed references to fuzzy terms in a query. We have assumed a special naming convention for parameters corresponding to the particular fuzzy terms. For example:

[FfA_FV *fuzzy value name*]   will be interpreted as a fuzzy value
[FfA_FQ *fuzzy quantifier name*] will be interpreted as a fuzzy quantifier

This maintenance of dictionaries of fuzzy terms defined by users strongly supports our approach to data summarization. In fact, the package comes with a set of predefined fuzzy terms but the user may enrich the dictionary too.

When the user initiates the execution of a query, it is automatically transformed and then run as a native query of Access. The transformation consists primarily of the replacement of parameters referring to fuzzy terms by calls to functions that secure a proper interpretation of these fuzzy terms. Then, the query is run by Access as usually.

In our approach, the interactivity, i.e. user assistance, is in the definition of summarizers (indication of attributes and their combinations). This proceeds via a user interface of a fuzzy querying add-on.

We use "natural" linguistic terms, i.e. (7±2!) exemplified by: *very low, low, medium, high, very high*, and also "comprehensible" fuzzy linguistic quantifiers as: *most, almost all*, etc.

In Kacprzyk and Zadrożny (1994a - 1995c), a conventional DBMS is used, and a fuzzy querying tool is developed. Basically, the so-called *quantified queries*, introduced by Kacprzyk and Ziółkowski (1986) and Kacprzyk, Zadrożny and Ziółkowski (1989), are used. They make it possible to express complex concepts as, e.g., a "*serious water* pollution" may well be equated with, say, "*almost all* of the *relevant* pollution indicators *considerably* exceed pollution limits (maybe imprecisely specified)". Zadeh's (1983) calculus of linguistically quantified propositions is use. Notice that the quantified queries are what we do need for implementing the linguistic summaries.

FQUERY for Access is embedded in the native Access's environment as an add-on. Definitions of attributes, fuzzy values etc. stored in proper locations, and a mechanism for putting them into the Query-By-Example (QBE) sheet (grid) are provided. All the code and data is put into a database file, a *library*, installed by the user. Parameters are used, and a query transformation is performed.

FQUERY for Access provides its own toolbar. There is one button for each fuzzy element, and the buttons for declaring attributes, starting the querying, closing the toolbar and for help (cf. Figure 3).

Generally, the user interacts with FQUERY for Access, by pressing a button, in order to:

- declare attributes,

- define fuzzy elements and put them automatically into the QBE sheet,

- start the querying process,

The above fuzzy querying add-on was the extended by Kacprzyk and Zadrożny (1996) to fuzzy querying over the Internet. The query is defined by using a WWW browser (more specifically, Microsoft Explorer or Netscape Navigator), and the user interface is similar as in Figure 3, with a WWW-browser-like toolbar. The definition of fuzzy values, fuzzy relations, and linguistic quantifiers is via Java applets. Basically, a query is sent to the WWW server, the searching program decodes the query, and, if fuzzy elements exist, a HTML page is sent back to specify their membership functions, search is done sequentially, and yields a matching degree. The results are sent back as a HTML document to be displayed. So, the interface consists of: HTML pages for query formation, membership functions specifications, list of records and the content of a selected record, a searching program, and a reporting program. The use of such a fuzzy querying add-on employing a WWW browser can greatly enhance the scope and perspectives of linguistic summarization.

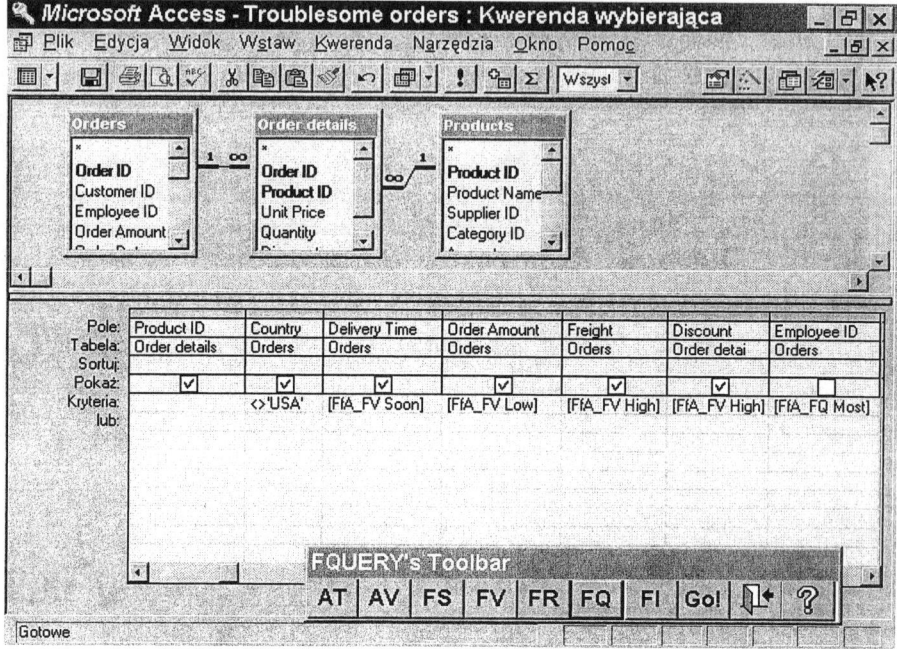

Figure 3. Composition of a fuzzy query

6 Linguistic Summaries via FQUERY for Access

FQUERY for Access, which extends the querying capabilities of Microsoft Access by making it possible to handle fuzzy terms, may be viewed as an interesting tool for the generation of linguistic summaries. The simplest method of data mining through ad-hoc queries becomes much more powerful by using fuzzy terms. Nevertheless, the implementation of various types of summaries mentioned before seems to be worthwhile, and is fortunately enough relatively straightforward.

We employ dictionaries of fuzzy terms maintained and extended by users during the subsequent sessions. To easily generate linguistic summaries we adopt context-free definitions of particular fuzzy terms. Hence, looking for a summarizer we may employ any term in the context of any attribute. Thus, we get a summarizer building blocks at hand and what is needed is an efficient procedure for their composition, compatible with the rest of the fuzzy querying system.

In case of Type 1 summaries only the list of defined linguistic quantifiers is employed. The query S is provided by the user and we are looking for a linguistic quantifier describing in a best way the proportion of records meeting this query. Hence, we are looking for a fuzzy set in the space of linguistic quantifiers such that

$$\mu_S(Q) = \text{truth}(QS(X)) = \mu_Q(\sum_{i=1}^{N}\mu_S(x_i)/m) \qquad (7)$$

is maximized.

FQUERY for Access processes the query, additionally summing up the matching degrees for all records. Thus, the sum in (7) is easily calculated. Then, the results are displayed as a list of records ordered by their matching degree. In another window, the fuzzy set of linguistic quantifiers sought is shown. We only take into account quantifiers defined by the users for querying, for efficiency. Currently, FQUERY for Access does not support fuzzy filters. As soon as this capability is added, also summaries of Type 1.1 will be available.

Type 2 summaries require more effort and the redesigning of the results' display. Now, we are given the quantifier and the whole query, but without some values. The user may leave out some values in the query's conditions and request the system to find a best fit for them. Obviously, such computations are time-consuming and are practically feasible only for one placeholder in a query. On the other hand, the case of searching for typical or exceptional values is the most useful form of a Type 2 summary. It is again fairly easy to embed Type 2 summaries in the existing fuzzy querying mechanism.

7 An Implementation: Linguistic Summarization of a Sales Database in a Computer Retailer

The linguistic data summarization procedure was implemented for a sales database of a small-to-medium size computer retailer (ca. 15 employees) located in the Southern part of Poland.

The database is characterized by:

Number of records:	8743
Number of attributes:	14
Number of transaction documents:	4000
Number of suppliers and customers:	3000
Number of products carried:	3000

and these numbers obviously vary over time.

The basic structure of the database is shown in Table 1.

Table 1. The basic structure of the database (in the "dbf" type format)

Attribute name	Attribute type	Description
Date	Date	Date of sale
Time	Time	Time of sale transaction
Name	Text	Name of the product
Amount (number)	Numeric	Number of products sold in the transaction
Price	Numeric	Unit price
Commission	Numeric	Commission (in %) on sale
Value	Numeric	Value = amount (number) x price; of the product
Discount	Numeric	Discount (in %) for transaction
Group	Text	Product group to which the product belongs
Transaction value	Numeric	Value of the whole transaction
Total sale to customer	Numeric	Total value of sales to the customer in fiscal year
Purchasing frequency	Numeric	Number of purchases by customer in fiscal year
Town	Text	Town where the customer lives or is based

The process of deriving linguistic summaries proceeds by employing the fuzzy querying interface described above. Basically, by using that interface the user specifies the class of linguistic summary of interest in the sense of what relations (between which attributes) he is interested in.

We will now give a couple of examples. First, suppose that we are interested in what the relation between the commission and the type of goods sold is. We obtain the linguistic summaries shown in Table 2.

As we can see, the results can be very helpful in, e.g., negotiating commissions for various products sold.

Next, suppose that we are interested in relations between the groups of products and times of sale. We obtain the results shown in Table 3.

Table 2. Linguistic summaries expressing relations between the group of products and commission

Summary	Degree of appropriateness	Degree of covering	Degree of validity
	Degree of imprecision	Weighted average	
About ½ of sales of network elements is with a high commission	0.2329	0.4202	0.3630
	0.1872	0.3165	
About ½ of sales of computers is with a medium commission	0.2045	0.5498	0.4753
	0.3453	0.3699	
Many sales of accessories are with a high commission.	0.1684	0.5779	0.5713
	0.4095	0.3919	
Much sales of components is with a low commission	0.1376	0.7212	0.6707
	0.5837	0.4449	
About ½ of sales of software is with a low commission	0.1028	0.4808	0.4309
	0.5837	0.3162	
About ½ of sales of computers is with a low commission	0.0225	0.5594	0.4473
	0.5837	0.3202	
A few sales of components is without commission	0.0237	0.0355	0.0355
	0.2745	0.2346	
A few sales of computers is with a high commission	0.1418	0.0455	0.0314
	0.1872	0.1881	
Very few sales of printers is with a high commission	0.1288	0.0585	0.0509
	0.1872	0.1820	

Notice that in this case the summaries are much less obvious than in the former case expressing relations between the group of product and commission. It should also be noted that the weighted average is here very low but this, by technical reasons, should not be taken literally as these values are mostly used to order the summaries.

Table 3. Linguistic summaries expressing relations between the groups of products and times of sale

Summary	Degree of appropriateness	Degree of covering	Degree of validity
	Degree of imprecision	Weighted average	
About 1/3 of sales of computers is by the end of year	0,0999	0,3009	0,2801
	0,2010	0,1274	
About ½ of sales in autumn is of accessories	0,0642	0,4737	0,4790
	0,4095	0,1143	
About 1/3 of sales of network elements is in the beginning of year	0,0733	0,2857	0,1957
	0,2124	0,0982	
Very few sales of network elements is by the end of year	0,0833	0,1176	0,0929
	0,2010	0,0980	
Very few sales of software is in the beginning of year	0,0768	0,1355	0,0958
	0,2124	0,0929	
About ½ of sales in the beginning of year is of accessories	0,0348	0,4443	0,4343
	0,4095	0,0860	
About 1/3 of sales in the summer is of accessories	0,0464	0,3209	0,3092
	0,2745	0,0853	
About 1/3 of sales of peripherals is in the spring period	0,0507	0,3032	0,2140
	0,2525	0,0809	
About 1/3 of sales of software is by the end of year	0,0446	0,2455	0,2258
	0,2010	0,0768	
About 1/3 of sales of network elements is in the spring period	0,0458	0,2983	0,2081
	0,2525	0,0763	
Very few sales of network elements is in the autumn period	0,0485	0,1471	0,0955
	0,1956	0,0692	
A few sales of software is in the summer period	0,0402	0,1765	0,1765
	0,1362	0,0691	

Finally, let us show in Table 4 some of the obtained linguistic summaries expressing relations between the attributes: size of customer, regularity of customer (purchasing frequency), date of sale, time of sale, commission, group of product and day of sale. This is an example of the most sophisticated form of linguistic summaries supported by the system described.

Table 4. Linguistic summaries expressing relations between the attributes: size of customer, regularity of customer (purchasing frequency), date of sale, time of sale, commission, group of product and day of sale

Summary	Degree of appropriateness	Degree of covering	Degree of validity
	Degree of imprecision	Weighted average	
Much sales on Saturday is about noon with a low commission	0,3843 0,2748	0,6591 0,3863	0,3951
Much sales on Saturday is about noon for bigger customers	0,3425 0,4075	0,7500 0,3648	0,4430
Much sales on Saturday is about noon	0,3133 0,4708	0,7841 0,3564	0,4654
Much sales on Saturday is about noon for regular customers	0,3391 0,3540	0,6932 0,3558	0,4153
A few sales for regular customers is with a low commission	0,3882 0,5837	0,1954 0,3451	0,1578
A few sales for small customers is with a low commission	0,3574 0,5837	0,2263 0,3263	0,1915
A few sales for one-time customers is with a low commission	0,3497 0,5837	0,2339 0,3195	0,1726
Much sales for small customers is for nonregular customers	0,6250 0,1458	0,7709 0,5986	0,5105

8 Conclusions

We considered the derivation of linguistic summaries of databases as an example of a new paradigm shift from computing with measurements (numbers) to computing with perceptions (words). We presented a realistic, interactive approach to linguistic summaries of large sets of data that, with ahuman intercation, made it possible to derive complex, "intelligent" and human-consistent linguistic summaries. Results of an implementation at a computer retailer are very encouraging, and prove that the new paradigm shift is viable and yields a new quality.

References

1. Bosc P. and J. Kacprzyk, Eds. (1995) Fuzziness in Database Management Systems. Physica-Verlag, Heidelberg,
2. George R. and R. Srikanth (1996) Data summarization using genetic algorithms and fuzzy logic, In: F. Herrera and J.L. Verdegay (Eds.): Genetic Algorithms and Soft Computing. Physica-Verlag, Heidelberg and New York, pp. 599 - 611.
3. Kacprzyk J. (1999) An interactive fuzzy logic approach to linguistic data summaries, In R.N. Dave and T. Sudkamp (Eds.), Proceedings of NAFIPS'99 – 18th International Conference of the North American Fuzzy Information Processing Society. IEEE Press, Piscataway, NJ, 1999, pp. 595 – 599.
4. Kacprzyk J. and P. Strykowski (1999) Linguistic data summaries for intelligent decision support, Proceedings of EFDAN'99 – 4th European Workshop on Fuzzy Decision Analysis and Recognition technology for Management, Planning and Optimization (Dortmund, 1999), 1999, pp. 3 – 12.
5. Kacprzyk J. and S. Zadrożny (1994) Fuzzy querying for Microsoft Access. Proceedings of the Third IEEE Conference on Fuzzy Systems (Orlando, USA), Vol. 1, pp. 167-171.
6. Kacprzyk J. and S. Zadrożny (1995a) FQUERY for Access: fuzzy querying for a Windows-based DBMS. In: P. Bosc and J. Kacprzyk (Eds.) Fuzziness in Database Management Systems, Physica-Verlag, Heidelberg, pp. 415 - 433.
7. Kacprzyk J. and S. Zadrożny (1995b) Fuzzy queries in Microsoft Access v. 2, Proceedings of 6th IFSA World Congress (Sao Paolo, Brazil), Vol. II, pp. 341 - 344.
8. Kacprzyk J. and S. Zadrożny (1997a) Fuzzy queries in Microsoft Access v. 2. In: D. Dubois, H. Prade and R.R. Yager (Eds.): Fuzzy Information Engineering - A Guided Tour of Applications, Wiley, New York, 1997, pp. 223 - 232.
9. Kacprzyk J. and S. Zadrożny (1997b) Implementation of OWA operators in fuzzy querying for Microsoft Access. In: R.R. Yager and J. Kacprzyk (Eds.) The Ordered Weighted Averaging Operators: Theory and Applications, Kluwer, Boston 1997, pp. 293 - 306.
10. Kacprzyk J. and S. Zadrożny (1997c) Flexible querying using fuzzy logic: An implementation for Microsoft Access. In: T. Andreasen, H. Christiansen and H.L. Larsen (Eds.): Flexible Query Answering Systems, Kluwer, Boston, 1997, pp. 247-275.

11. Kacprzyk J. and S. Zadrożny (1998a), "Data Mining via Linguistic Summaries of Data: An Interactive Approach", in T. Yamakawa and G. Matsumoto (eds.): Methodologies for the Conception, Design and Application of Soft Computing" (Proc. of 5th IIZUKA'98), Iizuka, Japan, 1998, pp. 668-671.
12. Kacprzyk J. and S. Zadrożny (1998b), "On sumarization of large datasets via a fuzzy-logic-based querying add-on to Microsoft Access", in: Intelligent Information Systems VII (Proceedings of the Workshop - Malbork, Poland), IPI PAN, Warsaw, 1998, pp.249-258.
13. Kacprzyk J. and S. Zadrożny (1999) "On interactive linguistic summarization of databases via a fuzzy-logic-based querying add-on to Microsoft Access", In B. Reusch (Ed.): Computational Intelligence – Theory and Applications (Proceedings of 6th Fuzzy Days, Dortmund, 1999). Springer, Berlin,pp. 462-472.
14. Kacprzyk J., Zadrożny S. and Ziółkowski A. (1989) FQUERY III+: a 'human consistent` database querying system based on fuzzy logic with linguistic quantifiers. Information Systems 6, 443 - 453.
15. Kacprzyk J. and Ziółkowski A. (1986) Database queries with fuzzy linguistic quantifiers. IEEE Transactions on Systems, Man and Cybernetics SMC - 16, 474 - 479.
16. Rasmussen D. and R.R. Yager (1997) Fuzzy query language for hypothesis evaluation. In Andreasen T., H. Christiansen and H. L. Larsen (Eds.) Flexible Query Answering Systems. Kluwer, Boston/Dordrecht/London, pp. 23-43.
17. Yager R.R. (1982) A new approach to the summarization of data. Information Sciences, 28, 69 – 86.
18. Yager R.R. (1988) On ordered weighted avaraging operators in multicriteria decision making. IEEE Trans. on Systems, Man and Cybern. SMC-18, 183-190.
19. Yager R.R. (1991) On linguistic summaries of data. In: W. Frawley and G. Piatetsky-Shapiro (Eds.): Knowledge Discovery in Databases. AAAI/MIT Press, pp. 347-363.
20. Yager R.R. and Kacprzyk J., Eds. (1997) The Ordered Weighted Averaging Operators: Theory and Applications. Kluwer, Boston.
21. Yager R.R. and Kacprzyk J. (1999) Linguistic summarization od databases: a perspective. Proceedings of IFSA'99 – World Congress of the International Fuzzy Sets Association (Taipei, Taiwan), Vol. 1, pp. 44 – 48.
22. Zadeh L.A. (1983) A computational approach to fuzzy quantifiers in natural languages. Computers and Maths with Appls. 9, 149-184.
23. Zadeh L.A. (1985) Syllogistic reasoning in fuzzy logic and its application to usuality and reasoning with dispositions. IEEE Transaction on Systems, Man and Cybernetics SMC-15: 754--763.
24. Zadeh L.A. (1996) Fuzzy logic = computing with words. IEEE Trans. On Fuzzy Systems FS-4, 103 – 111.
25. Zadeh L.A. (1999) Fuzzy logic = computing with words. In:Zadeh L.A. and J. Kacprzyk, Eds. Computing with Words in Information/Intelligent Systems. Vol 1: Foundations, Physica-Verlag, Heidelberg and New York, pp. 3 - 23.
26. Zadeh L.A. and J. Kacprzyk, Eds. (1992) Fuzzy Logic for the Management of Uncertainty, Wiley, New York.
27. Zadeh L.A. and J. Kacprzyk, Eds. (1999a) Computing with Words in Information/Intelligent Systems. Vol 1: Foundations, Physica-Verlag, Heidelberg and New York.

28. Zadeh L.A. and J. Kacprzyk, Eds. (1999b) Computing with Words in Information/Intelligent Systems. Vol. 2: Applications, Physica-Verlag, Heidelberg and New York.

Chapter 17. Hybrid Intelligent Decision Support Systems and Applications for Risk Analysis and Discovery of Evolving Economic Clusters in Europe

N. Kasabov[1], L. Erzegovesi[2], M. Fedrizzi[2], A. Beber[2], and D. Deng[1]

[1] Dept. of Information Science, Univ. of Otago, Dunedin, New Zealand.
   E-mail: nkasabov, ddeng@infoscience.otago.ac.nz
[2] Dept. of Informatics and Faculty of Economics, Univ. of Trento, Italy

Abstract. Decision making in a complex, dynamically changing environment is a difficult task that requires new techniques of computational intelligence for building adaptive, hybrid intelligent decision support systems (HIDSS). Here, a new approach is proposed based on evolving agents in a dynamic environment. Neural network and rule-based agents are evolved from incoming data and expert knowledge if a decision making process requires this. The agents are evolved from methods included in a repository for intelligent connectionist based information systems RICBIS (http://divcom.otago.ac.nz/infosci/kel/CBIIS.html) with the use of financial market data collected in an on-line mode, and with the use of macroeconomic data published monthly in the European Central Bank Bulletin. RICBIS includes different types of neural networks, including MLP, SOM, fuzzy neural networks (FuNN), evolving fuzzy neural networks (EFuNN), evolving SOM, rule-based systems, data pre-processing techniques, standard statistical and financial techniques. A case study project on risk analysis of the European Monetary Union (EMU) is considered and a framework of a system EMU-HIDSS is presented, which deals with different levels of information and users, e.g. the whole world, Europe, clusters of nations, a single nation, companies/banks. It combines modules for final decision making, global and national economic development, exchange rate trend prediction, stock index trend prediction, etc. Some experimental results on real data are presented.

Keywords. Intelligent decision support systems, risk analysis, connectionist-based information systems.

1 Introduction

Complex decision making problems very often require considering enormous amount of information distributed among many variables. The decision support systems (DSS) built to solve these problems should have advanced features such as [19]:

- Good explanation facilities, preferably presenting the decision rules used;
- Dealing with vague, fuzzy information, as well as with crisp information;
- Dealing with contradictory knowledge, e.g., two experts predict different trends in the stock market;
- Dealing with large databases with a lot of redundant information, or coping with lack of data.
- Having hierarchical organisation, as decision making is usually not a single-level process, but involves different levels of processing, comparing different possible solutions, using alternatives, sometimes in a recurrent way.

Techniques of computational intelligence, such as artificial neural networks, fuzzy logic systems, genetic algorithms, advanced statistical methods, hybrid systems, along with traditional statistical and financial analysis methods, have been widely applied on different problems from finance and economics. Here some of them are references and explained.

1.1 Specialised and Advanced Statistical and Econometric Models

In [45] a model for risk analysis on crashes of currencies is introduced. It considers a crash situation when a currency has been devaluated more than 10% within a month when compared to the US dollar. Four parameters are used to evaluate the likelihood of a crash (a value of 0 or 1) in the following month: measure of currency overvaluation; expected domestic output growth; ratio of federal reserves versus external debt; measures of international financial contagion. The latter is measured in risk appetite (a change in investors preferences before crashes occur), and clusters (whether crashes of currencies from the same block of countries have recently occurred). Other models are presented in [6].

The methods in this group require a reliable knowledge on the underlying rules of the financial and economic systems and can not be easily adjusted to a new economic or financial situation. They are applicable when crises occur according to recurrent and known patterns.

1.2 Symbolic and Fuzzy Rule-Based Systems

Rule-based systems and fuzzy rule-based systems in particular have been used in financial and economic decision making (see papers in [12][13]). The main advantage of rule-based systems is that their functioning is based on human expert rules. The main disadvantage is that these systems are not flexible enough to react to changes in the reality. Changing the rules usually is difficult, and nobody can articulate the perfect set of rules that do not have to change in a future time.

1.3 Artificial Neural Networks (ANN)

Extended studies of using ANN for financial DSS have been done in several books (see for example: [12][13][49][54]). Generally, the so called non-parametric models proved to outperform the statistical methods, especially when the underlying rules were not known, or they change over time. Different types of ANN were used, mainly multi-layer perceptrons (MLP), radial basis functions (RBF), and self-organising feature maps (SOM).

A MLP model and a RBF model on option pricing [16] proved to outperform several other models, that include a direct application of Black-Scholes formula, ordinary least squares, and projection pursuit. The test data was the daily S&P500 futures and option prices. Two input variables – the ratio stock price/strike price and time to maturity, and one output variable - option price, were used. This model was further extended into a multi-modular ANN model with the use of hints [10].

In [5] a substantial study of applying SOM to financial markets is presented. In [3] emerging and new markets are mapped into a SOM that shows groups (clusters) of similar economies.

The so far developed and used ANN models proved to be efficient, but they do not allow easily for dynamic on-line training, for changing the parameters in an on-line mode, for combining data and knowledge (rules) into one system.

1.4 Genetic Algorithms

Genetic algorithms (GA) are heuristic models that are based on generating possible solutions to a problem and evaluating their "goodness" based on a goodness (fitness) function (e.g. goal function) that has to be specified in advance. GA use terminology from the natural selection and evolution of species. There are several studies of using GA for economic and financial decision making (see also chapters in this volume). Software based on GA simulation, that works directly on data in an Excel format has been produced [8].

The main advantage of GAs is that they do not require much knowledge on the underlying rules, formulas, etc., but a goodness function to evaluate how good solutions are. The main disadvantages of GAs are: (i) they are computationally slow; (ii) they do not necessarily provide with the best solution as they are heuristically based; (iii) they do not work in on-line and real time modes.

1.5 Models Based on Dynamic System Analysis

In [50] the stock market is modelled as a complex dynamic system that can be in one of four states projected in a two-dimensional space of group thinking-fundamental bias: random walk, chaotic market, coherent bull market, coherent bear market. In [38][39] a stock index prediction problem is considered as equivalent to the prediction of a chaotic process with different characteristics at different time scales and a high frequency index prediction (e.g. daily prediction) being attempted after the low frequency one (e.g., monthly, or annual prediction) is performed.

1.6 Hybrid Systems

Hybrid systems combine several of the above methods into one system [19][20]. They achieve this combination either in a "loose" way, e.g. different modules in the same system use different methods (see examples of such systems in [12][13]), or in a "tight" way - methods are mixed at a low level, e.g. fuzzy neural networks ([18][29][42]). These methods are the most promising among the methods discussed above, as the hybrid systems integrate the advantages of all the methods combined, e.g. dealing with both data and expert rules, using both statistical formulas and heuristics or hints.

1.7 What is Missing in the Methods from Above?

The six groups of approaches presented above have been partially successful, with major problems as listed below:

- They do not consider the complexity of the problem in a whole with many hierarchical levels for decision making that include applying low-level processing and higher level expert knowledge; or
- They do not consider uncertainties at different levels of information processing and combining them or propagating them in a task dependent way to the final decision making block; or
- They do not apply sufficient variety of techniques and choose the most appropriate for each sub-task; or
- They do not offer adjustment of variable sets, optimisation criteria, rules, even if the real situation changes over time.

2 From DSS to HIDSS

Hybrid systems combine different techniques of computational intelligence with traditional statistical methods. Hybrid systems are especially suitable for building DSS.

Stock market index prediction is a good example of a complex problem that requires a hybrid system, as it is shown on the case study of the NZSE40 stock index in [19][38][39]. Several modules are included in the DSS system presented there as there are several tasks within the global one: data pre- processing (e.g. normalisation, moving averages calculation, etc); predicting the next value for the index; predicting longer-term values for the NZES40, final decision making that takes into account rules on the political and economic situation; extracting trading rules from the system - see Fig.1. A neural network is used to predict the next value of the index based on the current and the previous-day values. The predicted value from the neural network module is combined with expert rules on the current political and economic situation in a fuzzy inference module. These two variables are fuzzy by nature. The final decision is produced as a fuzzy one, and as a crisp one

- after a defuzzification process. Another module in the DSS from Fig.1 is devoted to extracting fuzzy trading rules, which are used to explain the current behaviour of the market.

An environment, called FuzzyCOPE/1, that can be used to create hybrid systems, is described in [19] and available from internet URL http://kel.otago.ac.nz/. It consists of the following modules that have compatible interfaces and can be connected in a DSS as a decision making sequence that represents the logic of the real decision making process: data processing modules (normalisation, fuzzification, filtering, etc.); multi-layer perceptron (MLP); self-organising map (SOM); fuzzy neural network (FuNN) as introduced by Kasabov in [18][19][28]; fuzzy logic inference engine (FLIE) based on simple fuzzy rules of Zadeh-Mamdani type (Zadeh, 1965); production rule-based system CLIPS and FuzzyCLIPS in particular. The Fuzzy-COPE/1 environment has been extended to FuzzyCOPE/3 with the inclusion of new MLP and SOM learning modes, and new modes for learning, rule extraction and rule insertion in FuNNs. Examples of hybrid systems built with the use of Fuzzy-COPE are given in [19]. The two environments FuzzyCOPE/1 and FuzzyCOPE/3 are available from the following WWW site and can be used for building hybrid DSS: http://divcom.otago.ac.nz/infosci/kel/CBIIS.html (Software → FuzzyCOPE).

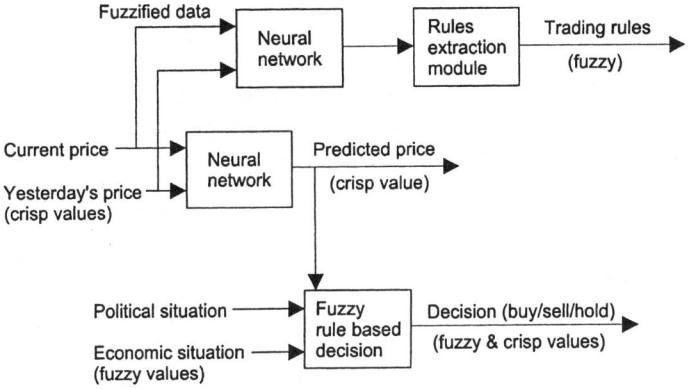

Fig. 1. An example of a hybrid DSS for stock trading (from [19])

The hybrid system environments developed so far, and the hybrid systems built with them, have been very useful techniques, but the complexity and the dynamics of the real-world problems, especially in finance and economics at present, require even more advanced and sophisticated methods and tools for building hybrid intelligent decision support systems (HIDSS). Such systems should be able to change as they operate, always updating their knowledge, and to refine the model through interaction with the environment. Some major requirements to the present day intelligent systems (IS), and to the HIDSS in particular are given in [21]-[25].

A framework for building adaptive intelligent systems, called ECOS (evolving connectionist systems) has been recently introduced in [21]-[25], along with its architecture, learning and evolving algorithms, rule extraction and rule insertion algorithms, of an evolving fuzzy neural network EFuNN [23]-[25]. EFuNNs can learn in an incremental, adaptive way through one-pass propagation of any new data examples. EFuNNs are much faster than FuNNs and MLPs and can learn data in an on-line mode. EFuNNs do not have a fixed structure, on the contrary – they start evolving/learning with no rule (hidden) nodes and they grow as data is presented to them. Pruning of nodes and node reduction is achieved with the use of fuzzy pruning rules, e.g.:

IF a node is not much used in a defined period, AND
it is old, THEN probability to prune the node is high.

EFuNNs have the following advantages when compared with traditional MLP or SOM networks ([24]:

1. they can learn in an on-line mode any new data as they are made available over time;
2. they can work in a complex environment with changing dynamics, e.g. a stock index system can be in a random walk state, then it moves to a chaotic state, and then - to quasi periodic state, and an EFuNN that predicts future stock values, learns all the time the new behaviour without any human intervention for parameter adjustment.
3. they can be used to mix expert rules and data as there are algorithms for rule insertion and rule extraction;
4. they can cluster data in an on-line mode without pre-defining the number of clusters, or the dimensionality and the size of the problem space;
5. they can be used for both supervised and unsupervised learning modes; as supervised systems they can be used to predict future values of the output variables.

Examples of using FuNNs and EFuNNs for adaptive, intelligent decision support systems for stock prediction and loan approval are given in [26]. Other examples include: image recognition [34]; speech and language recognition [24][33]; mobile robot control [24].

Simulators of EFuNN are available from http://divcom.otago.ac.nz/infosci/kel/CBIIS.html (Software).

Another ECOS algorithm, called Evolving Self-organizing Map (ESOM) [4], is proposed as a variation of the SOM algorithm based on the ECOS principles. ESOM uses a learning rule similar to SOM, but its network structure is evolved dynamically from input data. Simulations have shown that ESOM learns faster than SOM with a smaller quantisation error for feature vectors.

ECOS-based modules such as EFuNNs and ESOM are part of the New Zealand Repository of Intelligent Connectionist-Based Systems (RICBIS), which also integrates modules from the FuzzyCOPE environments, a Java version of rule-based system CLIPS (JESS), and a platform independent interface running as a Java applet, which allows for dynamic creation of new modules during the operation of an ECOS, or an HIDSS in particular. RICBIS is available from the following URL: http://divcom.otago.ac.nz/infosci/kel/CBIIS.html (Software).

A new expert system architecture called Adaptive Intelligent Expert Systems (AIES), based on dynamic generation of interconnected modules (agents) from the RICBIS, is explained in [34]. It is in sharp contrast to the conventional expert systems and DSS that usually have a fixed structure of modules and a fixed rule base. Although traditional expert systems and DSS have been successful in some specific and restricted areas, there was no, or little flexibility left for the expert system to adapt to changes required by the user, or by the dynamically changing environment in which the expert system and the DSS respectively operated.

AIES, or HIDSS in particular, consist of a series of modules which are agent-based and are generated "on the fly" as they are needed. Fig.2 shows a general architecture of an AIES [34]. The user specifies the initial problem parameters and tasks to be solved. The AIES then creates Modules that may initially have no rules or may be set up with rules from the Expert Knowledge Base. The Modules combine the rules with the data from the Environment. The Modules are continuously trained with data from the Environment. Rules may be extracted from trained FuNNs or EFuNNs for later analysis, or for the creation of new Modules. The Modules dynamically adapt their rule sets as the environment changes since the number of rules is dependent on the data presented to the Modules. Modules (agents) are dynamically created, updated and connected in an on-line mode. They can be removed if they are no more needed at a later stage of the operation of the AIES.

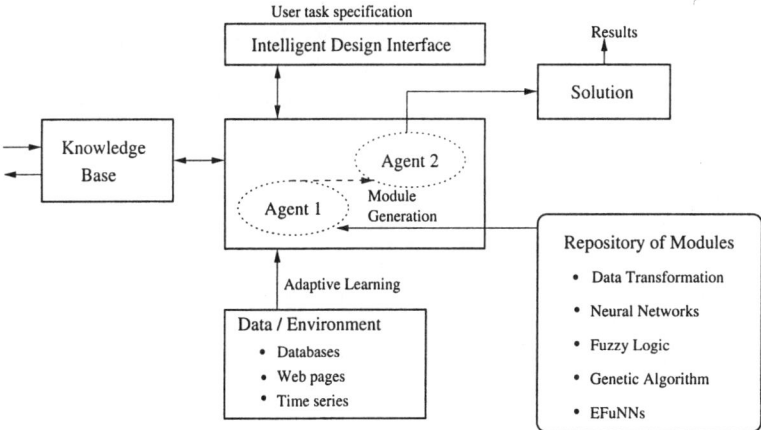

Fig. 2. A block diagram of an agent-based, adaptive intelligent decision support system – HIDSS that uses the architecture of an AIES.

A very complex problem of risk analysis of the European Monetary Union (EMU) system, established in 1999 to unify the currency and the economic development of eleven European countries, is the problem discussed and handled here. The rest of the material here presents first the problem and then a framework of a HIDSS for its solution. It then develops some specific modules and discusses some preliminary experimental and implementation issues.

3 The Problem of Risk Analysis in the European Monetary Union

Since its conception, there have been a lot of materials published on different issues concerning EMU. Global policy requirements exist, such as for each participating country to have a deficit less than 3% of its GDP and external debt less than 60% of the GDP. In the EMU framework, the European Central Bank (ECB) is in charge of the monetary policy, with priority and responsibility for inflation control. The ECB publishes a monthly bulletin containing a rich set of real, monetary and financial data regarding the EMU economies, other countries that would be members of the EMU, USA and Japan, in order to follow the evolution of the EMU in a world-wide context. Important financial and economic parameters are recorded and analysed monthly, quarterly, and annually, e.g.: Reserves and assets (gold, foreign exchange, other); Liabilities; Stock market indexes of the EMU, each country separately and the major world indexes (Dow Jones EURO STOXX, S&P500 – USA, Nikkei225 – Japan); Interest rates; Exchange rates Euro/US$, Euro/JY; Government bond yields (2, 3, 5, 7 and 10 years); Index of consumer prices; Industry and commodity prices; GDP; Employment/unemployment; Saving, deficit/surplus ratio (as a % of GDP); Gross nominal consolidate debt (as a % of GDP); Balance of payments (goods, services, income, capital account).

Of a particular interest is the analysis of the EMU as a dynamic cluster of economies in terms of volatility, variations, change, tendencies, and prediction (e.g., monthly, quarterly, yearly). There are smaller sub-clusters that evolve with the economic development of different groups of European countries and world economies that should be also modelled and predicted in relation to the EMU cluster. All these clusters move quickly in a dynamically changing problem space, thus making the problem of their prediction and risk analysis extremely difficult.

The problem this paper is dealing with as a case study, can be described as risk analysis of the EMU system. Here, more details are given.

With the EMU in effect, countries sharing the Euro as a common currency should overcome the risk of currency crises within the Euro area. However, the European monetary unification has not ruled out possible episodes of financial instability in Europe. The Maastricht Treaty imposes rigid constraints on public budget deficits. These constraints are aimed at preventing an excessive debt burden on national governments, which could lead to a weaker Euro. On the other side, EMU governments could put pressure in order to ease such constraints in the presence of external shocks, such as the crisis in the Balkans. Moreover, European financial markets are not immune from shocks originating in the world economy.

In an extreme scenario, risk of unilateral withdrawal by weaker member countries (otherwise called breakaway risk) cannot be excluded a priori, since the political costs associated with respect of rigid fiscal and monetary constraints in an anchored regime could make withdrawal imperative. Any withdrawal would be a disruptive event, anticipated and/or followed by instability and crashes in credit and asset markets. Credibility of EMU membership will be assessed and priced by financial markets. In the EMU, expectations of breakaway, unlike expected realignments or withdrawals in the Exchange Rate Mechanism operating from 1979 to 1998 under the European Monetary Systems, will no longer translate into wider interest rate differential among currencies, but into variation of credit spreads applied to sovereign debt from different countries. Holders of financial assets will bear a new sort of macro risk, which will be different from plain currency risk and more difficult to identify and measure. Studies on currency crises (e.g. [6]) must be reinterpreted and extended to the new context. Early warning systems used by central banks and speculators, fed with signals of real and financial dis-equilibrium, must be redesigned.

The goal of this project is to develop a computational model for analyzing and anticipating signals of abrupt changes of volatility in financial markets. The system will be aimed at assessing the possibility of speculative attacks against specific EMU member countries, prospective EMU members or the EMU area as a whole. Potential users of the system include monetary authorities, asset managers, traders on money, debt, currency and stock markets and corporate financial managers.

The conceptual model underlying the computational model will be derived from a representation of financial markets as complex dynamic systems, whose stochastic behavior is influenced by exogenous shocks and endogenous uncertainty, the latter caused by interaction among market participants (degree of consensus and tendency to crowd behavior). Inspiration for this approach came from a paper by Tonis Vaga ([50]). The system will be fed with information from different sources, namely:
- macroeconomic and macro-financial indicators, such as the real exchange rate of the Euro against US Dollar and Japanese Yen, inflation differentials, Government deficit, aggregate liquidity and solvency measures (debt/asset ratio, reserve/debt ratio);
- risk spread on debt securities issued by sovereign and private borrowers in EMU countries;
- risk appetite of investors in securities, measured on the basis of correlation between returns in "risky" and "safe" markets (risk appetite is high when riskier markets are rallying versus "safe" markets and the correlation is highly positive);
- returns and historical volatility in financial markets (stock, currency, bond, money, derivatives);
- signals of trend, reversal and change of regime from technical analysis of financial prices (moving averages, resistance and support levels, relative strength indicators, etc.); these signals will serve as proxy variables for endogenous uncertainty;
- implied volatility in option markets and expected distributions extracted from them;
- recent episodes of instability in other currency areas that can exert a contagion effect on the EMU area.

The logic of the system will be designed as an extension of existing event risk models applied to foreign exchange markets ([45]). The system will be tested on a set of recent episodes of market crash, and then extended taking care of unique features of the new EMU monetary regime. The system will produce a rich informative output, consisting of descriptive reports and warning signals.

Firstly, the system will provide an intelligent interface to information currently analyzed by economists and traders. Users will be able to navigate through a rich set of economic and financial data presented in tables and graphs. The presentation will focus on phenomena pertaining to the economic performance in the EMU area, emphasizing divergences among countries and sustainability of Maastricht constraints both at the national and the EMU level.

Secondly, the system will provide signals and indicators reflecting the likelihood of a crash in EMU financial markets. An extensive set of symptoms of financial fragility will be monitored in credit, bond and stock markets. New events will be checked against typical patterns of evolution of financial crises.

The system will provide a valuable support to analysts and decision makers in two ways: (1) selecting relevant information to be subsequently analyzed by human experts and (2) extracting and synthesizing signals from a vast array of information sources.

4 Collecting Relevant Data for Risk Analysis on the EMU. The Trento Financial Data Dictionary

The data collection is an important part of the research project, because the results and the reliability of the hybrid system depend crucially on the quality of the raw data. In fact a huge amount of data is necessary to monitor the so-called breakaway risk in the EMU, in order to detect the solidity of the EMU system, the degree of asymmetry and the potential external shocks.

Huge amount of data is being collected that includes different level of information: macroeconomic information; financial data. Macroeconomic data is available from many sources including EUROSTAT, the European Central Bank Monthly Bulletin, OECS and IMF. A vast amount of macroeconomic data is made available through Datastream. Financial data is available from different on-line sources, such as DataStream and Reuter. Different time-scales of data are present in the data repository (e.g., yearly, quarterly, monthly, daily, irregular time intervals, etc)

It is not easy to deal with this relevant amount of data without preparing at least a general initial structure. For this reason we have built two main prospects: the "sources dictionary" and the "data dictionary".

The first prospect has a three dimensional structure. In the x-axis we have placed the countries relevant for the EMU risk analysis; we have chosen the eleven participating countries, the potential entrants, the big world players, and some emerging countries to proxy for any contagion effect. In the y-axis we have set all the types of data which could be useful for the research project. The third dimension is concerned with the source of the gathered information; every single variable related to every involved country has been collected by a specific source. This aspect is important to evaluate mainly the reliability of the data, but also the degree of homogeneity in the data set. In fact, choosing the data source which covers the most part of the involved countries allows to achieve high degree of similarity in the modelling procedure and in the updating timing and method. Considering the specific purpose of the analysis, we rely upon official statistics from Eurostat for most of the required data.

The second prospect, named "data dictionary", is formed as a table with columns reporting the specific features of each data series; the most important ones are the first available date, the frequency of collection, and the mnemonic code, which is an alphanumeric expression useful to automate the information downloading process. We have distinguished two broad categories of data that are being collected. The first level of information is concerned with macroeconomic variables; the second level is more specific and regards financial market data. A third level of information, connected with qualitative knowledge, for example the political situation or particular news, is not formally considered in this first model implementation. The typical feature of the first group of variables – macroeconomic data – is the low frequency of collection, which is monthly, quarterly or even yearly, depending on the specific sector; another important feature is the potential variety of sources for the same kind of data. A third property is the potential different calculus and updating procedure for the same sort of indices. The most relevant macroeconomic data are generally available either as historical time series, or as a consensus forecast.

The financial market variables are collected more frequently, even on a daily basis, and are usually released officially by the exchanges. In our research project we have tried to consider not only the past evolution of the various financial series, but also the market expectations implied in option prices; there is a growing literature dealing with this aspect with two main approaches. The simpler approach is related to the calculation of implied volatility at different monetary degrees as a forecast for expected volatility. In the second approach the underlying risk neutral density function is extracted from option market prices; it is very useful subsequently to monitor the evolution of the various moment of the probability distribution.

5 The EMU-HIDSS: Architecture and Functionality

5.1 A General Framework of the EMU-HIDSS

Here a preliminary design of the EMU-HIDSS is presented (see Fig.3). It is a multi-level, multi-modular structure where many neural network modules (denoted as NNM), rule-based modules and other modules are connected with inter-, and intra-connections. EMU-HIDSS does not have a clear multi-layer structure, but rather a modular, "open" structure. It is an evolving hybrid connectionist-based system ([22]).

The main parts of the EMU-HIDSS are described below.
(1) Feature selection part. It performs filtering of the input information, feature extraction and forming input vectors. Typical features extracted from the input data either in an on-line mode or from the already stored data in files are:

- Basic statistical parameters;
- Probability distribution and cluster information;

- Moving averages;
- Wavelet transformations;
- Power spectrum and FFT frequency characteristics;
- Main frequencies;
- Lyapunov coefficients;
- Fractal dimensions;
- First and second derivatives;
- Skewness measures.

The above transformations are performed in the pre-processing (feature extraction) modules of the system applied to certain information input streams.

(2) Learning and memory part, where information (patterns) are stored. It is a multi-modular, evolving structure of neural network modules (NNM). These modules can be built with the use of MLP, SOM, ESOM, FuNNs, EFuNNs, etc. There are several levels of processing in these modules in terms of timing:

- Daily updated modules, these are modules that deal with daily financial prediction and daily input data, e.g. MIB30 prediction, Euro/US$ exchange rate prediction, etc.
- Weekly updated modules
- Monthly updated modules
- Yearly updated modules, e.g. long trend prediction, and in terms of the produced results that are passed to the next level higher decision modules:
- One day ahead prediction results
- Monthly prediction results
- Yearly prediction results
- Longer term predicted results

This part of the system will include several modules to deal with different levels and scales of prediction for each of the European countries, the big economies and the emerging economies. Different modules will deal with:

- Predicting values (differences in values)
- Predicting short term trends, e.g. one week trend if a stock value will be going up extremely high, or moderately down.
- Predicting long term trends, e.g. one-year trend if a stock value will be going up extremely high, or critically down.

(3) Higher-level decision part that consists of several modules, each taking decision on a particular problem. The modules receive input from the NNMs, inputs form other variables in the data, qualitative input from users, and make decisions on possible critical situations that might occur in the EMU. These modules can send a feedback to the NNM and the feature extraction part of the system in terms of requiring more information, different scenarios to be explored, different features extracted, etc. The modules here are mainly rule based with the use of production

systems, flat fuzzy rules, FuNNs and EFuNNs that can represent both fuzzy rules and data. There are several groups of in this part that interact between each other, for example:

(a) A group of modules that deal with a global risk evaluation problems, e.g.:

- Module evaluating the degree of stability in the EMU;
- Module evaluating the degree of symmetry / asymmetry between the economies within the EMU;
- Module evaluating the political sustainability in the EMU;
- Module evaluating the degree of suitability of a new country joining the EMU;
- Module evaluating the degree of instability in the EMU based on external factors (USA; Asia, Japan, India, Russia, wars, etc.)
- others

(b) A group of modules that deal with important economic factors and their results can be used either separately, or by the modules of type (1) above:

- GDP
- Rate of unemployment
- Internal debt
- External debt
- Short term global economic trends
- Long term global economic trends
- Solvency ratio of households, business, banks and government
- Indication of reallocation of investments by global asset managers
- Sharp movements of a certain commodity in a short or in a long term pattern
- Sharp falls in short term or long term trends
- Evaluating and indication of consecutive phases happening over a period of time, for example an economy has been in three consecutive phases that signal a critical situation for this economy and will be influential for the EMU

(4) Action modules, that take the output from the higher-level decision modules and produce output results or send output (control) information in an on-line or in an off-line mode to institutions that should be alerted on a critical situation.

(5) Self-analysis, and rule extraction modules. This part extracts compressed abstract information from the NNMs and from the decision modules in different forms of rules, abstract associations, etc. Here FuNNs' and EFuNNs' rule extraction capabilities will be utilised.

Initially the EMU-HIDSS will have a pre-defined structure of modules and very few connections between them defined through prior knowledge. Gradually, the system will become more and more "wired through self-organisation, and through creation of new NNM and new connections.

Each of the modules in the system are built, or automatically generated from the agent modules available from RICBIS, e.g.: data processing modules (e.g. normalisation, moving averages, FFT, filtering, wavelet transformation, fractal analysis, chaos analysis, etc); production rules in JESS; fuzzy inference rules, MLP, SOM, ESOM, FuNNs, EFuNNs, Hidden-Markov Models, etc.

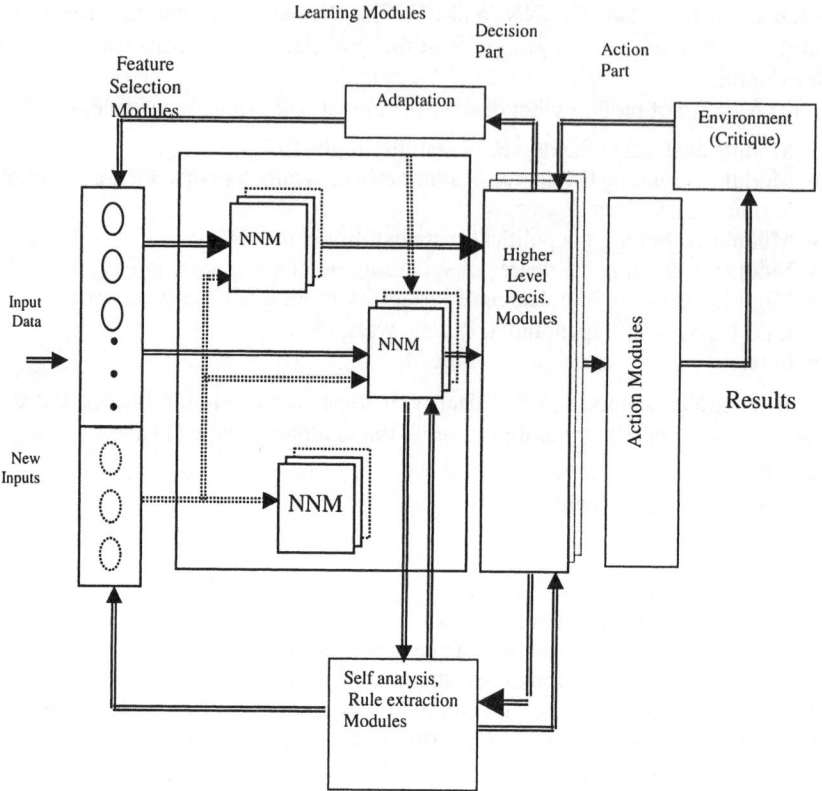

Fig. 3. A block diagram of the general framework of the EMU-HIDSS as an evolving connectionist-based system (adapted from [22]).

5.2 EMU-HIDSS/1

Here, the first version of the EMU-HIDSS is presented that includes a small number of modules and groups connected between each other as described below.

Group 1: Modules for higher-level risk analysis

Here are modules that make decision on the discrepancy/risk of a single country to develop in a direction away from the expected development of the EMU. Fuzzy production rules are used for the implementation of these modules, such as:
IF a country is politically unstable, or in war, AND the trend of its macroeconomic development in the last two periods is away from the centre of the EMU cluster, THEN the risk that this country will go even further away from the EMU is high.

Group 2: Modules for discovering trends in the macroeconomic development of the EMU cluster related to the development of other clusters and other countries.

Here the concept of EMU cluster is introduced based on the EMU aggregated data projected from a multidimensional space into a two (or a three) dimensional topological map with the use of self-evolving, self-organising maps. The vectors of the used 8 parameters for the last 5-6 year of all the EMU countries, the other European countries, some emerging market countries, and also the USA and Japan, are mapped into one SOM or ESOM.

The neuron (the point in the two-dimensional map space) where the current-period EMU vector is projected is considered to be the center of the EMU cluster. The cluster incorporates all points where the data of the individual EMU countries are mapped. The form of the cluster and its movement from one period to the next one can be observed on the maps and the information will be quantified based on the distance between the points. The shape of the EMU cluster and its dynamics can suggest further political and economic development in the EMU.

The movement of the center of the EMU cluster can be compared with the movement of the points where the different countries are mapped. A quantitative measure on the difference (the distance in the topological space) between different countries from the EMU, and outside the EMU, is evaluated that is used as a separate information results as well as an input information to the higher- level decision making modules.

Different clusters are formed on a SOM: the EMU cluster; the emerging economies cluster (e.g., Poland); the cluster of the non-European developed countries (e.g. the USA, Japan, Canada, Australia, New Zealand); the cluster of underdeveloped countries; the cluster of the developed non-EMU countries (e.g. the UK); the cluster of the developing non-EMU countries (e.g., Bulgaria, Romania). A vector of fuzzy membership degrees to which each country belongs to each of the clusters is calculated and traced over time.

Modules from group 2 cover different time-scales: annual macroeconomic mapping, quarterly macroeconomic map and monthly macroeconomic mapping.

The following variables describe the macroeconomic state of a country in a given period and all the vectors for all the relevant periods (years, quarters, months) are used in the unsupervised training: GDP, debt, deficit, inflation rate, interest rate, unemployment, balance of payment, production gap.

Group 3: Modules for evaluating trends in the exchange rate Euro/US$

The main module here predicts the trend of the exchange rate between Euro and the US dollar, but other modules deal with national currencies that are not part of the EMU.

The following 10 input variables for example can be used to predict the rate $R(t+1)$ of Euro/US$, where t is the current period: $R(t)$, $R(t-1)$, Euro/JY(t), Euro/JY(t-1), ratio inflation rate in EMU/inflation rate in the USA for both (t) and (t-1) periods; ratio interest rates in EMU/interest rates in USA for both (t) and (t-1) periods; ratio GDP in EMU/GDP in USA for both (t) and (t-1) periods.

Two types of models are used - FuNNs and EFuNNs. The two models use different techniques for extracting rules and the meaning of the extracted rules is different.

In the rules extracted from FuNNs the condition and conclusion elements have importance factors attached to them pointing to the importance of the different parts of a rule. The rules extracted from EFuNNs are also fuzzy, but they point to the clusters in the input and the output space that are linked together in the rule. EFuNNs require less time for training and can be updated very quickly with new data in an on-line mode.

The predicted trends of the exchange rates can be used either as separate output results, or as input values for the higher-level decision modules for both quarterly and monthly trends prediction.

Group 4: Modules for evaluating trends in major stock indexes and stock markets

Here a map of the different states of a stock market according to [50] is created. The transition between random walk state, chaotic state, or a coherent state will be modelled by the use of different techniques, that include: hidden Markov models; evolving fuzzy neural networks EFuNNs; production rules.

Another module from this group evaluates the volatility of monthly trends in the Dow Jones Euro STOXX50 index (DJE). Other modules evaluate weekly trends and daily values. The following 14 input variables can be used to evaluate the DJE$(t+1)$, where t is the current time period: DJE(t); DJE$(t-1)$; $S\&P500(t)$; $S\&P500(t-1)$; Euro/US\$$(t)$; Euro/US\$$(t-1)$; Euro/JY$(t)$; Euro/JY$(t-1)$; Inflation rate (t); Inflation rate $(t-1)$; GDP(t); GDP$(t-1)$; Interest rate (t); Interest rate $(t-1)$.

Other modules evaluate the trends in major European stock markets, such as the Italian MIB30 index (Milano).

6 Implementation and Current Experimental Results with the EMU-HIDSS

The implementation of the conceptual model of the EMU-HIDSS from section 5 is a very complicated task and a long-term objective. Here, different modules from the EMU-HIDSS/1 conceptual model that follow the general description and the logical links presented in the previous section, are developed and results are explained.

6.1 Group 1 Module for Statistically-Based Higher-Level Risk Analysis

This module takes dynamic input information from the cluster maps of the previous level of processing and calculates the Euclidean distance between each country's representation vector and the center of the EMU cluster over consecutive periods of time (years, quarters). In this way the countries that are moving away from the EMU cluster are indicated along with the speed at which they are moving. For example the distance between the cluster center of the main EMU countries and Italy for 1997 can be evaluated as 3.8 and for 1998 as 3.3, while the same distance between IT and JP for the same periods can be evaluated as 1.2 and 0.7 respectively (see Fig.4).

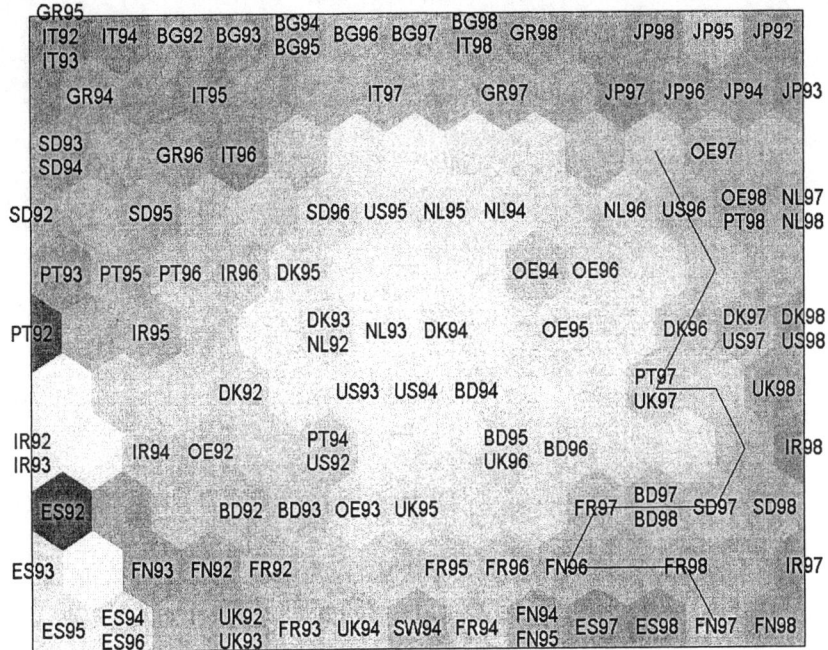

Fig. 4. The annual map of the 15 countries according to 5 characteristics (DBT/GDP, DEF/GDP, Inflation rate, Interest rate, Unemployment). The contour shows the center of the EMU cluster for 1998.

6.2 Group 2, SOM Modules for Visual Exploration of the Annual and Quarterly Macroeconomic Development of the EMU Cluster Related to the Development of Other Clusters and Other Countries

The following 5 variables describe the annual macroeconomic state of a country: DBT/GDP, DEF/GDP, Inflation rate, Interest rate, Unemployment. The SOM model was trained on 15 countries data from 1992 till 1998. It is seen how the points of the map of the main EMU countries and USA are moving from left to right. Fig.4 shows also the contour surrounding the centre of the EMU cluster for 1998. It is obvious that the following EMU countries are within the cluster: OE, NL, DK, IR, UK, SD, BD, FR in addition to the USA and the UK. But four countries are outside it (IT, BG, GR, ES) with only ES moving into the right direction towards the EMU cluster center.

Fig.5 shows the direction in which Italy (IT) is moving over the years from 1992 till 1998.

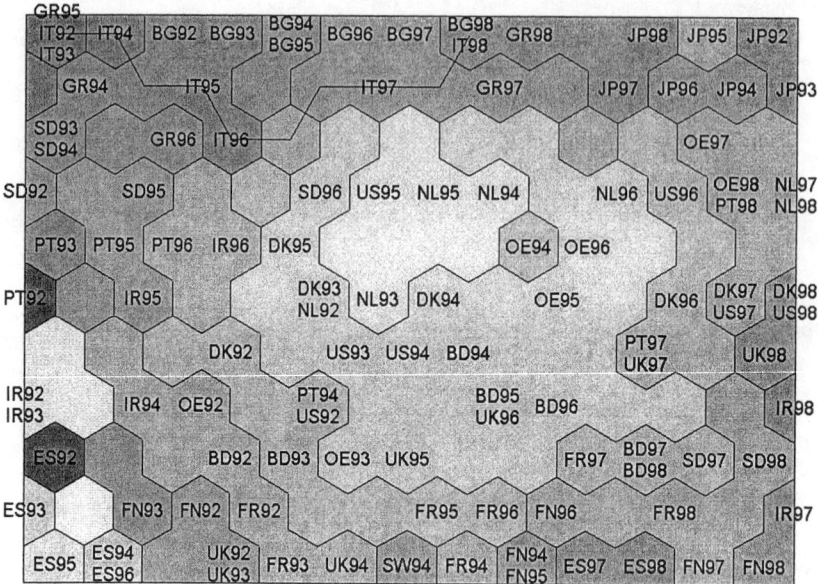

Fig. 5. Iso-graphs of the 15 countries annual development map and the direction of the development of Italy.

Another SOM module is trained on quarterly data of the following three variables: Inflation rate, Interest rate, Unemployment. Fig.7 shows the map and the line of the quarterly development of Germany (BD).

The SOM modules are suitable for visual exploration as SOMs provide with an efficient tool for vector quantisation, i.e. turning an n-dimensional space usually into two dimensional space preserving the similarity between the input vectors across all the attributes as topological distance between points of the map. SOMs have some difficulties, mainly: (a) they have a fixed structure; (b) they can not tolerate missing values; (c) learning usually takes a long time.

In order to overcome the limitations of the SOMs, here evolving SOM modules are also used.

6.3 Group 2, ESOM Module for Dynamic Mapping of Macroeconomic Analysis

Another annual map is evolved with the same data set as in the SOM modules and is shown in Fig.8. The map is first clustered using ESOM algorithm, and then projected onto a two-dimensional plane for visualisation using Sammon's algorithm. Hence both clustering and data structure are available within the map. The layout of labelled nodes is very similar to that of Fig.4, but the ESOM map gives more

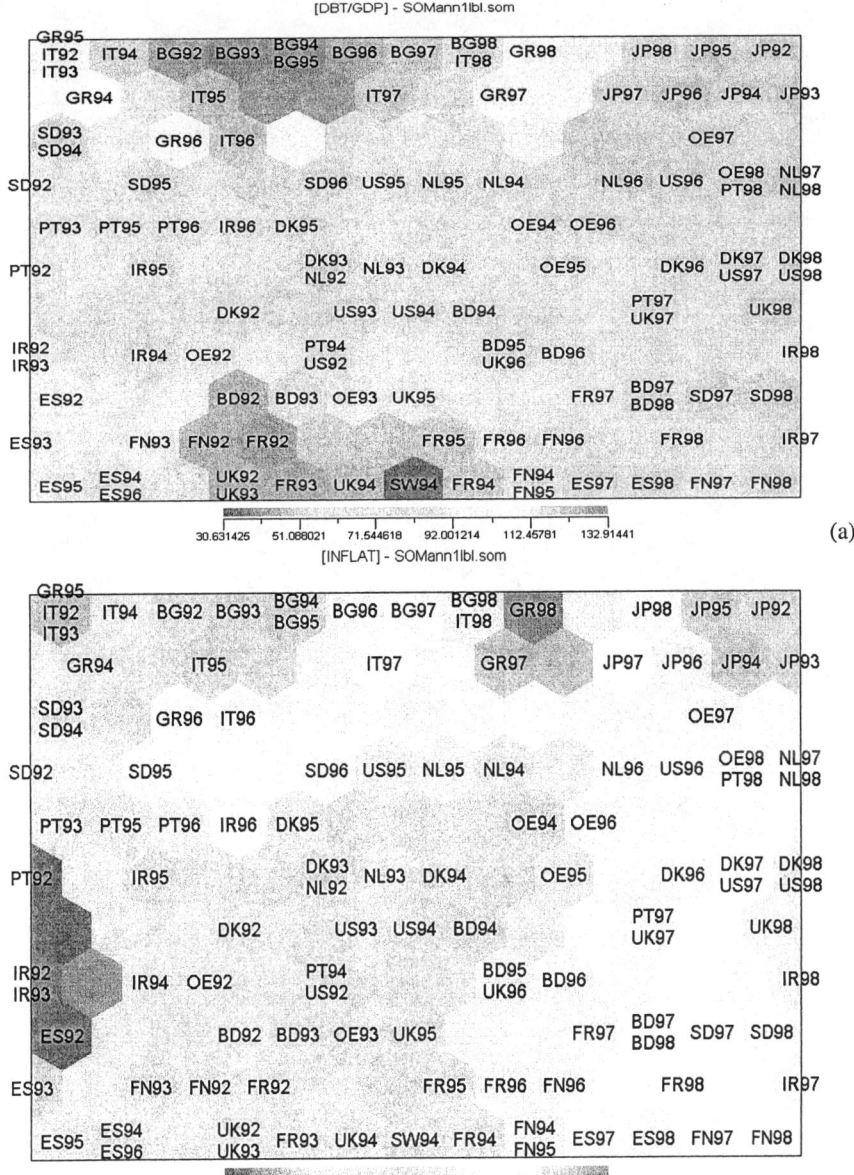

Fig. 6. Component analysis: (a) The first component (DBT/GDP) from the map of figs.4 and 5 show that in this respect BG, IT and GR are in a similar position. SD97 and SD98 form an "island" in the EMU cluster with a good tendency. (b) The Inflation component shows a dramatic increase in Greece from 1996 to 1997 and 1998.

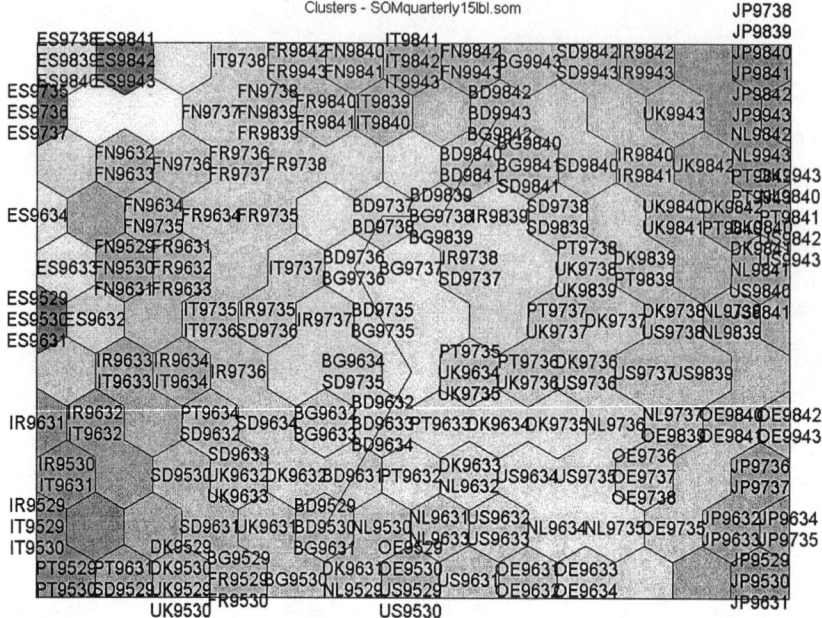

Fig. 7. The complete map of the quarterly development of the 15 countries from 1995 till 1999 and the line of the development of Germany (DB).

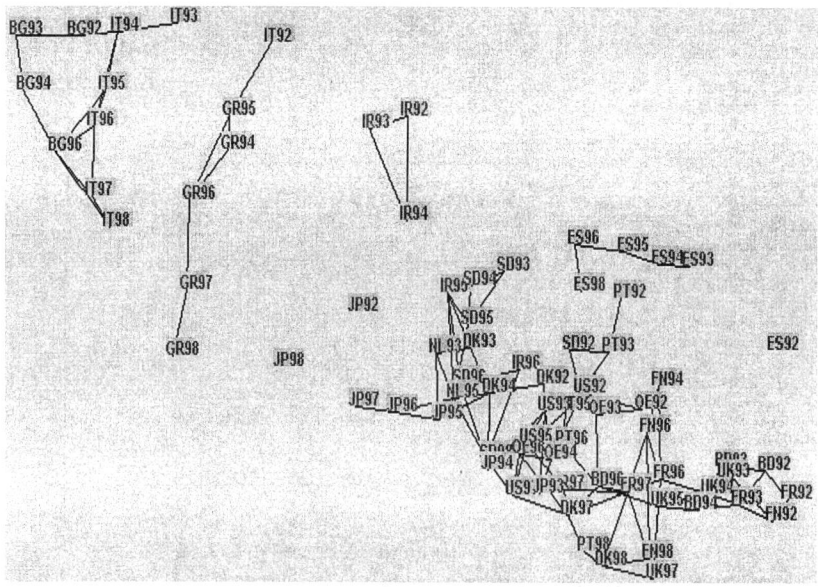

Fig. 8. The ESOM clusters of the macroeconomies of EMU countries.

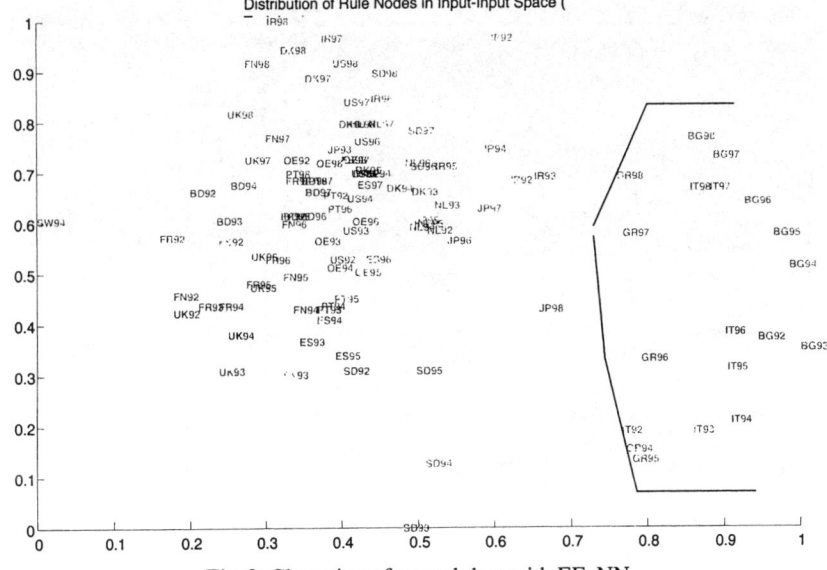

Fig. 9. Clustering of annual data with EFuNN.

explicit information on similarity of country performance which can be represented by distance between the nodes.

By clipping weak connections using a distance threshold, as shown in Fig.8, clusters in the annual macroeconomic performance data can be revealed. Here we find two major clusters, the EMU cluster with countries like FR, BD, FN, PT etc. plus UK and US, and the fall-out cluster with GR, IT, and BG.

6.4 Group 2, EFuNN Based Module for Prediction and Clustering of the Annual and Quarterly Economic Development of the 15 Countries

Here EFuNNs are used to develop modules that can predict values for all the five selected attributes in the annual development and the three selected attributes for the quarterly development. Such modules are the annual module and the quarterly module. The first one takes two input vectors each of 5 variables (at the time moment t and $t-1$) and calculates one output vector of 5 elements predicting what the values for these variables will be. Fig.9 shows the clustering of the annual data (displayed with the first two variables DBT/GDP and Deficit/GDP) achieved in the rule nodes of an evolved EFuNN. It is clear that IT, GR and BG form a cluster with high DBT/GDP value and DEF/GDP running from low to high values. The clustering of data samples are quite similar to that of the SOM module, but it is much more quickly learned in the one-pass learning EFuNN module.

Fig.10 shows the annual EFuNN predictor for the 15 countries run in an on-line mode to predict the DBT/GDP values annually.

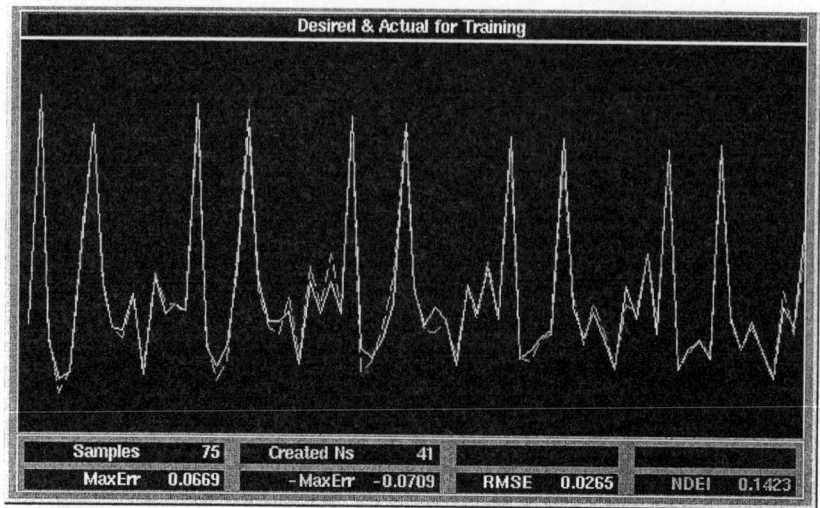

Fig. 10. Annual EFuNN predictor for prediction of annual DBT/GDP values.

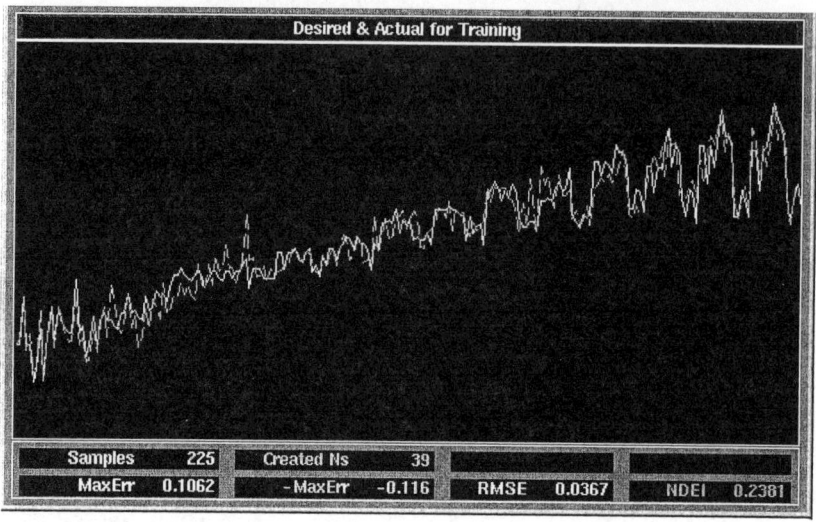

Fig. 11. Quarterly predictor for the 15 countries and the on-line prediction of the first variable (Inflation rate).

Rules can be extracted from EFuNNs that are fuzzy, and they point to the clusters in the input and the output space that are linked together in the rule. EFuNNs require less time for training and can be updated very quickly with new data in an on-line mode.

Fig.11 shows the 15 countries quarterly predictor, which can predict values of the three selected variables Inflation rate, Interest rate, Unemployment for any of the countries in the following quarter given the data for the current and previous quarter and also the annual DBT/GDP and the previous year DBT/GDP are entered. The first output variable is shown as predicted in an on-line mode. The system improves its prediction over time.

6.5 Group 4, EFuNN Module to Evaluate Trends in the Exchange Rate Euro/US$

The main module here predicts the monthly trend of the exchange rate between Euro and the US dollar. The following 10 input variables are used in order to predict the rate $R(t+1)$ of Euro/US$, where t is the current period: $R(t)$, $R(t-1)$, Euro/JY(t), Euro/JY$(t-1)$, ratio inflation rate in EMU/inflation rate in the USA for both (t) and $(t-1)$ periods; ratio interest rates in EMU/interest rates in USA for both (t) and $(t-1)$ periods; ratio GDP in EMU/GDP in USA for both (t) and $(t-1)$ periods.

6.6 Group 4, EFuNN Module to Evaluate Trends in the DJE501 Major Stock Index

This module evaluates the monthly trends in the Dow Jones Euro STOXX50 index (DJE). The following 14 input attributes are used to evaluate the DJE$(t+1)$, where t is the current time period: DJE(t); DJE$(t-1)$; $S\&P500(t)$; $S\&P500(t-1)$; Euro/US(t); Euro/US$(t-1)$; Euro/JY(t); Euro/JY$(t-1)$; Inflation rate (t); Inflation rate $(t-1)$; GDP(t); GDP$(t-1)$; Interest rate (t); Interest rate $(t-1)$.

Besides the modules listed above, several other modules are currently under development.

7 Conclusions and Directions for Further Research

A framework of hybrid intelligent decision system is presented in the paper. By applying a repository of intelligent information processing modules implemented in an agent-based architecture, a case study system EMU-HIDSS is built for risk analysis and prediction of evolving economic clusters in Europe.

The EMU-HIDSS is designed to be used at different levels of analysis and decision making about the EMU and about the relevant changes in the economic clusters of Europe and the world, that includes: the European Union level; the global world economies level; national level; company and bank level. Data and some of the developed models are available from internet URL http://divcom.otago.ac.nz/infosci/kel/CBIIS.html (Software - Financial Risk Analysis and Prediction).

Acknowledgements

This work is supported by a research grant of the Department of Computer and Management Sciences, University of Trento, and also by a grant PGSF-UOO808 funded by the New Zealand Foundation for Science, Research and Technology (FRST).

References

1. Amari, S., Kasabov, N., Eds. (1998) Brain-like Computing and Intelligent Information Systems. Springer Verlag.
2. Bollacker, K., Lawrence S., Giles L. (1998) CiteSeer: An autonomous Web agent for automatic retrieval and identification of interesting publications. Proc. of the 2nd International ACM conference on autonomous agents. ACM Press, 116-123.
3. Deboeck, G. (1999) Investment maps for emerging markets. In: N.Kasabov and R.Kozma Eds. Neuro-fuzzy techniques for intelligent information systems. Physica Verlag (Springer), 373-395.
4. Deng, D., Kasabov, N. (1999) Evolving Self-organizing Map and its Application in Generating A World Macroeconomic Map. Proc. of ICONIP/ANZIIS/ANNES'99 International Workshop, University of Otago, 7-12.
5. Deboeck, G., Kohonen, T. (1998) Visual exploration in finance with self-organizing maps, Springer Verlag.
6. Eichengreen, B., Rose, A., Wyplosz, C. (1995) Exchange market mayhem: the antecedents and aftermath of speculative attacks. Economic Policy, 251-312.
7. Monthly Bulletin (1999) European Central Bank.
8. Evolver software package for Excel, http://www.palisade.com/html/evolver_body.html.
9. Farmer, J.D., Sidorowitch J. (1987) Predicting chaotic time series. Phys. Rev. Lett., 59, 845-848.
10. Garcia, R., Gencay R. (1997) Pricing and hedging derivative securities with neural networks and a homogeneity hint. Technical Report, Department of Science and Economics, University of Montreal, Canada.
11. Goodman, R., Higgins, C.M., et. al. (1992) Rule-based neural networks for classification and probability estimation. Neural Computation, 14, 781-804.
12. Goonatilake, S., Khebbal S. Eds. (1995). Intelligent Hybrid Systems. John Wiley & Sons London.
13. Goonatilake, S., Trelevan P. (1995) Intelligent Systems for Finance and Business. John Wiley & Sons.
14. Hashiyama, T., Furuhashi, T., Uchikawa, Y. (1992) A decision making model using a fuzzy neural network. Proceedings of the 2nd International Conference on Fuzzy Logic & Neural Networks, Iizuka, Japan, 1057-1060.
15. Heskes, T.M., Kappen B. (1993) On-line learning processes in artificial neural networks. Mathematical foundations of neural networks, Elsevier, Amsterdam, 199-233.
16. Hutchinson, J., Lo A., Poggio T. (1994) A nonparametric approach to pricing and hedging derivative securities via learning networks. The Journal of Finance, vol.XLIL, No.3, 851-890.

17. Ishikawa, M. (1996) Structural learning with forgetting. Neural Networks, 9, 501-521.
18. Kasabov, N. (1996) Adaptable connectionist production systems. Neurocomputing, 13(2-4) 95-117.
19. Kasabov, N. (1996) Foundations of Neural Networks, Fuzzy Systems and Knowledge Engineering. The MIT Press, CA, MA.
20. Kasabov, N. (1996) Learning fuzzy rules and approximate reasoning in fuzzy neural networks and hybrid systems. Fuzzy Sets and Systems, 82(2), 2-20.
21. Kasabov, N. (1998) The ECOS Framework and the ECO Learning Method for Evolving Connectionist Systems. Journal of Advanced Computational Intelligence, 2(6), 1-8.
22. Kasabov, N. (1998) ECOS: A framework for evolving connectionist systems and the ECO learning paradigm. Proc. of ICONIP'98, Kitakyushu, Japan, IOS Press, 1222-1235.
23. Kasabov, N. (1998) Evolving fuzzy neural networks - algorithms, applications and biological motivation. In Yamakawa and Matsumoto Eds., Methodologies for the Conception, Design and Application of Soft Computing. World Scientific. 271-274.
24. Kasabov, N. (1999) Evolving connectionist and fuzzy connectionist systems for on-line adaptive decision making and control. In: Advances in Soft Computing - Engineering Design and Manufacturing, R. Roy, T. Furuhashi and P.K. Chawdhry Eds. Springer, London.
25. Kasabov, N. (1999) Evolving fuzzy neural networks for adaptive, on-line intelligent agents and systems. In: O. Kaynak, S.Tosunoglu and M.Ang Eds. Recent Advances in Mechatronics. Springer, Berlin.
26. Kasabov, N. Fedrizzi, M. (1999) Fuzzy neural networks and evolving connectionist systems for intelligent decision making. Proc. of the 8th International Fuzzy Systems Association World Congress, Taiwan, 17-20.
27. Kasabov, N., Kozma, R. (1999) Multi-scale analysis of time series based on neuro-fuzzy-chaos methodology applied to financial data. In Refenes, A., Burges, A. and Moody, B. Eds. Computational Finance 1997. Kluwer Academic.
28. Kasabov, N., Kim J.S. et.al. (1997) FuNN/2- a fuzzy neural network architecture for adaptive learning and knowledge acquisition. Information Sciences, 101(3-4), 155-175.
29. Kasabov, N., Kozma, R. (1997) Neuro-fuzzy-chaos engineering for building intelligent adaptive information systems. In: Intelligent Systems: Fuzzy Logic, Neural Networks and Genetic Algorithms. Da Ruan Eds. Kluwer Academic Boston/London/Dordrecht, 213-237.
30. Kasabov, N., Israel, S., Woodford, B. (1999) Methodology and evolving connectionist architecture for image pattern recognition. In: Pal, Ghosh and Kundu Eds. Soft Computing and Image Processing, Physica-Verlag(Springer) Heidelberg, accepted.
31. Kasabov, N., Song, Q. (1999) Dynamic, evolving fuzzy neural networks with 'm-out-of-n activation nodes for on-line adaptive systems. Technical Report TR99-04, Department of Information Science, University of Otago.
32. Kasabov, N., Watts M. (1997) Genetic algorithms for structural optimisation, dynamic adaptation and automated design of fuzzy neural networks. In: Proc. of the Inter. Conf. on Neural Networks (ICNN97), IEEE Press, Houston.
33. Kasabov, N., Watts, M. (1999) Spatial-temporal evolving fuzzy neural networks STEFuNNs and applications for adaptive phoneme recognition, Technical Report TR99-03, Department of Information Science, University of Otago.
34. Kasabov, N., Woodford, B. (1999) Rule insertion and rule extraction from evolving fuzzy neural networks: algorithms and applications for building adaptive, intelligent expert systems. Proc. of Int. Conf. FUZZ-IEEE, Seoul.
35. Kawahara, S., Saito, T. (1996) On a novel adaptive self-organising network. Cellular Neural Networks and Their Applications, 41-46.

36. Kohonen, T. (1990) The Self-Organizing Map. Proceedings of the IEEE, 78, 1464-1497.
37. Kohonen, T. (1997) Self-Organizing Maps, second edition. Springer.
38. Kozma, R., Kasabov, N. (1998) Chaos and fractal analysis of irregular time series embedded into connectionist structure. In: Brain-like Computing and Intelligent Information Systems. S. Amari, N. Kasabov Eds. Springer Singapore, 213-237.
39. Kozma, R., Kasabov, N. (1999) Generic neuro-fuzzy-chaos methodologies and techniques for intelligent time-series analysis. In: Soft Computing in Financial Engineering, R. Ribeiro, R. Yager, H.J. Zimmermann, J. Kacprzyk (Eds.), Physica-Verlag Heidelberg.
40. Krogh, A., Hertz, J.A. (1992) A simple weight decay can improve generalisation. Advances in Neural Information Processing Systems 4, 951-957.
41. Le Cun, Y., Denker J.S., Solla, S.A. (1990) Optimal Brain Damage. In Touretzky, D.S. Ed. Advances in Neural Information Processing Systems, 2, 598-605.
42. Lin, C.T., Lee C.S.G. (1996) Neuro Fuzzy Systems. Prentice Hall.
43. Mitchell, M.T. (1997) Machine Learning, MacGraw-Hill.
44. Moody, J., Darken, C. (1989) Fast learning in networks of locally-tuned processing units. Neural Computation, 1, 281-294.
45. Persaud, A. (1998) Global foreign exchange research. In: Event Risk Indicator Handbook, JP Morgan, London, 44-171.
46. Saad, D. Ed. (1999) On-line Learning in Neural Networks. Cambridge University Press.
47. Sankar, A., Mammone, R.J. (1993) Growing and pruning neural tree networks. IEEE Trans. Comput., 42(3), 291-299.
48. Swope, J.A., Kasabov, N., Williams, M. (1999) Neuro-fuzzy modelling of heart rate signals and applications to diagnostics. In: Szepanjuk Ed., Fuzzy Systems in Medicine, Physica Verlag, accepted.
49. Trippi, R., Turban E. Eds. (1993) Neural Networks in Finance and Investing. Irwin Professional Publications, New York.
50. Vaga, T. (1990) The coherent market hypothesis. Financial Analysts Journal, November-December, 36-49.
51. Watts, M., Kasabov, N. (1998) Genetic algorithms for the design of fuzzy neural networks. In: Proc. of ICONIP'98, Kitakyushu, Japan, 21-23.
52. Watts, M., Kasabov, N. (1999) Neuro-genetic tools and techniques. In: Neuro-Fuzzy Techniques for Intelligent Information Systems, N.Kasabov and R.Kozma, Eds., Physica Verlag Heidelberg, 97-110.
53. Woldrige, M., Jennings, N. (1995) Intelligent agents: Theory and practice. The Knowledge Engineering Review, 10.
54. Zirilli, J. (1997) Financial Prediction Using Neural Networks. Thomson Computer Press, London.

Chapter 18. Intelligent Resource Management Through the Constrained Resource Planning Model

David Y. Y. Yun

Laboratory of Intelligent and Parallel Systems (LIPS)
Holmes 492, 2540 Dole Street, Honolulu, HI 96822
College of Engineering, University of Hawaii (UH)
Phone: (808) 956-7627; Fax: (808) 941-1399; Email: dyun@hawaii.edu

Abstract. Objective: *Introduce a resource management methodology, known as the Constrained Resource Planning (CRP), together with a flexible and powerful computing "engine", suitable for most planning and scheduling applications under stringent solution requirements, tightly interacting constraints, as well as restricted resource availability and utilization.* Methodology: *By the effective utilization of two domain-independent guiding principles (i) the most-constrained strategy for task identification and the least-impact strategy for solution selection (ii) the algorithmic procedure of CRP allows widely varying problems to be mapped and solved without changing the underlying problem solving mechanisms – the CRP engine.* Quality: *Unlike most heuristic approaches that can be trapped into locally- or non-optimal solutions, CRP has been shown to produce provably optimal solutions for a variation of the classic, NP-complete traveling salesman problem, known as the diamond lattice problem.* Applicability: *Over 40 resource allocation and activity scheduling problems have been solved using the CRP methodology and computing engine. Capitalizing on CRP's ability to delicately integrate and balance complementary and conflicting objectives, CRP-solved applications have demonstrated a consistency to achieve quality solutions and display surprising intelligence for well known difficult problems, such as multiprocessor scheduling, 3D model discovery, and DNA folding, in addition to providing intelligent solutions for most challenging and NP-hard problems as maximal common subgraph.* Significance: *Due to its broad applicability, solution quality and computational efficacy, CRP is offered both as a general, problem-solving methodology for tackling difficult problems and as an executable computing engine capable of achieving solutions even beyond human intelligence. CRP also guides a problem solver to tackle resource confined decision problems with simultaneous objectives and conflicting constraints through a set of disciplined strategies and balanced principles.*

Keywords. *Resource management, constrained resource planning, constraint satisfacttion, optimization,planning and scheduling*

1 Introduction

Resource management is an intellectual process permeating daily human activities and important business decisions [84]. *Planning* for resource utilization and *scheduling* interrelated efforts are the necessary steps for solving these problems in order to achieve predefined goals under predetermined conditions. These resource management problems can, and often do, reach such complexity that require exceptional skills for judicious use and clear tracking of the resources. The key lies in the sharing of available resources to satisfy all the competing demands and restrictions on their use. Due to the diverse range of these problems, they are usually tackled one problem at a time, with little attention given to shared characteristics (among problems), generic techniques (across code modules), or common approaches (of analytic thinking) for this important class of problems.

A broadly applicable *planning and scheduling* methodology supported by a computer executable "*engine*" is presented here. The primary aim is to assist problem solvers with not only the general guidelines for characterizing their problems into systematic types and patterns but also transforming them into a general resource management paradigm where an embeddable, executable, work-horse module can assist to yield solutions. The *Constrained Resource Planning (CRP)* model, originating from work of Keng and Yun [42, 40, 93], has already been firmly established as a broadly applicable technique to solve numerous resource management problems (Tables 1 and 2). These problems all deal with *managing* a combination of *resources* categorized as *time, space, people,* and *material*. Common to these problems are six essential concepts of the CRP model: *resource, task, constraint, solution, criticality,* and *cruciality*.

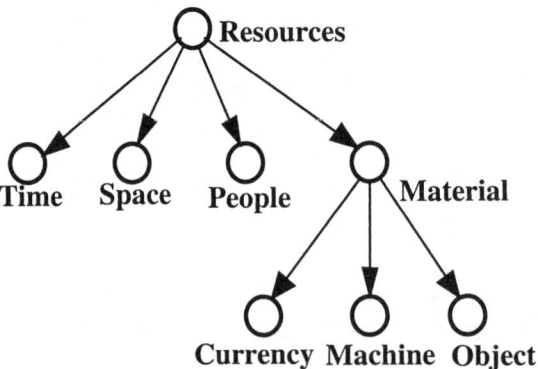

Figure 1. Resource categorization

Resource is the collection of usable elemental units to perform the tasks of a given problem. Resources can be organized into four basic categories: *time, space, people,* and *material,* which can be further separated into objects, machines, cost, etc. *Tasks* are work to be done in order to accomplish the overall goal of a resource management problem, using a specific amount of available, but often constrained, resource units. The tasks are usually interrelated due to competition among multiple tasks over the scarcity of resources. During the process of CRP problem solving, the active tasks are those still to be performed using the remaining resources. The set of active tasks is called the *task agenda*. The identification of the most appropriate task to perform first is a critical decision.

A *solution* to a task is a collection of resource units assigned to perform that task. There may be alternative resources suitable for the performance of a specific task. All possible solutions of a task form the *solution space* for that task. The selection of available resource units among alternatives is the crucial resource management decision that determines the most appropriate solution. Relationships among different resources, solutions, and tasks are specified by *constraints* in resource management.

2 The Constrained Resource Planning (CRP) Model

CRP model provides two domain-independent guiding principles for the identification of tasks and for the selection of solutions [36]. The tasks are evaluated by a *criticality* function in order to apply the *most-constrained strategy*. To the first order of approximation, the criticality function estimates the number of possible, valid solutions for each task. The most-constrained strategy operates by identifying the task with the least number of possible solutions, or the most critical task. This strategy is a natural generalization of the simple heuristic: always tackle the task with only one possible solution first, lest all needed resource units are consumed by performing other tasks and thereby reducing its own chance of completion. Thus, the most-constrained task is one that has the least flexibility for delay, hence has the highest priority - criticality.

The other guiding principle in the CRP model is used to direct the selection of a solution from the solution space. This domain-independent strategy selects the solution, which allows the most flexibility for the other tasks still to be have a solution. Called the *least-impact strategy*, it selects the solution of the current task by minimizing the impact to other tasks, thus maximizing the feasibility of completing all remaining tasks. The impact of a solution on other tasks is measured by the *cruciality* function, which calculates the reduction to the number of valid solutions for other tasks. Selecting the solution with the least cruciality allows maximum flexibility for other solutions to satisfy other tasks.

Based on these six fundamental concepts, the operational mechanism of the CRP model is the *Four-Corner Loop* iterative process (of Figure 2), constituting an *executable engine* that transforms one *state* (corner box) to the next in accordance to the *actions* (specified by edge labels).

The CRP Four-Corner Loop has been repeatedly demonstrated to achieve superb efficacy by *balancing* the needs of *using and sharing resources* simultaneously under the requirement of completing all the tasks for a given problem. The task identification and the solution selection steps both compete over the available resources, thereby incorporating the natural compromising and economizing principles of resource sharing and utilization. Such *global considerations* at the *local execution* levels help to achieve the desirable delicate *balance* between "*task demand pull*" and "*resource supply push*". This balance allows the CRP engine to exhibit an uncanny ability to avoid backtracking (a common pitfall for most heuristic problem solvers) [44], so that at least one sub-problem is solved through each loop. When executing without backtracking, the CRP model achieves linear efficiency with respect to the number of tasks in the problem. Such economy, such efficiency, and such precision distinguishes CRP from many traditional AI (Artificial Intelligence) planning methods which could spend all the computational efforts "worrying" about the complex interactions among the sub-problems to an extent that nothing useful gets done. Thus, CRP proves its ability to serve as an intelligent planning and scheduling engine, creating suggested solutions to its users. Some CRP applications also keep score on the resulting plans and schedules, so as to provide a balanced user satisfaction index that allows further performance enhancement.

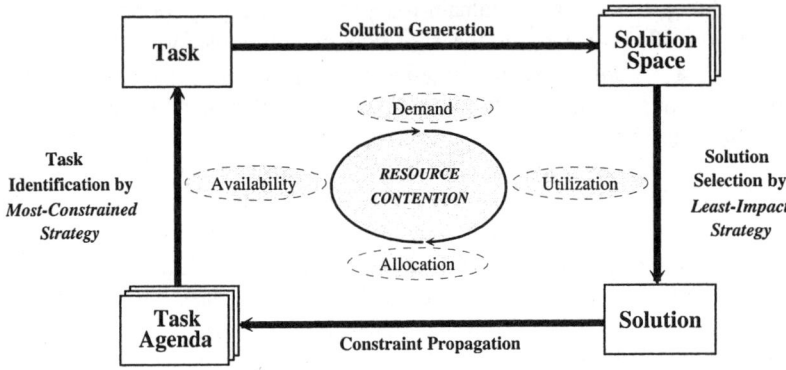

Figure 2. CRP Four-Corner-Loop

3 Applications of the CRP Model

In the area of *job-shop scheduling,* CRP [48] has already become an acknowledged *performance benchmark* [69, 72]. The research community in business administration recently promotes theories and models of *"resource based firm management"* [3, 4]. However, theories and models are not operational unless supported by an *executable engine,* as offered by CRP. The CRP model is a general decision support engine [46, 47, 48, 56, 60, 61], which possesses built-in intelligence through the criticality and cruciality heuristics, applicable to numerous domains. A prototype system incorporating the *CRP Intelligent Engine*, was on exhibit at COMDEX '92 and was a nominee of a Byte Magazine best new software award.

Though still a heuristic-based method, CRP has demonstrated consistent achievement of *solution optimality* in certain domains. One striking example arises from the well-known NP-complete Traveling Sales Problem (TSP) where solution optimality is difficult to determine [21]. One sub-problem of TSP, called Diamond Lattice Connections (DLC), has provably optimal solutions that are simple to check [88]. The CRP-based algorithm for solving the general TSP was directly applied to DLCs and produced optimal solutions for each size [88]. This further demonstrates CRP's inherent robustness to attain optimal solutions, although it is well known that no heuristic algorithm can guarantee optimality.

Recently, significant progress has been made not only to unify all graph matching problems and solve them by CRP but also to address the effective solutions of NP-hard problems [14]. The *Maximal Common Sub-graph* (MCS) problem is usually associated with the effort of finding hidden common patterns. Thus, solving these problems opens an entirely new spectrum of *pattern discovery* applications. As it turns out, the MCS problem can be tackled by solving a transformed problem of *Maximum Clique* (MC). Furthermore, the close mathematical relationship between the MC and the *Graph Coloring* (GC) problems are exploited and solved separately by CRP. Both MC and GC solutions mutually feedback in a tightly-coupled iterative loop in order to converge to a solution for MCS. The CRP-based graph matching algorithm has been demonstrated to discover common patterns. In Protein Common Sub-structure extraction problem, the tertiary structure of proteins are transformed into graphs and the common sub-structure of a protein pair is detected as the MCS [95]. The MCS problem is also applied to the 3D object model registration where two 3D models are registered/aligned by detecting their maximal common components [12].

Pattern No.	Resource Type = Application Pattern	*Time*	*Space*	*People*	*Material*
#1	Production Optimization	X			X
#2	Space Utilization		X		X
#3	Occupancy Planning		X	X	
#4	Inventory Distribution			X	X
#5	Facility Reservation	X	X		
#6	Work-Shift Scheduling	X		X	

Table 1. CRP application patterns

Table 2 shows numerous solved resource management problems (see relevant papers) and classifies them according to their similarities to the application patterns of Table 1.

Application Problem \ Application Pattern No.	#1 TM	#2 SM	#3 SP	#4 PM	#5 TS	#6 TP
Job-shop Scheduling	X					
Machine Utilization	X					
Income and Expense Management	X					
Process (Chemical, Production, ...) Planning	X					
Multi-processor Scheduling	X					
Floorplan Design (2D Packing)		X				
Pentomino (game)		X				
Steel Mill Production Planning		X				
VLSI Chip Placement and Wire Routing		X				
Shipping Container (3D) Packing		X				
Line-drawing Labeling		X				
3D Image Feature/Object Recognition		X				
Store Shelf Loading and Optimization		X				
Graph Matching (SGI, MCS, MOS)		X				
Block-based Image Coding		X				
N-Queens and General Max Clique Finding			X			
Map Four-Coloring, K-Coloring & Min Coloring			X			
Room (Hotel, Meeting, Class) Allocation			X			
3D Model Reconstruction from 2D Views			X			
Protein Common Tertiary Substructure Identification			X			
Workflow Model Discovery			X			
Inventory Control				X		
Goods (Food, Tools, etc.) Distribution				X		
Weapon-Target Assignment				X		
Traveling Salesman Problem					X	
Itinerary Planning					X	
Telescope or Satellite Antenna Steering					X	
Diamond Lattice Connections					X	
Transportation (Bus) Dispatching					X	
Power Generator Maintenance Scheduling					X	
Nurse Scheduling and Shift Scheduling						X
Course (Teaching) Assignment						X
Work-load Assignment	X					X
Vehicle Path (Trajectory) Planning	X	X				
Supply Allotment and Scheduling	X	X				
Sheet Metal Cutting		X			X	
Robot Motion Planning		X			X	

Airport Gate Assignment			X		X	
Railway Train Scheduling	X				X	
Network Management	X				X	
Ocean Resource Management	X			X	X	
Conference Scheduling			X		X	X
Factory Automation Design and Control	X	X		X		X
War Games	X	X	X	X	X	X
Mission Planning	X	X	X	X	X	X

Table 2. Resource management problems and classification

Differentiating Factors	CRP	SS	DB	ES	OR
1. Generic vs. Domain Specific	Generic	DomSp	DomSp	DomSp	DomSp
2. Solution Intelligence	Yes	No	No	Yes	No
3. What-if Scenarios	Yes	Yes	No	No	No
4. Planning/Scheduling Tool	Yes/Yes	Yes/No	Could	Could	Could
5. Ability to Optimize	Yes	No	No	Could	Yes
6. End-User must be Expert	No	No	Yes	No	Yes
7. User Mod of Constraints	Yes	Yes	No	No	Yes
8. Batch vs. Dynamic	Dynamic	Batch	Batch	Dynamic	Batch
9. Demand/Supply Balancing	Yes	No	No	Could	Could
10. Backtracking Reduced	Yes	n.a.	n.a.	No	n.a.

Table 3: Capability/technology differentiation comparisons

Since resources utilized in different problems are primarily time, space, people, material, or combinations thereof, any new resource management problem can be mapped into a pattern of representative problems already solved with the CRP model. Table 1 identifies six representative application patterns, each involving two categories of resources. Each application pattern is a well-known pattern of planning and scheduling problem where typical resources characterize all applications within that pattern. Of course, since material has many distinct sub-types, it is possible to form an application pattern from two material types alone, such as cost and machine, without involving the other three types. Tables 1 and 2, though, focus on combinations of the top level categories of resources to give an already rich collection of application examples, illustrating the benefit of characterizing applications by their resource types.

These solved problems, all using the CRP engine as the embedded work horse, demonstrates the inherent breadth of the CRP approach and the generic strength of this planning and scheduling tool for resource management. Comparable techniques, some of which already have marketable, computer-executable tools while others require more human intervention, include spreadsheets (SS), databases (DB), expert systems (ES), and operations research (OR) optimization techniques. Comparing several differentiating factors among these techniques with CRP, Table 3 offers some insights into the various differentiating factors when a user chooses a problem solving methodology together with an executable engine for solving resource management problems.

The methodology underlying CRP has also provided useful guidance for planners/schedulers faced with complex resource management problems [77, 78, 1, 88]. Usually, these problems involve limited resources, stringent requirements, conflicting constraints, and overwhelming data that confuse the problem solver. The experience of working through many different problems (in Table 2), dealing with all six application patterns (of Table 1), has revealed that a CRP-based way of thinking (approach) to sort out the complexities can be very helpful. CRP's resource categorization (Fig. 1), the six essential concepts (resource, task, constraint, solution, criticality, and cruciality), and the Four-Corner Loop process (Fig. 2) provide a useful conceptual structure to think through a complex problem in an organized manner. Often, intricacies in the problem can be clarified, confusing details can be organized, resource utilization and interrelationship can be simplified, conflict can be more easily resolved, and even useful heuristics can be generated and tested. Indeed, this captures the spirit of resource-focused management and problem solving.

4 Conclusions

The CRP model succeeds over traditional planning and scheduling techniques due to its balance between resource sharing and utilization, its separation of tasks and solutions, its sensitivity to interactions among them [44, 48], as well as its domain-independent control strategies that combine to effect both local execution objectives and global resource considerations. The result is a general methodology both for "thinkers" by conceptually guiding the problem solving and for "doers" armed with an executable engine, together with the demonstrably improved effectiveness. The model is embodied as a planning/scheduling engine and the methodology is supported by a set of tools that, collectively, offer a development environment available for industrial applications [93].

References

1. M. R. Akhter, D. Y. Y. Yun and H. Tian, (1991), "A Resource Management Approach to Schedule Manufacturing Processes", University of Hawaii MS thesis
2. Bowers, M. R. and J. P. Jarvis, (1992), "A Hierarchical Production Planning and Scheduling Model," Decision Sciences, 23, 144-159.

3. J. Barney, (1991) "The Resource-Based Model of the Firm: Origins, Implications, and Prospects", Journal of Management, Vol. 17, No. 1, pp. 97-98.
4. J. Barney, (1991) Firm Resources and Sustained Competitive Advantage, Journal of Management, Vol. 17, No. 1, pp. 99-120.
5. P. S. Bender, (1983), Resource Management - An Alternative View of the Management Process, John Wiley & Sons
6. Blazewicz, J., M. Dror, and J. Weglarz, (1991), Mathematical programming formulations for machine scheduling: A survey, European Journal of Operational Research, 51, 283-300.
7. T. Cheng and D. Y. Y. Yun, (1991), "Reasoning Three-Dimensional Relationships from Two-Dimensional Drawings by Planning," *Proceedings of OCEANS'91*, Honolulu, HI.
8. W. K. Chou and D. Y. Y. Yun, (1992), "Logic Minimization: A Planning Approach", *Proc. the 1992 International Computer Symposium* (ICS'92), Taiwan.
9. T. Cheng and D. Y. Y. Yun, (1994), Drawing Labeling by a Cascaded Planning Model - CCRP, *Proc. 3rd International Conf. on Automation, Robotics and Computer Vision* (ICARCV '94), SPICIS'94, pp.483-488
10. T. Cheng and D. Y. Y. Yun, (1994), Canonical Modeling and Multiview Indexing for 3D Object Recognition, *Proc. 3rd Intern'l Conf. on Automation, Robotics and Computer Vision* (ICARCV '94), pp.2000-2004
11. C. W. Chen and D. Y. Y. Yun, (1997), Variable Block-size Image Coding by Resource Planning, *Proc. International Conference on Imaging Science, Systems, and Technology*, Las Vegas, NV.
12. C. W. Chen and D. Y. Y. Yun, (1998), Automation of 3D Object Modeling from Multiple Line-Drawing Views, *1998 Symposium on Image, Speech, Signal Processing, and Robotics,* Hong Kong, China.
13. H. M. Chen and D. Y. Y. Yun, (1996), Sharing Computational Resources and Medical Images via ACTS-linked Networks, *Proc. 18th Annual Pacific Telecommunication Conference*, Honolulu, HI; also published in Pacific Telecommunication Review Vol. 17, No. 4, June 1996, pp. 15-28.
14. C. W. Chen, (1999), Algorithms for Maximal Common Subgraph Problem using Resource Planning, Ph.D. Dissertation, (Advisor: D. Y. Y. Yun), U. Hawaii
15. T. Cheng, (1994), Automatic Labeling, Modeling and Recognition for Line-Drawing Interpretation, Ph.D. Dissertation, (Advisor: D. Y. Y. Yun), U. Hawaii
16. H.-M Chen, D. Y. Y. Yun, Y. Ge and J. Khan, (1996), "Computer-Aided Resource Planning and Scheduling for Radiological Services," *Proc. PACS Design and Evaluation: Engineering and Clinical Issues, Medical Image*, CA
17. Elmaghraby, S. E., (1993), Resource Allocation Via Dynamic Programming in Activity Networks, European Journal of Operational Research, 64, 199-215.
18. S. M. Feng, B. G. Qiu and D. Y. Y. Yun, (1994), A Graphic Simulator for Autonomous Underwater Vehicle Docking Operations, *Proc. of OCEANS '94*, Sept. 13-16, Brest, France.
19. X. M. Feng and D. Y. Y. Yun, (1992) Scheduling of Steerable Antenna on Satellite by Resource management, *Proc. Advanced Communications Technology Satellite Conf.*
20. M. M. Flood, (1956), The Traveling-Salesman Problem, Operation Research, 4, 61-75.
21. K. C. Gilbert and R. B. Hofstra, "A New Multi-period Multiple Traveling Salesman Problem with Heuristic and Application to a Scheduling Problem," Decision Sciences, 23 (1992), 250-259

22. Gu, J. and X. Huang, (1994), Efficient Local Search With Search Space Smoothing: A Case Study of the Traveling Salesman Problem, IEEE Transactions on Systems, Man, and Cybernetics, 24, 5, 728-735.
23. D. G. Gattrysse and L. N. V. Wassenhove, (1992), A Survey of Algorithms for the Generalized Assignment Problem, European J. of Operational Research, Vol. 60, pp. 260-272.
24. H. M. C. Garcia and D. Y. Y. Yun, (1995), Intelligent Distributed Medical Image Management, *Proc. SPIE Medical Imaging*, San Diego, CA.
25. Y. Ge and D. Y. Y. Yun, (1996), "Simultaneous Compression of Makespan and Number of Processors Using CRP," *Proceedings of 1996 International Parallel Processing Symposium*
26. Y. Ge and D. Y. Y. Yun, (1996), "A Method that Determines Optimal Grain Size and Inherent Parallelism Concurrently," *Proceedings of International Symposium on Parallel Architecture, Algorithms, and Networks* (ISPAN)
27. Y. Ge and D. Y. Y. Yun, (1996), Simultaneous Determination of Optimal Grain Size and Inherent Parallelism for Parallel Programming Environments, *Proc. of IEEE Second International Conference on Algorithms and Architectures for Parallel Processing*, Beijing, China.
28. Y. Ge and D. Y. Y. Yun, (1996), A General Planning and Scheduling Methodology for Solving Task Scheduling with Multiple Objectives, *Proc. of IASTED International Conf. on Artificial Intelligence, Expert Systems and Neural Networks*
29. Y. Ge, (1996), A Simultaneous Compression Model for Multiprocessor Scheduling, Ph.D. Dissertation (Advisor: D. Y. Y. Yun), Department of Electrical Engineering, University of Hawaii
30. H. M. C. Garcia, Y. Ge, and J. Khan and D. Y. Y. Yun, (1996), Computer-Aided Resource Planning and Scheduling for Radiological Services, *Proc. SPIE workshop on PACS Design and Evaluation: Engineering and Clinical Issues*
31. J. J. Gray, D. McIntire and H. J. Doller, (1990), Preferences for specific work schedules: Foundation for an expert-system scheduling program, *Dallas Veterans Administration Hospital Technical Report*
32. Held, M. and R. M. Karp, (1970 and 1971), The travelling Salesman Problem and Minimum Spanning Trees, Part I, Operational Research, 18, 1138-1162; and Part II, Mathematical Programming, 1, 6-26.
33. Y. H. Hu and S. J. Chen, (1990), "GM_Plan: a gate matrix layout algorithm based on artificial intelligence planning techniques", IEEE Transactions on Computer-Aided Design, Vol. 9, No. 8
34. P. W. Huang and D. Y. Y. Yun, (1987), "A Representation Model and Inference Procedure for Intelligent Agents", *Proceedings of the 2nd International Conference on Applications of Artificial Intelligence in Engineering*, Cambridge. MA.
35. Y. H. Hu and D. Y. Y. Yun, (1991), "Model-Based Expert System for Digital System Design", IEEE Computer Society Press, Tutorial: Knowledge-Based Systems: Fundamentals and Tools, O. N. Garcia and Y-T Chien, eds.,.
36. Ho, W. P-C., D. Y. Y. Yun, and Y. H. Hu, Planning Strategies for Switchbox Routing, *Proc. IEEE International Conf. on Computer Design*, (1985), 463-467.
37. Hsu, W.-L., M. J. Prietula, and G. L. Thompson, "A Mixed-Initiative Scheduling Workbench: Integrating AI, OR and HCI," Decision Support Systems, 9 (1993), 245-257

38. P. W. Huang, The Formation and Application of an Intelligent Agent Model, Ph.D. Dissertation (Advisor: D. Y. Y. Yun), Computer Science and Engineering Department, SMU, 1989.
39. Jeffcoat, D. E. and R. L. Bulfin, "Simulated Annealing for Resource-Constrained Scheduling," European *Journal of Operational Research*, 70, 1993, pp. 43-51.
40. N. P. Keng and D. Y. Y. Yun, "A Planning/Scheduling Methodology for the Constrained Resource Problem", *Proc. of the Eleventh International Joint Conf. on Artificial Intelligence*, Aug. 1989, pp. 20-25.
41. Y. H. Kuo and D. Y. Y. Yun, "A CRP-based Planning and Scheduling Shell for Resource Management", EE Technical Report, University of Hawaii, Jan. 1991
42. N. P. Keng, Planning and Scheduling with the Constrained-Resource Model (CR-Model), Ph.D. Dissertation (Advisor: D. Y. Y. Yun), Computer Science and Engineering Department, SMU, 1988.
43. N. P. Keng, W. P-C. Ho and D. Y. Y. Yun, (1987), "The Manual of the Switchbox Router", SMU Computer Science and Engineering, TR87-CSE-2
44. N. P. Keng, W. Ho, and D. Y. Y. Yun, Interaction-Sensitive (Intelligent) Backtracking in a Planning System for Switchbox Routing, *Proc. of Applications of Artificial Intelligence in Engineering 87*.
45. Kirkpatrick, S. et al, Optimization by Simulated Annealing, Science, 220 (1983), 671.
46. G. Klein, I. C. Liu, and D. Y. Y. Yun, (1988), "Artificial Intelligence in the Architecture of Decision Support Systems", *Proceedings of the Conference on Impact of Artificial Intelligence on Business and Industry*
47. G. Klein, I. C. Liu, and D. Y. Y. Yun, (1990), "An Agent for Intelligent Model Management", Journal of Management Science
48. N. P. Keng, D. Y. Y. Yun, and M. Rossi, (1988), "Interaction-Sensitive Planning System for Job-Shop Scheduling", Expert Systems and Intelligent Manufacturing, M. Oliff (ed.), N. Holland, pp.57-69.
49. Lin, S. and B. W. Kernighan, "An Effective Heuristic Algorithm for the Travelling Salesman Problem," Operational Research, 21 (1977), 498-516.
50. Lawrence, S. R. and T. E. Morton, (1993), "Resource-Constrained Multi-Project Scheduling with Tardy Costs: Comparing Myopic, Bottleneck, and Resource Pricing Heuristics," European Journal of Operational Research, 64, 168-187.
51. Leclerc, M. and F. Rendl, (1989), "Constrained Spanning Trees and the Travelling Salesman Problem," European Journal of Operational Research, 39, 96-102.
52. Ling, Z. and D. Y. Y. Yun, (1996), "Solving Subgraph Isomorphism Problem by a New Planning Methodology", *Proc. IASTED International Conf. on Artificial Intelligence, Expert Systems and Neural Networks*, Honolulu, Hawaii
53. Ling, Z. and D. Y. Y. Yun, (1996), "An Efficient Subcircuit Extraction Algorithm by Resource Management Approach", *Proceedings of Second International Conference On ASIC*, Shanghai, P. R. China, 21-24
54. T. Y. Lu and D. Y. Y. Yun, (1997), Optimal Iso-Surface Triangulation, *Proc. International Conf. on Imaging Science, Systems, and Technology*, Las Vegas, NV.
55. Leung, J., "A New Graph-Theoretic Heuristic for Facility Layout," Management Science, 4, 38 (1992), 594-607
56. I. C. Liu, An Intelligent Agent to the Decision Support Model/Knowledge Base, Ph.D. Dissertation (Advisor: D. Y. Y. Yun), Computer Science and Engineering Department, SMU, 1989.

57. T. Y. Lu, J. Khan and D. Y. Y. Yun, (1996), An Optimizing Algorithm for 3D Object Surface Triangulation, *Proc. of IASTED International Conf. on Artificial Intelligence, Expert Systems and Neural Networks*, Honolulu., HI.
58. Lawler, E. L., J. K. Lenstra, A. H. G. R. Kan and D. B. Shmoys, (1985), *The Travelling Salesman Problem: A Guided Tour of Combinatorial Optimization*, John Wiley & Sons, NY
59. Luss, H., "Minimax Resource Allocation Problems: Optimization and Parametric analysis," European Journal of Operational Research, 60 (1992), 76-86.
60. I. C. Liu, D. Y. Y. Yun and G. Klein, (1989), "An Intelligent Agent Consulting System for Production and Operations Management", *Proc. Third International Conference on Expert Systems and the Leading Edge in Production and Operations Management*, Hilton Head, SC
61. I. C. Liu, D. Y. Y. Yun and G. Klein, (1990), "An Agent for Intelligent Model Management," Journal of Management Information Systems, 7, 1, 101-122.
62. Magirou, V. F. and J. Z. Milis, An Algorithm for the Multiprocessor Assignment Problem, Operations Research Letter, 8, 1989, 351-356.
63. Miller, D. L. and J. F. Perkny, Exact Solution of Large Asymmetric Traveling Salesman Problems, Science, 251, 1991, 754-761.
64. Prentis, E. L. and B. M. Khumawala, (1989), "Efficient Heuristics for MRP Lot Sizing with Variable Production/Purchasing Costs," Decision Sciences, 20, 439-450
65. Padman, R. and D. E. Smith-Daniels, "Early-Tardy Cost Trade-Offs in Resource Constrained Projects with Cash Flows: An Optimization-Guided Heuristic Approach," European Journal of Operational Research, 64 (1993), 295-311.
66. Padmanabhan, G., "Analysis of Multi-item inventory systems under resource constraints: A Non-linear Goal Programming Approach," Engineering Costs and Production Economics, 20, 2 (1990), 121
67. B. G. Qiu and X. M. Feng, (1991), "Trajectory Planning of Underwater Vehicle using Dynamic CRP Model", *Proceedings of OCEANS'91*, Honolulu, HI.
68. M. A. Rossi, Application of the Constrained-Resource Model (CR-Model) toward Flexible Job-shop Scheduling, MS Thesis, (Advisor: D. Y. Y. Yun), Computer Science and Engineering Dept., SMU, 1989.
69. N. Sadeh and M. Fox, "Variable and Value Ordering Heuristics for Activity-based Job-shop Scheduling", School of Computer Science Report, Carnegie Mellon University, 1990.
70. Shen, C. C. and W. H. Tsai, "A graph matching Approach to Optimal Task Assignment in Distributed Computing Systems Using a Minimax Criterion," IEEE Transactions on Computers, 34 (1985), 197-203.
71. E. Sacerdoti, (1974), Planning in a Hierarchy of Abstraction Spaces, Artificial Intelligence, 5, 115-35.
72. N. Sadeh, (1990), "Generating Advice for Hard Constraint Satisfaction Problems: an Application to Job-Shop Scheduling", *Computer Science Report*, Carnegie Mellon University
73. Sun, T., P. Meakin and T. Jossang, A Fast Optimization Method Based on Hierarchical Strategy for the Travelling Salesman Problem, PHYSICA, (1993).
74. Sofianopoulou, S., "The Process Allocation Problem: a Survey of the Application of Graph-Throretic and Integer programming Approaches," also published in the Journal of Operational Research Society, 43, 5, 1992, pp. 407-413.

75. Stone, H. S., "Multiprocessor Scheduling With the Aid of Network Flow Algorithms," IEEE Transactions on Software Engineering, 3 (1977), pp. 85-93
76. Sim, S. K., K. T. Yeo and W. H. Lee, An Expert Neural Network System for Dynamic Job Shop Scheduling, International Journal of Production Research, 32, 8 (1994), 1759.
77. H. Tian and D. Y. Y. Yun, (1991), "A Requirements/Knowledge Acquisition Methodology", *Proc. AAAI'91 Workshop on Automatic Software Design*, Anaheim
78. H. Tian and D. Y. Y. Yun, (1991), "A Heuristic Methodology for Oceanic Resource Management", *Proceedings of OCEANS'91,* Honolulu, HI.
79. H. Tian and D. Y. Y. Yun, (1991), "A More Complete Knowledge Acquisition Methodology", *Proceedings of The World Congress on Expert Systems,* Orlando
80. Tsai, W. H., Nonlinear Multiproduct CVP Analysis With 0-1 Mixed Integer Programming, Engineering Costs and Production Economics, 20, 1 (1990), 81
81. R. L. Tsai, A Heuristic for Static Task-to-Processor Assignment using a Constrain Satisfaction Approach, Ph.D Dissertation (Advisor: D. Y. Y. Yun), Computer Science and Engineering Dept., SMU, 1988.
82. Warkkentin, M. and P. Evans, AI in Business and Management, PC AI, 7, 2 (1993), 35.
83. J. G. Wu, Y. H. Hu, D. Y. Y. Yun, and W. P. C. Ho, (1990), "A Model-based Expert System for Digital System Design", IEEE Design and Test of Computers, pp. 24-41.
84. Wilkins, D. E., *Practical Planning: Extending the Classical Planning Paradigm*, Morgan Kaufmann Publishers, Inc., (1988).
85. Wilson, J. M., "Approaches to Machine Load Balancing in Flexible Manufacturing Systems," OR: the Journal of Operational Research Society, 43, 5 (1992), 415-423.
86. Yun, D. Y. Y. and X. Feng, (1993), "Solving 2-Dimensional Packing Problem by Constraint Resource Planning Model with Efficient Symmetry Detection," *UH Technical Report*
87. D. Y. Y. Yun and H. Tian, (1991), "A Heuristic Methodology for Oceanic Resource Management", *Proceedings of OCEANS'91,* Honolulu, HI
88. D. Y. Y. Yun, B. G. Qiu and Z. L. Chen, (1992), "Solving Travelling Salesman Problems by Planning", *Proceedings of the International Conference on Intelligent Information Processing & Systems*, (Keynote Paper), Beijing, China, Oct..
89. S. Yih, B. Shirazi and D. Y. Y. Yun, (1989), "Neural Network for Control System Modeling and Design," *Proceedings of Twentieth Annual Pittsburgh Conference on Modeling and Simulation*
90. L. Yin, M. Tanik, and D. Y. Y. Yun, (1989), "Intelligent Interactive Interfaces for Laboratory Equipment", *Proceedings of Thirteenth Annual Computer Science Conference*, Irving, Texas
91. L. Yin, M. Tanik, and D. Y. Y. Yun, (1989), "Design Activity Agent: A Knowledge-Based Software Design Environment", *Proc. International Conf. on Software Engineering and Knowledgee Engineering*
92. D. Y. Y. Yun, (1990), "The Evolution of Computing Software Tools" (Keynote), Computing Tools for Scientific Problem Solving, Academic Press, pp. 7-22.
93. D. Y. Y. Yun, (1992), CRP - an Intelligent Engine for Resource and Production Management, exhibition brochure for COMDEX '92, nominated for Byte Magazine best new software award.

94. Y. Zong, T. Yang and J. P. Ignizio, (1993), "An Expert System Using an Exchange Heuristic for the Resource-Constrained Scheduling Problem," Expert Systems with Applications, 6, 3, 327.
95. C. W. Chen and D. Y. Y. Yun, (1999), "Knowledge Discovery for Protein Tertiary Substructures", *Proc. 7th workshop on Rough Sets, Fuzzy Sets, Data Mining, and Granular-Soft Computing,* also to be published in the Lecture Notes in Artificial Intelligence series, Springer-Verlag.

Chapter 19. Evaluative Studies of Fuzzy Knowledge Discovery through NF Systems

Arthur Ramer[1], Maria do Carmo Nicoletti[1] & Sam Yuan Sung[2]

[1]School of Computer Science and Engineering, University of New South Wales, Sydney 2052, Australia

[2]Department of Computer Science, National University of Singapore, Singapore

[1]ramer@cse.unsw.edu.au, [2]marian@pop3.cse.unsw.edu.au

Abstract. *Recent years witnessed a rapid growth in fuzzy modelling using neuro-fuzzy systems (NF Systems or NFS for brief). Typically neural net structures would be extended to permit fuzzy inputs and to allow the outputs to be interpreted as fuzzy. The last step would defuzzify these outcomes producing a classifier, which would consist of a family of threshold ("partitioning") values. This chapter focuses on fuzzy modelling as a formalism for representing and inferring knowledge. It is mainly concerned with the investigation of the use of a neuro-fuzzy system for inducing fuzzy knowledge in the form of fuzzy if-then rules from a set of preclassified instances. We describe the experimental studies which we conducted in a few different knowledge domains. The motivation of our work is to evaluate the possibilities offered by this approach, to learn about its contributions and to establish situations where its use is appropriate. We close by discussing lines of further reerach into performance of such systems.*

Keywords. *Neuro-fuzzy systems, fuzzy classification, machine learning*

1 Introduction

The simplest possible representation of a conventional expert system comprises three main modules – domain knowledge base, inference engine and user interface. It is augmented by a structured working memory to hold a collection of case-specific data, and an explanatory unit to provide the user with explanation and justification of actions. The knowledge base consists of the knowledge that is specific to the domain of application; the inference engine is the mechanism that uses the knowledge in order to make inferences and the user interface is the module in charge of providing the communication between user and system.

Generally the knowledge contained in the knowledge base is represented by inference rules of the form

if condittion_1 and condition_2 and... and condition_n then conclusion

The fact that the if-then rule is the most successful and widely used representation does not prevent the adoption of other forms of potentially powerful representations (such as connectionist expert systems where the knowledge base is represented by a neural network). On top of solving problems in a given domain, an expert system is expected to explain the problem-solving process and it is also expected to be capable of handling uncertain and incomplete information. According to Zadeh [27] "Fuzzy logic provides a natural framework for the management of uncertainty in expert systems because its main purpose is to provide a systematic basis for representing and inferring from imprecise knowledge. In effect, in fuzzy logic everything is allowewd to be a matter of degree."

In a *fuzzy expert system* the knowledge base contains knowledge pertaining to the problem domain, generally but not necessarily, represented by a set of *fuzzy rules*. These rules commonly have the form "if X then Y" where X and Y are fuzzy sets. The inference engine of a fuzzy expert system operates on a set of fuzzy rules and makes fuzzy inferences (a process usually referred to as approximate reasoning). The notion of a fuzzy expert system was first introduced in [27]. Quoting Vadiee and Jamshidi [25] "The goal of a fuzzy expert system is to take in subjective, partially true facts that are randomly distributed over a sample space, and build a knowledge-based expert system that will apply certain reasoning and aggregation strategies to make useful decisions. These decisions are again approximate, and have partial degrees of truth and likelihood; the decisions and derived facts are reliable to the best of our available knowledge."

It is becoming increasingly evident that there are two central problems in fuzzy expert systems: the acquisition and maintenance of knowledge and the design of inference engines. As can happen with any expert system, be it fuzzy or not, the construction of the knowledge base can be a difficult task which generally takes a long time and which can be conducted in many different ways using a broad range of different strategies and approaches. These vary from totally manual (interview based) to totally automatic (learning). There is no general methodology for creating fuzzy rule bases but it is well known that they are hard to create and that it is difficult to ensure their consistency. There is also agreement about the difficulty of defining representative and appropriate membership functions. An additional problem is that, generally, fuzzy rules are not portable from one system to another.

In order to deal with the knowledge acquisition problem, some expert systems and their counterpart fuzzy have a module generally known as *knowledge acquisition module*, which is responsible for collecting/generating and sometimes maintaining the knowledge that constitutes their base. In spite of the large variety of methods that could be implemented by this module we are concerned here with those that make use of machine learning techniques. We are particularly interested in investigating those known as neuro-fuzzy, which have been developed for many different purposes.

In this chapter[1] we will be focusing on the neuro-fuzzy system NEFCLASS, trying to exploit its capabilities as a machine learning system for acquiring fuzzy knowledge. In order to acomplish this goal we have organized this chapter as follows: Section 2 presents and discusses the main ideas that support the neuro-fuzzy model. Section 3 introduces the 3-layer fuzzy perceptron and its specialisation, the system NEFCLASS, as proposed in [3,4,9,13,14,16,17,19]. Section 4 describes a few of the experiments conducted and discusses the results. Finally we present our conclusions and highlight future perspectives for this line of research.

2 Neuro-Fuzzy Systems

Over the past few years there have been an increasing number of systems which combine, in many different ways and with many different purposes, fuzzy techniques and neural networks. There are plenty of justifications in literature for these combinations, the main one being that of profiting from the capabilities of both paradigms. One of the most successful of these combinations is implemented by systems that use neural networks to produce a set of fuzzy rules (which can be further incorporated into the knowledge base of a *fuzzy expert system*). In this approach neural networks are generally part of a bigger system and are conveniently adapted (fuzzified) in order to be used as a learning technique that can deal with connection weights, inputs and concepts represented as fuzzy numbers. These systems are generally known as neuro-fuzzy systems.

Neuro-fuzzy systems can provide a "translator" which in the next phase converts the trained fuzzy neural net into a set of fuzzy rules. The inclusion of a "translator" in these system is justified in the same way as some of the existing symbolic-connectionist systems are justified: the need for an interpretation of the learned knowledge in order to make it understandable by a human expert. As the knowledge embedded in a trained neural net is not symbolically expressed because it is represented by the topology of the net, its connections and weights, an option for making this knowledge "interpretable" is translating the trained neural net into a set of rules. In the case of neuro-fuzzy systems this is particularly supported by the linguistic representation of fuzzy variables.

In [15] three advantages for using a fuzzy classifier are listed: (1) vague knowledge can be used, (2) the classifier is interpretable in the form of linguistic

---

[1] This chapter is a revised and extended version of [21].

rules, and (3) from an applicational point of view the classifier is easy to implement, to use and to understand. Whilst partially agreeing with the authors we believe these advantages are not restricted to neuro-fuzzy systems. Many symbolic-fuzzy classifiers also share these advantages, with the additional benefit of not needing a translation in order to obtain an "interpretable" knowledge. Table 1 extracted from [8] shows all the different combinations that can be considered in the process of fuzzifying a neural net. As commented in the above reference "The last three cases in Table 1 are not realistic. In Case 5, outputs are always real numbers because both inputs and weights are real numbers. Therefore, weights should be fuzzified for handling fuzzy targets in Case 5. In Cases 6 and 7, the fuzzyfication of weights is not necessary because targets are real numbers."

The main concern in this paper is with NEFCLASS, which is a tool for learning fuzzy rules and fuzzy sets by supervised learning. NEFCLASS can be considered a fuzzy rule-based system that is represented by a network architecture and trained as such. According to Table 1, NEFCLASS can be considered a special Case 6 since, depending upon the connection, the corresponding weight can be fuzzy or real.

Table 1. Degrees of fuzzification of neural nets

	weights	inputs	targets
conventional NN	real	real	real
Fuzzy NN: case 1	real	fuzzy	real
Fuzzy NN: case 2	real	fuzzy	fuzzy
Fuzzy NN: case 3	fuzzy	real	fuzzy
Fuzzy NN: case 4	fuzzy	fuzzy	fuzzy
Fuzzy NN: case 5	real	real	fuzzy
Fuzzy NN: case 6	fuzzy	real	real
Fuzzy NN: case 7	fuzzy	fuzzy	real

Generally, systems which combine neural network and fuzzy techniques employ a multilayer feedforward neural network. It has been proved that feedforward networks are universal approximators of any continuous mapping [5]. However there have been attempts to use other kinds of architectures, such as the one described in [26], where a learning method for neuro-fuzzy systems, based on Kohonen's learning algorithms self-organizing map and learning vector quantification 2.1 is proposed. In what follows, however, we will be focussing on a feedforward neural architecture, particularly on a 3-layer perceptron applied to classification domains.

The advantage of this model is that it allows the extraction of symbolic knowledge from the neural net and also allows the user to feed symbolic knowledge into the neural net. This model is implemented in the system NEFCLASS that we briefly present and discuss next. The concepts and definitions related to NEFCLASS have been extracted from the various references about this system and are shown at the end of this chapter.

3 The NEFCLASS System

NEFCLASS *(NEuron Fuzzy CLASSification)*, NEFCON (NEuro Fuzzy CONtrol), NEFPROX *(NEuro Fuzzy function ApPROXimation)* and FCLUSTER are all part of a collection of software made available by the University of Magdeburg as tools for fuzzy processing. The first three systems are based on a model known as generic 3-layer fuzzy perceptron, defined next and described in detail in [10]. These tools can be downloaded from http://fuzzy.cs.uni-magdeburg.de/. Each tool is generally available for different computational platforms.

The 3-layer fuzzy perceptron architecture can be approached as a 3-layer perceptron that has some degree of "fuzziness" attached to it. Only the weights (sometimes only part of them), the net inputs and the activation of the output units are modelled as fuzzy sets. The first layer is called input layer, the hidden layer is the rule layer and the third and last layer, the output layer. As stated in [10] "A fuzzy perceptron is like a usual perceptron used for function approximation. The advantage is to interpret its structure in the form of linguistic rules, because the fuzzy weights can be associated with linguistic terms. The network can also be created partly in the whole out of fuzzy if-then rules".

Definition 1 A 3-layer fuzzy perceptron is a 3-layer feedforward neural network (U,W,NET,A,O,ex) with the following specifications:

1) $U = \bigcup_{i \in \{1,2,3\}} A_i$ is a non-empty set of units, where U_1 is the input layer, U_2 the rule layer (hidden layer) and U_3 the output layer. For all i,j∈ {1,2,3}, $U_i \neq \emptyset$ and $U_i \cap U_j = \emptyset$ for i≠j

2) The structure of the network (connections) is defined as W: U × U → F(R), such that there are only connections W(u,v) with u ∈ U_i, v ∈ U_{i+1}(i ∈ {1,2}) (F(R) is the set of all fuzzy subsets of R (real numbers))

3) A defines for each u ∈U an activation function A_u, to calculate the activation a_u

   3.1) For input and rule units u ∈ $U_1 \cup U_2$, A_u: R → R, $a_u = A_u(net_u) = net_u$

   3.2) For output units u ∈ U_3, A_u:F(R) → F(R), $a_u = A_u(net_u) = net_u$

4) O defines for each u ∈ U an output function O_u, to calculate the output o_u

   4.1) For input and rule units $u \in U_1 \cup U_2$, $O_u: R \to R$, $o_u = O_u(a_u) = a_u$

   4.2) For output units $u \in U_3$, $O_u: F(R) \to R$, $o_u = O_u(net_u) = DEFUZZ_u(net_u)$, where $DEFUZZ_u$ is a suitable defuzzification function

5) NET defines for each unit u ∈ U a propagation function NET_u to calculate the net input net_u

   5.1) For input units $u \in U_1$, $NET_u: R \to R$, $net_u = ex_u$

   5.2) For rule units $u \in U_2$, $NET_u: (R \to F(R))^{U_1} \to [0,1]$, $net_u = T_{u' \in U_1} (W(u',u)(o_{u'}))$, where T is a t-norm.

   5.3) For output units $u \in U_3$, $NET_u: ([0,1] \times F(R))^{U_2} \to F(R)$, $net_u: R \to [0,1]$, $net_u(x) = \perp_{u' \in U_2} (T(o_{u'}, W(u',u)(x)))$, where $\perp$ is a t-conorm.

   If the fuzzy sets $W(u',u)$, $u' \in U_2$, $u \in U_3$ are monotonic in their support, and $W^{-1}(u',u)(\tau) = x \in R$ such that $W(u',u)(x) = \tau$ holds, then the propagation function net_u of an output unit $u \in U_3$ can alternatively be defined as

   $$net_u(x) = \begin{cases} 1 & \text{if } x = (\sum_{u' \in U_2} o_{u'} \cdot m(o_{u'})) / \sum_{u' \in U_2} o_{u'} \\ 0 & \text{otherwise} \end{cases}$$

   with $m(o_{u'}) = W^{-1}(u',u)(o_{u'})$. To calculate the output o_u in this case $o_u = x$, with $net_u(x) = 1$ is used instead of (4.2)

6) ex: $U_1 \to R$ defines for each input unit $u \in U_1$, its external input $ex(u) = ex_u$. For all other units ex is not defined.

Definition 2 A NEFCLASS system is a 3-layer fuzzy perceptron with the following specifications:

1) $U_1 = \{x_1,...,x_n\}$, $U_2 = \{R_1,...R_k\}$, $U_3 = \{c_1,...,c_m\}$

2) Each connection between units $x_i \in U_1$ and $R_r \in U_2$ is labelled with a linguistic term $A_{j_r}^{(i)}$ ($j_r \in \{1,...,q_i\}$)

3) $W(R,c) \in \{0,1\}$ holds for all $R \in U_2$ and $c \in U_3$

4) Connections coming from the same input unit x_i and having identical labels, bear the same weight at all times. These connections are called *linked connections*, and their weight is called *shared weight*

5) Let $L_{x,R}$ denote the label of the connection between the units $x \in U_1$ and $R \in U_2$. For all $R, R' \in U_2$ holds

$$(\forall (x \in U_1)\ L_{x,R} = L_{x,R'}) \rightarrow R = R'$$

6) For all rule units $R \in U_2$ and all units $c, c' \in U_3$ we have

$$(W(R,c) = 1) \wedge (W(R, c') = 1) \rightarrow c = c'$$

7) For all output units $c \in U_3$, $o_c = a_c = net_c$ holds

8) For all output units $c \in U_3$ the net input net_c is calculated by:

$$net_c = \frac{\sum_{R \in U_2} W(R,c).o_R}{\sum_{R \in U_2} W(R,c)}$$

NEFCLASS implements a supervised learning procedure where the net is trained with a set of pre-classified training examples (training set), each of them belonging to one of a number of distinct crisp classes. Its first layer of units represents the pattern features – there should be as many units as the number of features that describe the training instances. The hidden layer is composed of rule units that represent the fuzzy rules. The number of output units in the third layer is given by the number of different classes in the training set.

Note that a NEFCLASS system is a specialization of a 3-layer fuzzy perceptron, where: 1) the weights associated with connections between rule units and output units can only be 0 or 1; 2) a rule unit should only be connected to one output unit and 3) there are no two rule units connected to input units that have the same connection weights. In order to make the next algorithm clearer, we have slightly changed the original notation as presented in [14,16].

NEFCLASS Learning Algorithm

Consider a NEFCLASS system with n input units $x_1,...,x_n$, $k <= k_{max}$ rule units $R_1,...,R_k$ and m output units $c_1,...,c_m$. Also given is a learning task

$$L = \{(p_{11}, p_{12}, p_{13},...,p_{1n}, t_{11}, t_{12},...,t_{1m}),...,(p_{s1}, p_{s2}, p_{s3},...,p_{sn}, t_{s1}, t_{s2},...,t_{sm})\}$$

of s patterns, each consisting of an input pattern $p \in R^n$ and a target pattern $t \in \{0,1\}^m$. The *rule learning algorithm* that is used to create the k rule units of the NEFCLASS system consists of the following steps:

1) Select the next pattern $(p_{k1}, p_{k2}, p_{k3}, ..., p_{kn}, t_{k1}, t_{k2}, ..., t_{km})$ from L

2) For each input unit $x_i \in U_1$ find the membership function $\mu_{j_i}^{(i)}$ such that

$$\mu_{j_i}^{(i)}(p_{ki}) = \max_{j \in \{1,...,q_i\}} \{\mu_j^{(i)}(p_{ki})\}$$

3) If there are still less than k_{max} rule nodes and there is no rule node R with
$$W(x_1,R) = \mu_{j_1}^{(1)},\ldots, W(x_n,R) = \mu_{j_n}^{(n)}$$
then create such a node and connect it to the output node c_q if $t_{kq} = 1$

4) If there are still unprocessed patterns in L and $k < k_{max}$ then proceed with step 1) otherwise stop.

In the classification problem "Glass Identification" which will be discussed in Section 4, the training instances are described by 9 numerical attributes; each of them belonging to one out of six classes. The way an instance is presented to NEFCLASS is as a sequence of 9 real values followed by a sequence of six binary digits (only one of them being 1). For example, in the instance

1.51761   13.89   3.60   1.36   72.73   0.48   7.83   0.00   0.00   1 0 0 0 0 0

the pattern 1 0 0 0 0 0 indicates that this particular instance belongs to the first of the six classes.

At the beginning the net has n input units (n=number of attributes that describe the instances of the training set), m output units (m=number of existing classes in the training set) but has no rule units and no connections. The rule learning algorithm creates the rule units and establishes the connections between input and rule units and between rule and output units. Also, the algorithm assumes that associated to each input unit x_i there are q_i fuzzy sets and chooses the one that maximizes the membership function for the input value p_{ji} into x_i (step 2). Note in step 3 that a new rule unit is always created (provided the number of rules has not reached its maximum) if there is no rule unit connected to the input units with that particular combination of fuzzy weights. However, it may happen that some of these connections are aready present in the net; that gives rise to what is known as *linked connections-shared weights* (see Definition 2). This assures consistency, preventing the same fuzzy set evolving into two different fuzzy sets during the procedure described next. Once the 3-layer perceptron is created, in order to obtain a classification procedure that generalizes the classification embedded in the training set, NEFCLASS refines the fuzzy sets weights of connections between input and rule units by running the *fuzzy set learning algorithm* cyclically:

1) Select the next pattern (p,t) from L, propagate it through the NEFCLASS system and determine the output vector c.

2) For each output unit c_i determine the delta value $\delta_{c_i} = t_i - a_{c_i}$

3) For each rule unit R with $a_R > 0$

   3.1) Determine the delta value

$$\delta_R = a_R(1 - a_R) \sum_{c \in U_3} W(R,c)\delta_c$$

3.2) Find x′ such that

$$W(x',R)(a_{x'}) = \min_{x \in U_1}\{W(x,R)(a_x)\}$$

3.3) For the fuzzy set W(x′,R) determine the delta values for its parameters a, b, c using the learning rate $\sigma > 0$:

$$\delta_b = \sigma\delta_R(c - a)\,\text{sgn}(a_{x'} - b)$$

$$\delta_a = -\sigma\delta_R(c - a) + \delta_b$$

$$\delta_c = \sigma\delta_R(c - a) + \delta_b$$

and apply the changes to W(x′,R) if this does not violate a given set of constraints Φ. (Note: the weight W(x′,R) might be shared by other connections, and in this case might be changed more than once).

4) If an epoch was completed, and the end condition is met, then stop; otherwise proceed with step 1)

After the training, the resulting 3-layer perceptron can be interpreted as a set of fuzzy rules. For example, one of these rules in the "Glass Identification" domain (see next Section) is:

if V_1 is medium and V_2 is medium and V_3 is large and V_4 is small and V_5 is medium and V_6 is small and V_7 is small and V_8 is small and V_9 is small then Class_1

where V_1 up to V_9 are the attributes that describe the training instances; medium, large and small are fuzzy sets associated to the respective attribute and Class_1 is one of the six existing classes in this domain. NEFCLASS allows the user to name attributes and classes conveniently. For instance, instead of V_1 we could have used refractive_index, since V_1 stands for this index.

When operating NEFCLASS the user has first to define the number of initial fuzzy sets associated with each attribute that describe the training set and also provide the system with a number k, to limit the maximum number of rule nodes the perceptron will have. An interesting aspect that has not been explored in this work is that NEFCLASS also allows that a prior partial knowledge about the fuzzy concepts can be given and then, using a pre-classified training set, refines this knowledge.

4 Experiments

We have used NEFCLASS in a few knowledge domains aiming to experiment with the system in an exploratory way. The choice of these domains was basically determined by two characteristics they have: the type of attributes, since we were interested only in numerical attributes, and the absence of missing values for attributes. Since four of these domains (all but Excipients) are well-known datasets and part of the UCI Repository [1], we present only a short description of them, since any additional information can be found at this site.

Excipients

This is a pharmaceutical knowledge domain, with 170 instances, related to the industrial production of pharmaceutical drugs. In order to optimise drug delivery systems, a better understanding of excipients, their properties and limitations is required The excipient domain consists of data with 14 different characteristics associated to excipients: bulk density, tapped density, compressibility, angle of repose, relative filling, flow rate, particle size distributions (with percentage of retention on 840, 420, 250, 177, 149 and < 149 μm), moisture content and Hausner ration. The 170 training examples are distributed among 17 different excipients (10 training examples for each excipient). All the measurements were obtained in a laboratory. The excipients are nine different types of microcrystalline cellulose (Avicel PH-102®, Avicel PH-200®, Microcel MC-101®, Microcel MC-102®, Microcel MC-250®, Microcel MC-301®, Microcel MC-302®, Vivapur type 101® and Vivapur type 102®), Cellactose, Lactose, Encompress®, Ludipress®, Manitol powder and granules, Sodium Chloride and CornStarch. This domain has been previously used with the machine learning system CN2 and details on the experiments conducted can be found in [20].

Pima Indians Diabetes

This is a domain with 768 training instances, each of them described by eight numeric-valued attributes. The attributes are: number of times pregnant; plasma glucose concentration; diastolic blood pressure (mm Hg); triceps skin fold thickness (mm); 2-hour serum insulin (mu U/ml); body mass index (weight-kg)/(height-m)^2; diabetes pedigree function and age in years. The instances are classified into two classes, depending upon whether the patient shows (or not) signs of diabetes according to the World Health Organization.

Bupa Liver Disorders

This domain contains 345 training instances, each of them described by six numeric-valued attributes. The attribute are: mean corpuscular volume; alkaline phosphatase; alumina aminotransferase; aspartate aminotransferase; gamma-glutamyl transpeptidase and drinks (number of half-pint equivalents of alcoholic beverages drunk per day). The instances are classified into two classes: patients with or without liver disorders.

Glass Identification Database

This domain contains 214 instances of different types of glass – the correct identification of the type of a glass can help forensic services. The original dataset contains 10 attributes. Since the first one is the identification number of the instance, this attribute has not been considered in the experiments described in this chapter. The remaining nine attribute are: refractive index, sodium, magnesium, aluminium, silicon, potassium, calcium, barium and iron. All the nine attributes, with the exception of the refractive index, have been measured using as a unit of measurement the weight percentage in the corresponding oxide. The class attribute identifies the type of glass. Since no instance of class four is part of the original dataset, we renamed the classes. Consequently we worked with a dataset with six different classes.

Haberman's Survival Data

This dataset contains 306 cases from a study that was conducted between 1958 and 1970 at the University of Chicago's Billings Hospital on the survival of patients who had undergone surgery for breast cancer. The instances are described by three numerical attributes: age of patient at time of operation, patient's year of operation and number of positive axillary nodes detected. Each instance belongs to one of the two classes: 1: the patient survived 5 years or longer and 2: the patient died within five years.

For each of the five datasets, the experiments consisted of performing, ten times, the following sequence of steps: the original training set was randomly divided into two sets: a new training set – TrS (containing approximately 75% of the original) and a testing set – TeS (containing the remaining –approximately 25%). The new training set was then used by NEFCLASS for training the net and creating the corresponding set of fuzzy rules. Subsequently the associated testing set was used for evaluating the performance of the NEFCLASS system when classifying new examples. Table 2 shows the number of instances in the original dataset, the number of training instances and testing instances used in each domain, as well as the number of attributes and classes in each domain.

The results of the experiments were obtained using the defaults of the system. We used the option "same for all units" with the value 3, when defining the number of fuzzy sets associated with each input unit (attribute) because 3 seemed to be a reasonable value if "interpretability" is one of the goals the user has in mind. However the system allows the user to define an individual number of fuzzy sets for each attribute. We believe that the use of this option in a suitable way, depending on the domain, can be a difficult task because the user needs not only to be familiar with but also to have a very precise idea of the domain and the distribution of the values of each particular attribute over its range.

Table 2. Number of instances (original, training and testing sets), number of attributes and number of classes per domain

Domain	|Original|	|TrS|	|TeS|	|attributes|	|classes|
Excipients	170	136	34	14	17
Pima	768	580	188	8	2
Bupa	345	260	85	6	2
Glass	214	161	53	9	6
Haberman	306	230	76	3	2

In the tables that follow we adopted the notation *#:* number of the experiment; *NR:* number of rules created; *misc TrS:* number of misclassified instances in TrS and *misc TeS:* number of misclassified instances in TeS. Initially we used NEFCLASS with each of the ten pairs training set-testing set, for each of the five domains, limiting the maximum number of rules to k=30. Tables 3, 4, 5, 6 and 7 show the results for Excipients, Pima, Bupa, Glass and Haberman domains respectively.

Table 3. Excipients

		k=30				k=50			
#	NR	misc TrS	mean error	misc TeS	mean error	NR	misc TrS	mean error	misc TeS	mean error
0	30	15	0.2622	4	0.2735	31	3	0.2446	1	0.2570
1	29	3	0.2255	1	0.2461	29	1	0.2255	1	0.2461
2	30	19	0.2851	5	0.2852	32	3	0.2335	1	0.2262
3	30	16	0.2602	5	0.2512	31	3	0.2337	1	0.2336
4	30	12	0.2734	2	0.2679	32	0	0.2316	0	0.2365
5	30	3	0.2358	1	0.2531	30	3	0.2358	1	0.2531
6	30	3	0.2400	0	0.2261	30	3	0.2400	0	0.2261
7	30	3	0.2317	1	0.2408	30	3	0.2317	1	0.2408
8	30	16	0.2690	3	0.2452	32	3	0.2327	0	0.2376
9	29	3	0.2320	1	0.2440	29	3	0.2320	1	0.2440

Table 4. Indians diabetes (Pima) (k=30)

#	NR	misc TrS	mean error	misc TeS	mean error
0	30	174	0.4958	68	0.5395
1	30	214	0.6407	61	0.6270
2	30	185	0.4922	63	0.5225
3	30	184	0.5016	64	0.4803
4	30	179	0.4862	62	0.4724
5	30	187	0.4786	61	0.4764
6	30	196	0.5910	74	0.6323
7	30	181	0.4832	57	0.4858
8	30	194	0.4953	53	0.4706
9	30	200	0.5804	71	0.6148

Table 5. Liver disorders dataset (Bupa)

		k=30				k=100				
#	NR	misc TrS	mean error	misc TeS	mean error	NR	misc TrS	mean error	misc TeS	mean error
0	30	120	0.5470	40	0.5455	42	100	0.6105	33	0.6077
1	30	103	0.6366	36	0.6499	45	103	0.5740	39	0.5930
2	30	115	0.5740	44	0.5996	48	105	0.5744	41	0.6092
3	30	97	0.6114	42	0.6929	46	97	0.6065	43	0.6661
4	30	104	0.5860	30	0.5889	48	102	0.5983	26	0.5925
5	30	102	0.6751	38	0.7157	50	105	0.5922	39	0.5876
6	30	100	0.5615	35	0.5654	46	101	0.5687	35	0.5662
7	30	100	0.5697	42	0.6674	47	97	0.5698	42	0.6157
8	30	95	0.6756	44	0.7304	47	111	0.5763	41	0.5788
9	30	98	0.6491	43	0.6647	47	102	0.6123	45	0.6211

Table 6. Glass identification dataset (Glass) (k=30)

#	NR	misc TrS	mean error	misc TeS	mean error
0	30	64	0.6887	28	0.7228
1	30	76	0.7186	25	0.7016
2	30	68	0.6983	26	0.7321
3	30	74	0.6966	26	0.7430
4	30	73	0.7470	21	0.7241
5	30	77	0.7041	28	0.7340
6	30	70	0.6666	27	0.7148
7	30	71	0.7089	26	0.7517
8	30	68	0.7102	31	0.7930
9	30	74	0.7107	24	0.6980

Table 7. Haberman's survival dataset (k=30)

#	NR	misc TrS	mean error	misc TeS	mean error
0	19	66	0.5497	20	0.5964
1	18	62	0.5246	22	0.4725
2	18	64	0.5224	17	0.5035
3	16	62	0.5239	26	0.5695
4	18	61	0.5230	17	0.5061
5	18	60	0.5183	17	0.4922
6	19	65	0.5027	19	0.5463
7	19	60	0.5308	20	0.5209
8	16	60	0.5128	19	0.5207
9	19	61	0.4792	20	0.4714

As can be seen from the previous tables, the experiments conducted with Excipients (Table 3) were the ones where NEFCLASS had its best performance. We believe that part of this achievement is due to the nature of the dataset – the training instances that represent each of the 17 classes have the values associated with each attribute differing only slightly.

Considering the results of the Pima experiments shown in Table 4, the performance of NEFCLASS can be considered reasonable, with approximately 30% of misclassified instances in both training and testing dataset. As we were curious about the results a symbolic machine learning system would produce in this domain, we decided to use the machine learning system CN2 [2] with one of the 10 Pima training datasets (#0). For this particular dataset CN2 induced 44 rules (one of them the default rule) with an overall accuracy of 73.9% when evaluated in the corresponding testing set. Later on in this chapter we will return to CN2.

After running the experiments with k=30 we believed that the performance of the fuzzy rules could be improved by allowing the system to create a bigger number of rules (this however was not the case for the Haberman domain where the limit 30 was far beyond the number of rules the system created). As can be seen in Table 3 (k=30 and all experiments but 1 and 9), Table 4, Table 5 (k=30) and Table 6, the number of rules (NR) in those experiments reached the limit. So, we decided to run a few of the experiments again varying the limit on the number of rules. Table 3 (k=50), Table 5 (k=100), Table 8 (k=100, k=200) and Table 9 are the results obtained by increasing the limit on the number of rules in the Excipients, Bupa, Pima and Glass domains, respectively. As shown in Table 3, when the limit was set to 50, the number of rules the system created increased very little. This reflected positively on their accuracy since they slightly improved their performance in both the training and testing sets.

In the Bupa domain (Table 5) however, when the limit was set to 100 (actually 50 would have had the same impact on the learning since the largest number of rules created by NEFCLASS was 50 (#5)), the results were difficult to analyse comparatively to those obtained for k=30. As figures in Table 5 show, with the increase in the value of k, in some experiments the accuracy of the rules improved and in others it did not. Consider for instance, experiment #6, where the number of misclassifications in the training set has dropped from 100 to 97 (all the other values remained practically the same). Can this value be considered an "improvement" if the set of rules has increased from 30 up to 46? This is a difficult question to answer if we consider that interpretability of the rules is one of the issues that should be taken into account.

The Pima domain was one where we expected a considerable improvement in accuracy by increasing the limit on the number of rules. However, that did not happen. As can be seen by comparing the results shown in Table 4 and Table 8, the best accuracy of the set of rules was obtained when k was set to 200. We should keep in mind that this little improvement in accuracy on the training data was obtained at the expense of an approximately 50% increase in the number of rules. It also seems that the rules, particularly in this case, overfitted the training set and there was no improvement in their accuracy when classifying the instances from the testing set.

Table 8. Indians diabetes (Pima)

#	NR	misc TrS	mean error	misc TeS	mean error
k = 100					
0	100	154	0.5270	51	0.5552
1	100	160	0.5505	47	0.5463
2	100	132	0.5567	53	0.5770
3	100	152	0.5390	50	0.5247
4	100	170	0.5357	47	0.5066
k = 200					
0	137	139	0.5585	48	0.5829
1	152	148	0.6041	47	0.6145
2	148	132	0.5475	57	0.5561
3	148	148	0.6087	49	0.5884
4	154	147	0.6153	39	0.5941

Table 6 shows that the accuracy of the rules induced in the domain Glass, for k=30, to predict the class of instances of the training set, is below 50%, which makes them useless. Since the reason for such an inaccuracy could be the limit we imposed on the number of rules, we tried different values for k, as shown in Table 9. When running the experiments for k=50, in three of them (#=2,3,4) the system reached the limit 50 so we decided to run another set of tests with k=100. However, the results obtained were the same, except for experiment #=2, because 53 rules were created (note that their performance on the training set decreased a little but the results were still very similar to those for k=50). In the Glass domain by comparing the results of the five experiments (#=0,1,2,3,4) shows that by increasing the number of rules from 30 to 40, their accuracy slightly decreases in both the training and testing set. However when their number was increased from 50 to 100, the induced rules performed better in some of the experiments and worse in others.

In the Haberman domain (Table 7), the 10 experiments had a reasonable performance with approximately 25% of misclassified instances in both training and testing sets. It can be seen that the limit of 30 was also far beyond the number of rules NEFCLASS created (maximum of 19). This fact motivated us to explore smaller values for k as shown in Table 10. The results for k=20 and 30 are the same since the number of rules has not gone beyond 19. But surprisingly, considering the number of rules has dropped by almost half, the rules induced with the limit set to 10 gave, on average, a slightly better performance when classifying instances of the testing set, compared to those induced with the limit set to 20.

Table 9. Glass identification dataset (Glass)

#	NR	misc TrS	mean error	misc TeS	mean error
k = 40					
0	40	73	0.7128	30	0.7360
1	40	74	0.6692	25	0.6620
2	40	67	0.7106	27	0.7574
3	40	70	0.6483	26	0.6842
4	40	103	0.8222	34	0.8375
k = 50					
0	47	81	0.6963	30	0.7170
1	48	76	0.6539	26	0.6469
2	50	71	0.7123	26	0.7381
3	50	68	0.6284	24	0.6417
4	50	109	0.8238	35	0.8584
k = 100					
0	47	81	0.6963	30	0.7170
1	48	76	0.6539	26	0.6469
2	53	74	0.7020	26	0.7401
3	50	68	0.6287	24	0.6417
4	50	109	0.8238	35	0.8584

The results obtained with NEFCLASS in the Glass domain made us wonder how a conventional machine learning system would perform in this dataset. The motivation for using such a system was only to get an idea about the number of rules and number of conditions per rule it would create for this particular domain. We did not intend to establish patterns for comparing their performance simply because that would not take us anywhere. They are two different systems that, in spite of aiming for the same goal, serve different purposes, implement different strategies and deal with different types of data. We decided to run the ten experiments (ie. using the same ten pairs of training-testing sets) with the system CN2. The algorithm CN2 was proposed by Clark & Niblett [2] and the system that implements it, also called CN2 is available for download from many different sites (see, for example, http://www.cs.utexas.edu/users/pclark). CN2 is a rule-induction program for crisp data, which takes a set of examples (vectors of attribute-values and an associated class) and generates a set of rules for classifying them. As with NEFCLASS, it allows the user to evaluate the accuracy of a set of rules in terms of a set of pre-classified examples (testing set).

Table 10. Haberman's survival dataset

#	NR	misc TrS	mean error	misc TeS	mean error
k = 10					
0	10	64	0.4718	18	0.4973
1	10	59	0.4916	21	0.4720
2	10	65	0.4999	16	0.4404
3	10	59	0.4629	24	0.5285
4	10	63	0.4694	21	0.4799
k = 20					
0	19	66	0.5497	20	0.5964
1	18	62	0.5246	22	0.4725
2	18	64	0.5224	17	0.5035
3	16	62	0.5239	26	0.5695
4	18	61	0.5230	17	0.5061

The results obtained with CN2 in the Glass domain are shown in Table 11, where *ov_ac* stands for overall accuracy (the accuracy of the whole set of induced rules) and *def_ac* stands for default_accuracy. CN2 always generates a default rule which, during the classification process, is triggered when none of the previous rules is triggered; an instance that has not been classified by any rule will have the class given by the default rule. In Table 11, the default rule is not included in the value NR. The column *number rules-number of variables* shows the number of rules which have a certain number of test of variables in their condition part. This column is intended to give an idea of the volume of information the system generated. For example, in experiment #=0, CN2 generated 23 rules (plus a default rule). The 23 rules are distributed into: 5 rules which have in their condition part, 4 tests of variables; 6 rules with three such tests, 9 rules with two tests and 3 rules with one test. Also, CN2 created 6 rules for class_1, 7 for class_2, 4 for class_3, 2 for class_4, 1 for class_5 and 3 for class_6. This information can also be used to determine how "interpretable" this set of 23 rules can be. On average the 10 experiments generated 20 rules each; on average, 13 of them had at most, 3 tests of variables.

For this same pair of training-testing sets (#=0) NEFCLASS created 47 rules, all of them with 9 tests of variables in their condition part and distributed among the classes as: 9 rules for class_1, 23 for class_2, 0 for class_3, 5 for class_4, 3 for class_5 and 6 for class_6. We believe that a set of 23 rules that describe a particular class, whether or not it has been created by a conventional or a neuro-fuzzy system, is very hard to be fully understood by human beings. It should not be forgotten that, in the case of neuro-fuzzy learning, for the "interpretation" to be complete, there is still the additional "cost" of understanding the meaning of the linguistic terms that describe the fuzzy sets associated with the attributes. As shown in Table 11, the rules induced by CN2, in spite of having, on average, an overall accuracy of 85% in the training set, did not perform well on the testing set giving, on average, an overall accuracy of 65%. This made us wonder whether some important information that could be determinant for characterising the classes might be missing from the dataset

It is important to mention that NEFCLASS allows the user to know what rules are embedded within the network and also, to alter them. Through the dialog box "Fuzzy Rules" the user has the option of browsing, modifying, deleting or adding new rules. This is a very important feature of NEFCLASS that has not been explored in this work. Through this option the user can "tune" the induced set of fuzzy rules, changing its number and/or changing the number of fuzzy sets associated to variables. NEFCLASS helps the user in this task by providing some statistics about the data.

The learning process conducted by NEFCLASS can be controlled by the user through a few options the system offers in a dialog box named "Learning Control Parameters". It is the user's choice to stop this process after a pre-determined time and/or when a good classification result is obtained. The possible choices available are: *the number of epochs* (cycles through the training set); *stopping at error value* (stop learning when a specific error value is reached); *maximum number of errors* (maximum number of allowed misclassification); *stop if minimal error is not decremented for a given number of epochs.* According to the authors, the minimal error is usually the best criterion – to use it the user needs to provide a number of epochs and the system checks whether the learning process has to be stopped because the error does not become smaller during the last specified number of epochs.

Table 11. Glass identification dataset (Glass) - using CN2

#	NR	number rules-number of variables	misc TrS	ov_ac (%)	def_ac (%)	misc TeS	ov_ac (%)	def_ac (%)
0	23	5-4; 6-3;9-2;3-1	23	85.7	36.0	22	58.5	34.0
1	20	3-5;2-4;8-3;6-2;1-1	33	79.5	36.0	17	67.9	34.0
2	20	6-5;3-4;5-3;4-2;2-1	24	85.1	35.4	17	67.9	35.8
3	22	2-5;3-4;9-3;6-2;2-1	21	87.0	35.4	19	64.2	35.8
4	20	1-6;2-5;5-4;6-3;3-2;3-1	15	90.7	37.9	19	64.2	28.3
5	20	1-5;7-4;4-3;5-2;3-1	27	83.9	34.2	15	71.7	39.6
6	20	1-6;2-5;3-4;5-3;8-2;1-1	25	84.4	36.2	20	62.3	34.0
7	21	1-5;5-4;5-3;7-2;3-1	31	80.7	35.4	19	64.2	35.8
8	20	3-5;6-4;4-3;7-2	3	98.1	34.8	17	67.9	37.7
9	20	2-5;3-4;7-3;7-2;1-1	28	82.6	38.5	17	67.9	26.4

In all our experiments we chose to use as stopping criterion the number of epochs (which was set to 1000 in all of them). We were aware that in many of the experiments we would have obtained practically the same results in a much shorter time if the recommended criterion had been adopted. However, we wished later to explore the effects the number of epochs could have on the learning process.

We did notice that in all experiments the global error and the number of misclassification changed very little from the beginning of the training (around 100 epochs) up to the end, when reaching 1000 epochs. We conducted only a few experiments in the Glass and Haberman domains, with bigger values for epochs. Table 12 and Table 13 show the results we obtained for the first two pairs of datasets (#=0,1) of each domain, with 2000 and 3000 epochs respectively. As can be seen when comparing the results of these tables with the previous results obtained with 1000 epochs, they are practically the same, which means the learning process has stabilized.

As far as the options in the dialog box "Learning Parameters" are concerned, their default values were used in all experiments. So, in relation to the options listed in "Fuzzy Set Constraints", the following values were used: 1) *do not pass neighbours* (yes), meaning that during the learning process a fuzzy set must not pass its left or right neighbours; 2) *learning asymmetrically* (yes), meaning that during the learning process only one side of the triangular function is changed – where the input is located; 3) *FS intersect at 0.5* (no), meaning it is not necessary that two adjacent fuzzy sets must always intersect at the membership value of 0.5; 4) *fixed conclusions* (yes), meaning that the conclusion weights are fixed at 1.0; 5) *conclusions in [0,1]* – this option is only considered if fixed conclusions is set to 'no'. Also, in the dialog box "Learning Rates" the default value of 0.01 for sigma_a, sigma_b and sigma_c, which are the learning rates for the respective parameters of the triangular fuzzy sets, was not changed.

Table 12. Glass – 2000 and 3000 epochs

		2000 epochs and k=50				3000 epochs and k=50			
#	NR	misc TrS	mean error	misc TeS	mean error	NR	misc TrS	mean error	misc TeS	mean error
0	47	81	0.6964	30	0.7163	47	81	0.6973	30	0.7168
1	48	76	0.6538	26	0.6469	48	76	0.6540	26	0.6470

Table 13. Haberman – 2000 and 3000 epochs

		2000 epochs and k=50				3000 epochs and k=50			
#	NR	misc TrS	mean error	misc TeS	mean error	NR	misc TrS	mean error	misc TeS	mean error
0	19	66	0.5558	18	0.5795	19	66	0.5557	18	0.5794
1	18	62	0.5246	22	0.4722	18	62	0.5249	22	0.4724

The default value of 'Best' for *Rule Learning Procedure* was maintained, meaning that all patterns are propagated three times for finding the best rules and, by doing that, the rule learning procedure is made independent from the sequence of patterns. We would like to add that, in spite of not having used it in any of the experiments, another promising value for the *Rule Learning Procedure* is the 'Best per Class' which has similarities with 'Best'. With the value 'Best' it is possible that the best rules are only for one class and that the remaining classes are not considered. In 'Best per Class' the second evaluating step selects the best M rules for every class, where M = N div L3, where N = maximum number of rules and L=number of output units. The final option *Relearn the Rule Base* was kept set to 'no', meaning that the rule base is only learned if the user defines a new network (either without rules – which was the case in all experiments described in this chapter, or with prior rules entered by the user), after the user has changed the maximum number of rules or after the user has changed a fuzzy partitioning.

It is important to mention that the fuzzy if-then rules created by NEFCLASS always contain, in their condition part, all the attributes that describe the training instances. In most situations this can be very misleading because no matter how irrelevant an attribute is for describing a class, it will always be part of the rule that describes that class. Just by looking at a rule the user is not able to know the relevance of each attribute when categorizing the corresponding class. However, in order to deal with this problem, as mentioned earlier, NEFCLASS allows the user to alter the set of rules and fuzzy sets associated to variables. The editor is operated by the user who can conveniently refine and tune the set of rules. This feature of NECLASS requires its own investigation and will be subject of our studies next.

We also carried out an experiment using the dataset "rice taste data", with 105 instances, as described in [22]. In this dataset however, as the overall evaluation of each instance (class) is represented by a real number, we "transformed" that value into a discrete value. Probably because of this the results we obtained are not worth discussing. We also tried different partitions of the classes without any concrete result.

5 Conclusions and Future Research

One of the stated advantages that the use of a neurò-fuzzy system could give – that of producing an interpretable knowledge – is questionable. For instance, how interpretable is a set of 152 rules for discriminating between two classes where each rule is described by 8 fuzzy variables? Unless the user carefully edits the induced rules trying to maintain their representativeness, we believe that interpretability will be far from being achieved in many domains. The problem of "interpretability" of a set of induced rules occurs also with the rules induced by conventional machine learning systems. Feasibility of interpreting the derived rules has much to do with their number and the number of variables used in their context hence with the dimension of the domain. This interpretability issue arises whether a conventional or neuro-fuzzy system has been used. For these reasons we believe that a tool such as NEFCLASS can be of help when there are not too many rules and attributes; then the results it produces can help the user to understand more about the domain.

It is important to note that NEFCLASS has a large number of parameters which need be tuned by the user. We have not explored this issue in depth and confined our analysis to the study of criticality of dimensional aspects of the space of rules.

Further research will include the analysis of other aspects of NEFCLASS which have not been targeted by our experiments. An example is 'tuning' of the rules and the 'Best per Class' rule learning. We also want to explore the possibilities offered by NEFCLASS of learning the conclusion weights, despite the fact that this option is not normally used as it may lead to results that cannot be interpreted (see [11][18]). It is our aim also to investigate how the tool FCLUSTER [12] can help the learning process. The studies described in this chapter will continue with the investigation of other models of neuro-fuzzy learning, particularly the multilayer perceptron architecture known as FuNe I [6,7] and the Fuzzy RuleNet [23,24]. The results will form a series of separate reports. The comment about NEFCLASS found in [16] is very appropriate in this context:

"... it supports the user to find a desired fuzzy system based on training data, but it cannot do all the work. A good and interpretable fuzzy classifier can usually not simply be found by an automatic learning process... The user must be able to supervise and interpret the learning procedure in all its stages".

We expect to apply this paradigm in a future extension of our analyses.

Acknowledgments

The second author wishes to thank to Brenda Padgett for her comments and suggestions. FAPESP (Fundação de Amparo à Pesquisa do Estado de São Paulo) is acknowledged for the research scholarship awarded to Maria do Carmo Nicoletti (Proc. 98/05547-6).

References

1. Blake, C. L., Merz, C. J. (1998) UCI Repository of machine learning databases. Irvine, CA: University of California, Dept.of Information and Computer Science [http://www.ics.uci.edu/~mlearn/MLRepository.html]
2. Clark, P., Niblett, T. (1989) The CN2 induction algorithm. Machine Learning 3, pp 261-283
3. Detlef, N., Kruse, R. (1996) Designing neuro-fuzzy systems through backpropagation. In W. Pedrycz (ed.) Fuzzy Modelling: Paradigms and Practice, Kluwer, Boston, pp 203-228
4. Detlef, N., Detlef, U. and Kruse, R. (1996) Generating classification rules with the neuro-fuzzy system NEFCLASS. In Proc. Biennial Conference of the North American Fuzzy Information Processing Society NAFIPS'96, Berkeley, CA, June 1996
5. Funahashi, K. (1989) On the approximate realization of continuous mappings by neural networks. Neural Networks 2, pp 183-192

6. Halgamuge, S.K., Glesner, M. (1994) Neural networks in designing fuzzy fuzzy systems for real world applications. Fuzzy Sets and Systems 65, pp 1-12
7. Halgamuge, S.K., Poechmueller, W. and Glesner, M. (1995) An alternative approach for generation of menbership functions and fuzzy rules based on radial and cubic basis function networks. International Journal of Approximate Reasoning, 12, pp 279-298
8. Ishibuchi, H., Kwon, K. and Tanaka H. (1995) A learning algorithm of fuzzy neural networks with triangular fuzzy weights. Fuzzy Sets and Systems 71, pp 277-293
9. Klawonn, D. Nauck D., Kruse R. (1995) Generating rules from data by fuzzy and neuro-fuzzy methods. In: Proc. Fuzzy-Neuro-Systeme'95, Darmstadt, pp 223-230
10. Nauck, D. (1994) A fuzzy perceptron as a generic model for neuro-fuzzy approaches. In: Proc. Fuzzy-Systems'94, 2nd GI-Workshop, Munich, Siemens Corporation
11. Nauck, D. (1995) Beyond neuro-fuzzy: perspectives and directions. Proceedings of the Third European Congress on Intelligent Techniques and Soft Computing (EUFIT'95), Aachen, August 28-31, pp 1159-1164
12. Nauck, D., Klawoon, F. (1996) Neuro-fuzzy classification initialized by fuzzy clustering. Proc. Fourth European Congress on Intelligent Techniques and Soft Computing (EUFIT'96), Aachen
13. Nauck, D. (1997) Neuro-fuzzy systems: review and prospects. In: Proc. Fifth European Congress on Intelligent Techniques and Soft Computing (EUFIT'97), Aachen, Sep. 8-11, pp 1044-1053
14. Nauck, D., Kruse, R. (1995) NEFCLASS – a neuro-fuzzy approach for the classification of data. In: Proc. ACM Symposium on Applied Computing, Nashville, Feb. 26-28, ACM Press, pp 461-465
15. Nauck, D., Kruse, R. (1997) What are neuro-fuzzy classifiers? In: Proc. Seventh International Fuzzy Systems Association World Congress IFSA'97, Vol. IV, Academia Prague, pp 228-233
16. Nauck, D., Kruse, R. (1997) New learning strategies for NEFCLASS. In: Proc. Seventh International Fuzzy Systems Association World Congress IFSA'97, Vol. IV, Academia Prague, pp 50-55
17. Nauck, D., Kruse, R. (1997) A neuro-fuzzy method to learn fuzzy classification rules from data. Fuzzy Sets and Systems 89, pp 277-288
18. Nauck, D., Kruse, R. (1998) How the learning of rule weights affects the interpretability of fuzzy systems. In Proc. IEEE International Conference on Fuzzy Systems 1998 (FUZZ-IEEE'98), Anchorage, AK, May 4-9, pp 1235-1240
19. Nauck, D., Nauck, U. and Kruse, R. (1996) Generating classification rules with the neuro-fuzzy system NEFCLASS. Proc. Biennial Conference of the North American Fuzzy Information Processing Society (NAFIPS'96), Berkeley
20. Nicoletti, M. A., Nicoletti, M. C. and Magalhães, J. F. (1999) Machine learning system investigation in pharmaceutical development. Bollettino Chimico Farmaceutico, Febbraio, Vol. 138, 1999, pp. XXVI
21. Nicoletti, M. C., Ramer, A. (1999) Experiments in inducing fuzzy knowledge using a neuro-fuzzy system. In: Proceedings of the ICONIP/ANZIIS/ANNES'99 International Workshop, N. Kasabov & K. Ko (eds.) 22-23 Nov, Dunedin, New Zealand, pp 244-249
22. Nozaki, K., Ishibuchi, H. and Tanaka, H. (1997) A simple but powerful heuristic method for generating fuzzy rules from numerical data. Fuzzy Sets and Systems 86, pp 251-270

23. Tschichold-Gurman, N. (1995) Generation and improvement of fuzzy classifiers with incremental learning using fuzzy rulenet. In Proc. ACM Symposium on Applied Computing, Nashville, Feb. 26-28, pp 466-470
24. Tschichold-Gurman, N. (1997) The neural network model RuleNet and its application to mobile robot navigation. Fuzzy Sets and Systems 85, pp 287-303
25. Vadiee, N., Jamshidi, M. (1994) The promising future of fuzzy logic. IEEE Expert, August, pp 36-38
26. Vuorimaa, P. (1994) Fuzzy self-organizing map. Fuzzy Sets and Systems 66, pp 223-231
27. Zadeh, L.A. (1983) The role of fuzzy logic in the management of uncertainty in expert systems. Fuzzy Sets and Systems 11, pp 199-227

Druck: Strauss Offsetdruck, Mörlenbach
Verarbeitung: Osswald, Neustadt

Studies in Fuzziness and Soft Computing

Vol. 25. J. Buckley and Th. Feuring
Fuzzy and Neural: Interactions and Applications,
1999
ISBN 3-7908-1170-X

Vol. 26. A. Yazici and R. George
Fuzzy Database Modeling, 1999
ISBN 3-7908-1171-8

Vol. 27. M. Zaus
Crisp and Soft Computing
with Hypercubical Calculus, 1999
ISBN 3-7908-1172-6

Vol. 28. R. A. Ribeiro and H.-J. Zimmermann,
R. R. Yager and J. Kacprzyk (Eds.)
Soft Computing in Financial Engineering, 1999
ISBN 3-7908-1173-4

Vol. 29. H. Tanaka and P. Guo
Possibilistic Data Analysis for Operations Research,
1999
ISBN 3-7908-1183-1

Vol. 30. N. Kasabov and R. Kozma (Eds.)
Neuro-Fuzzy Techniques for Intelligent
Information Systems, 1999
ISBN 3-7908-1187-4

Vol. 31. B. Kostek
Soft Computing in Acoustics, 1999
ISBN 3-7908-1190-4

Vol. 32. K. Hirota and T. Fukuda
Soft Computing in Mechatronics, 1999
ISBN 3-7908-1212-9

Vol. 33. L. A. Zadeh and J. Kacprzyk (Eds.)
Computing with Words in Information/
Intelligent Systems 1, 1999
ISBN 3-7908-1217-X

Vol. 34. L. A. Zadeh and J. Kacprzyk (Eds.)
Computing with Words in Information/
Intelligent Systems 2, 1999
ISBN 3-7908-1218-8

Vol. 35. K. T. Atanassov
Intuitionistic Fuzzy Sets, 1999
ISBN 3-7908-1228-5

Vol. 36. L. C. Jain (Ed.)
Innovative Teaching and Learning, 2000
ISBN 3-7908-1246-3

Vol. 37. R. Słowiński and M. Hapke (Eds.)
Scheduling Under Fuzziness, 2000
ISBN 3-7908-1249-8

Vol. 38. D. Ruan (Ed.)
Fuzzy Systems and Soft Computing
in Nuclear Engineering, 2000
ISBN 3-7908-1251-X

Vol. 39. O. Pons, M. A. Vila and J. Kacprzyk (Eds.)
Knowledge Management in Fuzzy Databases, 2000
ISBN 3-7908-1255-2

Vol. 40. M. Grabisch, T. Murofushi and M. Sugeno
(Eds.)
Fuzzy Measures and Integrals, 2000
ISBN 3-7908-1255-2

Vol. 41. P. Szczepaniak, P. Lisboa and J. Kacprzyk
(Eds.)
Fuzzy Systems in Medicine, 2000
ISBN 3-7908-1263-4

Vol. 42. S. K. Pal, A. Ghosh and M. K. Kundu (Eds.)
Soft Computing for Image Processing, 2000
ISBN 3-7908-1217-X

Vol. 43. L. C. Jain, B. Lazzerini and U. Halici (Eds.)
Innovations in ART Neural Networks, 2000
ISBN 3-7908-1270-6

Vol. 44. J. Aracil and F. Gordillo (Eds.)
Stability Issues in Fuzzy Control, 2000
ISBN 3-7908-1277-3

Vol. 45. N. Kasabov (Ed.)
Future Directions for Intelligent Systems
on Information Sciences, 2000
ISBN 3-7908-1276-5

Vol. 46. J. N. Mordeson and P. S. Nair
Fuzzy Graphs and Fuzzy Hypergraphs, 2000
ISBN 3-7908-1286-2

Vol. 47. E. Czogała† and J. Łęski
Fuzzy and Neuro-Fuzzy Intelligent Systems, 2000
ISBN 3-7908-1289-7

Vol. 48. M. Sakawa
Large Scale Interactive Fuzzy Multiobjective
Programming, 2000
ISBN 3-7908-1293-5

Vol 49. L. I. Kuncheva
Fuzzy Classifier Design, 2000
ISBN 3-7908-1298-6

Vol. 50. F. Crestani and G. Pasi (Eds.)
Soft Computing in Information Retrieval, 2000
ISBN 3-7908-1299-4